T5-DGT-439

Extinction and Survival in the Fossil Record

Proceedings of a Symposium held in Durham, UK, September 1986

The Systematics Association
Special Volume No. 34

Extinction and Survival in the Fossil Record

Edited by

G. P. Larwood

Department of Geological Sciences, University of Durham, Science Laboratories, South Road, Durham City

Published for the SYSTEMATICS ASSOCIATION by
CLARENDON PRESS • OXFORD
1988

Oxford University Press, Walton Street, Oxford OX2 6DP

Oxford New York Toronto
Delhi Bombay Calcutta Madras Karachi
Petaling Jaya Singapore Hong Kong Tokyo
Nairobi Dar es Salam, Cape Town
Melbourne Auckland

and associated companies in
Beirut Berlin Ibadan Nicosia

Oxford is a trade mark of Oxford University Press

Published in the United States
by Oxford University Press, New York

British Library Cataloguing in Publication Data
Extinction and survival in the fossil record.
—(The Systematics Association special
volume; no. 34).
1. Extinction (Biology) 2. Paleontology
I. Larwood, G. P. II. Systematics
Association III. Series
575'.7 QE721.2.E97
ISBN 0-19-857708-7

Library of Congress Cataloging in Publication Data
Extinction and survival in the fossil record.
(The Systematcs Association special volume; no. 34)
"Proceedings of a symposium held in Durham, UK,
September 1986"—P.
Includes indexes.
1. Extinction (Biology)—Congresses. I. Larwood,
Gilbert P. (Gilbert Powell) II. Systematics Association.
III. Series.
QE721.2.E97E968 1987 575 87-13447
ISBN 0-19-857708-7

Filmset and Printed in Northern Ireland by The Universities Press (Belfast) Ltd.

Preface

The Systematics Association has published a number of Special Volumes relating to evolutionary change and process in various groups of plants and animals in general or in more detail. Volume 12 (1979) was concerned with the origins of major invertebrate groups, Vol. 18 (1981) reviewed the Ammonoidea, Vol. 28 (1985) dealt with the origins and relationships of lower invertebrates, and most recently, Vol. 32 (1986) considered coevolution and systematics.

In examining extinction and survival in the fossil record this present volume continues these studies over a very wide range of organisms. It was never intended that it should be a complete compilation of a data base nor does it set out to cover all groups as in the Geological Society Symposium *The Fossil Record* of 1967. Contributors were invited to comment on the current understanding of or problems in defining the character and causes of major changes in groups of organisms through time with some evaluation of the current data base and its biases. Experts on a wide range of fossil plants and animals considered their groups in relation to Raup and Sepkoski's original (1982) recognition of five major extinction events in the Phanerozoic, and in relation to their later (1986) assessment of a much higher number of mass extinction phases during this interval of time.

There was some measure of agreement that some groups were involved in extinctions at these times — foraminiferids, Palaeozoic corals, brachiopods, cephalopods, echinoderms, arthropods, graptolites, conodonts, and reptiles to some extent reflect these recognized phases of higher extinction rates. Other groups which were considered — notably plants and pollen, bryozoans, bivalves, agnathans, birds, and mammals — were involved in substantial extinction events which do not necessarily coincide with those major events recognized by Raup and Sepkoski nor with the 26-million-year periodicity pattern which they also propose. Smith and Patterson have more recently questioned this cyclicity in relation to fishes and echinoderms as partly a taxonomic artefact because true terminations of monophyletic clades are not being considered.

Contributions to the Symposium were invited individually, but in the event, some authors have combined their work into extended joint papers — for example, those on patterns of plant extinction, or on the extinction and fossil record of the arthropods. In all, there are 15 papers reflecting the authors' varied views on the topic in relation to their 'own' groups. Although nearly all agree that the data base leaves much to be desired in terms of taxonomic and stratigraphic refinement, the authors have produced reviews and commentaries which are themselves a valuable assessment and basis for continued study of the highly complex nature of extinction and survival in the fossil record.

I am most grateful to my colleagues for contributing papers and discussion at a relaxed, informative, and lively meeting, to the Head of Department of Geological Sciences in the University of Durham for permission to use department facilities, to the technical staff of the Department, and to the staff of Hatfield College, Durham, for their excellent support.

Durham
January 1987

G. P. L.

Contents

Contributors

D. V. AGER

Department of Geology, University College of Swansea, Singleton Park, Swansea SA2 8PP, UK.

R. J. ALDRIDGE

Department of Geology, University of Nottingham, University Park, Nottingham NG7 2RD, UK.

M. J. BENTON

Department of Geology, The Queen's University of Belfast, Belfast BT7 1NN, UK.

M. C. BOULTER

Palynology Research Unit, NE London Polytechnic, Romford Road, London E15 4LZ, UK.

M. D. BRASIER

Department of Geology, University of Hull, Cottingham Road, Hull HU6 7RX, UK.

D. E. G. BRIGGS

Department of Geology, University of Bristol, Wills Memorial Building, Queen's Road, Bristol BS8 1RJ, UK.

E. N. K. CLARKSON

Grant Institute of Geology, University of Edinburgh, West Mains Road, Edinburgh EH9 3JW, UK.

R. A. FORTEY

Department of Palaeontology, British Museum (Natural History), Cromwell Road, London SW7 5BD, UK.

A. HALLAM

Department of Geological Sciences, University of Birmingham, P.O. Box 363, Birmingham B15 2TT, UK.

L. BEVERLY HALSTEAD

Departments of Geology and Zoology, University of Reading, Whiteknights, P.O. Box 227, Reading RG6 2AB, UK.

M. R. HOUSE

Department of Geology, University of Hull, Cottingham Road, Hull HU6 7RX, UK.

G. P. LARWOOD

Department of Geological Sciences, University of Durham, Science Laboratories, South Road, Durham DH1 3LE, UK.

A. I. MILLER

Department of Geology, University of Cincinnati, Cincinnati, Ohio, USA.

C. R. C. PAUL

Department of Geology, University of Liverpool, Brownlow Street, Liverpool L69 3BX, UK.

R. B. RICKARDS

Department of Earth Sciences, University of Cambridge, Sedgwick Museum, Downing Street, Cambridge CB2 3EQ, UK.

R. J. G. SAVAGE

Department of Geology, University of Bristol, Wills Memorial Building, Queen's Road, Bristol BS8 1RJ, UK.

C. T. SCRUTTON

Department of Geology, The University, Newcastle upon Tyne NE1 7RU, UK.

R. A. SPICER

Life Sciences Department, Goldsmiths' College, Creek Road, London SE8 3BU, UK.

P. D. TAYLOR

Department of Paleontology, British Museum (Natural History), Cromwell Road, London SW7 5BD, UK.

B. A. THOMAS

Department of Botany, National Museum of Wales, Cathays Park, Cardiff CF1 1XL, UK.

D. M. UNWIN

Department of Zoology, University of Reading, Whiteknights, P.O. Box 228, Reading RG6 2AJ, UK.

1. Patterns of plant extinction from some palaeobotanical evidence

MICHAEL C. BOULTER

Palynology Research Unit, North East London Polytechnic, London, UK

ROBERT A. SPICER

Life Sciences Department, University of London Goldsmiths' College, London, UK

BARRY A. THOMAS

Botany Department, National Museum of Wales, Cardiff, UK

Abstract

The majority of plants fossilize as fragments of separate organs such as leaves, wood, and reproductive structures. These are produced and preserved in many different ways, and different taphonomic regimes influence the assemblages in which they are found. Within the scientific literature the names given to these diverse fossil plant parts have created a complex taxonomy. The variable abundance of particular groups at any one time and place add difficulties to this complex background. Thus, it is no surprise that evolutionary processes and rates are obscured within the data currently available.

This article summarizes our interpretation of the palaeobotanical data which we have consulted to examine the likely patterns within some of the existing records of plant history, including work on both megafossils and palynomorphs. Through the noise and confusion within the data due to the factors mentioned above we have detected that the fossil plant record indicates regional and apparently global changes in some parts of the

Extinction and Survival in the Fossil Record (ed. G. P. Larwood), Systematics Association Special Volume No. 34, pp. 1–36. Clarendon Press, Oxford, 1988.

geological column. The times of first and last appearance of some pollen and spore taxa have been taken from the literature and appear to follow a similar punctuated pattern to that of the megafossils, in which first appearances precede the last appearances of other taxa. The results of our surveys from megafossils and microfossils are compared with one another and set against some of the evidence from other sources.

We review the ample palaeobotanical evidence for a profound short-term ecological trauma at the Cretaceous–Tertiary boundary. Cretaceous vegetation was devastated in the middle latitudes with evergreen elements being most susceptible, suggesting a cold dark phase. Early Palaeocene leaves typically indicate a higher precipitation than do the latest Cretaceous fossils, suggesting a change in climate as well.

Introduction

Plant fossils pose many more taxonomic problems than do those of animals due to their more fragmentary nature and this, together with the complex processes leading to the formation of fossil assemblages, make interpretation of the fossil record for the purposes of evolutionary assessment very difficult. Discussion on either evolutionary change or extinction should be based upon the soundest knowledge possible of the various plant groups that have existed throughout geological history.

In our search for patterns within the evolutionary history of the plant kingdom we have been conscious of the formidable problems caused by depositional processes as well as the difficulties created by the inadequacy of the data available. This article summarizes the most well known peculiarities of the fossil plant record and warns that the palaeobotanical data currently available is not very well suited for the kind of analysis we are making to try and help understand evolutionary processes. Although much of the lengthy descriptive palaeobotany up to about 1950 has monographed apparently substantial amounts of taxonomic detail, the taxonomy is usually so unreliable that it cannot be used as a source of data for this kind of analysis. In compiling this analytical review we have used only the most recent sources of palaeobotanical evidence which we feel are the most reliable.

The plant fossil record has not been uniformly sampled for extinction studies. Palaeozoic plant fossils, partly because of their relatively simple and consistent morphology, and partly because of the economic importance of Carboniferous coal deposits and the attendant stratigraphy, have been studied in such a way that biostratigraphic data are readily available. Comparable information about megafossil plants for the Mesozoic and Tertiary is not so readily obtainable because compilations of fossil stratigraphic range data have less application. If such detailed compilations were to be undertaken they would have confusing meaning in view of the taxonomic problems that exist. These difficulties are summarized below.

The nature of the fossil plant record

1. Fossil plant taxa

Living plants may shed organs for a variety of reasons throughout their life: leaves or shoots may be abscissed naturally or lost through mechanical damage, spores, or pollen shed for reproductive purposes, and seeds or fruits lost for animal or wind dispersal. Then, on death through old age, fungal or animal attack, or mechanical damage, the last remains of the organism may be liberated into the surrounding environment. All of these organs, whether separated through natural abscission, mechanical damage, or death, may then find their way differently into an environment where they may or may not become preserved from further decay. The vast majority of organs do not become preserved and ultimately fossilized, being instead recycled in the environment through decay or animal ingestion and digestion. Indeed, many plants live in environments that are at sufficient distances from sites of potential preservation that virtually ensure that none of their larger organs will ever be represented in the fossil record.

Plant fossil assemblages therefore represent mixtures of plant organs which may have come from a variety of parent plants, while there is a virtual certainty that not all the organs of any one plant will be preserved in that assemblage, having been separated through aerial or fluvial sorting. Morphologists are then faced with the problem of trying to reconstruct the parent plants from such incomplete fragments, while taxonomists attempt to name and classify the remains in a repeatable and usable manner.

Palaeobotanists have always grappled with the problems of classifying fragements of plants and many different ways have been proposed. Some have tried to develop a biologically-based classification system, incorporating nearly every known plant fossil, while others have suggested that many plant organs cannot be used in any evolutionarily meaningful system. For these reasons the particularly palaeobotanical taxa of the form genus and the organ genus have been used; they are not meant to be taken as, or imply anything similar to, genera of whole plants. Nevertheless, classification systems are built up on fragmentary plant fossils as palaeobotanists do their best to integrate the evidence. We can therefore find within an overall classification families based on different plant organs; some may be based on vegetative organs while others are based on reproductive organs or a mixture of both. So when we try to use fossil plants to help interpret evolutionary rates and extinctions we are immediately faced with taxa that have a different nomenclatural status from living ones. Any calculations of evolutionary changes in the number of taxa or the number of first and last appearances must therefore be used very cautiously indeed.

Even with the use of new methods and techniques there can be major disagreements over the best way to classify some plant fossils. It is the total biology of the once living plants that is so vital to us in any discussion of plant extinctions. It is much easier to look for extinctions at the higher levels of plant classification than at the generic level, especially when considering the more unfamiliar extinct groups of pteridophytes and gymnosperms. These large groups, such as Orders, usually reach a period of maximum diversity about half way through their existence as shown by the familiar balloon-shaped representation in summarizing stratigraphic diagrams (for example, Stewart 1983). Hence, it becomes difficult when using data from the fossil record of large groups to monitor rates of evolution and to know at what time to record the extinctions. Within any such generalized graphic portrayal of the whole geological column, or of a large part of it, there will be many changes and extinctions of lower taxonomic units that are hidden by the necessary use of families or even orders. Thus, although genera may become extinct, families may persist. Similarly, in any portrayal of presence at the generic level, changes at the species level tend to be hidden. The rate of extinction of species within the large group is more useful than the time of the single last occurrence, and detailed data of rates are rarely if ever available.

2. *Differential evolutionary rates of plant organs*

There is another important feature of plant evolution which causes difficulty when trying to interpret what we know of the fossil record. When plants evolve the morphology of one or more of their organs may be altered. Vegetative organs seem potentially to be more stable than reproductive organs, especially in angiosperms, but even so may become independently affected by environmental pressures. The disappearance of an organ in the fossil record might then be the result of evolutionary change rather than the extinction of the plant as a whole. This is called heterobathmy or mozaic evolution, in which different organs of plants within a single lineage evolve at different rates.

The most abundant plant megafossils are leaves. They are produced in large numbers during the lifetime of an individual plant and, being relatively robust, are therefore readily fossilized. The morphology of leaves is, however, attuned to efficiency of food production and as such leaves exhibit a strong environmental, and in particular climatic, signal. Changes in palaeobotanical leaf taxa, therefore, not only may be indicative of biological species change, but also provide critical clues as to possible variations in the physical environment that may have brought about the biotic changes. This sensitivity of leaf morphology to environment, a characteristic particularly well developed in angiosperms, compounds the taxonomic difficulties encountered when dealing with detached organs, but fortunately, the detection and utilization of the inherent environmental signal is largely independent of taxonomic assignment and may be used to chart vegetational change.

3. *Taphonomy*

Terrestrial plants produce variable quantities of a variety of organs that have vastly differing predepositional transport and preservation potentials. These taphonomic effects inevitably give rise to biased assemblages and a highly distorted picture of ancient plant life (see Ferguson 1985, and Spicer and Greer, 1986, for recent reviews). A leaf may be one of several hundred thousand leaves produced by one individual tree throughout its lifetime, or a pollen grain one of several million from that same tree. An assemblage of fossil leaves is essentially the product of vegetation local to the depositional site although a variety of discrete communities may be represented. On the other hand, a pollen and spore assemblage may represent both local and regional vegetation, and therefore might be reasonably considered to present a more reliable image of vegetational status at a single moment. Unfortunately, understanding of the relationship between palynomorph assemblages and vegetation is extremely crude for any time-scale prior to the Quaternary because most palynomorphs cannot be related to the megafossils of their parent plant. Leaves and flowers cannot be reworked without obvious detection and are therefore amenable to studies involving high resolution stratigraphy. Fossil wood, fruits, and seeds, and pollen are frequently reworked and their stratigraphic ranges, and therefore patterns of extinction and survival, may become blurred.

4. *Plant migration*

The final major need for care when using plant fossils to monitor plant extinctions is shared by animals. The evolutionary process is strongly influenced by changing environments within the physical world. However, living organisms first have a much simpler way than having to evolve in order to respond to these changes — they can move around. So plant migration is a further factor which causes difficulty when monitoring the first and last appearances of fossil taxa. If a fossil plant taxon is absent from one region the plant is not necessarily extinct: it may have migrated to a more favourable place.

When rapid environmental changes (measured on a biological time-scale) occur it is usually the adaptable (or the fortuitously 'preadapted') rather than the adapted, that survive. This is particularly true of plants because of their limited capacity to migrate rapidly. Subsequent specialization of the survivors in the new environment may be reflected in morphological change which in the fossil record might be interpreted as an extinction of the earlier lineage followed by replacement by the new form. This is in fact a pseudo-extinction because the lineage has been preserved albeit in a modified form. In all but the most complete fossil sequences this event is difficult to demonstrate with the result that the fossil plant record is seen to be inherently punctuated.

The distinction between real extinctions and pseudo-extinctions is an important one. Real extinctions may be seen as evolutionary failures whereas pseudo-extinctions are evolutionary successes. Unfortunately, for biostratigraphic purposes this critical difference is usually overlooked and the disappearance of a particular morphology from the fossil record in favour of one more suited to the new environment is regarded as an extinction.

5. The origin of new groups and reticulate evolution

Reticulate speciation, as recognized in animals by Sylvester-Bradley (1977), typically is thought to occur during the initial radiation of groups of organisms. It is the result of high levels of successful hybridization during geographic expansion and it produces a large number of forms with intergrading morphologies.

In the early phase of plant diversification following the colonization of land in the late Silurian, there were great numbers of morphologically similar and genetically simple plants which must have been capable of hybridizing. Any genetic change could have resulted in relatively large scale morphological changes. Also, in the case of angiosperms the evolution of novel, relatively unspecialized, reproductive morphologies coupled with high levels of polyploidy resulted in extreme hybridization during the middle and late Cretaceous poleward spread of angiosperms. The possible early development of leaf polymorphy in response to novel habitat exposure during this expansive phase may have added to the melange of leaf forms that confused early attempts at angiosperm leaf taxonomy (Spicer 1986). Unfortunately, nomenclatural practices tend to perpetuate false assignments, obscure subtle but critical character combinations, and generate inflated numbers of organ taxa. Thus, rates of true speciation (as understood from studies of perhaps relatively stable extant organisms) and extinction are exaggerated. Although so far best demonstrated in angiosperms, reticulate evolution is likely to have occurred in other groups to a greater or lesser extent throughout the Phanerozoic, and at present the effect of this on the record of extinction and survival cannot be meaningfully assessed.

Tiffney (1981) attempted an overview of plant diversity through time based on species citations, mostly using data from North America. Each citation records the presence of one species at one locality at one point in time. Tiffney was able to relate diversity to samples from non-marine sediments laid down at any time. His data also show that plant diversity, as measured by the citation of species, is lower than might be expected for North American Permian, Triassic, and Jurassic rocks. Although there are fewer rocks of these ages than others exposed in North America the real reason for the drop in diversity might well be the lack of suitable depositional environments. Parrish *et al.* (1982) have argued that during these times when large areas of North America experienced arid climates there were few environments suitable for deposition. If a similar exercise were to be carried out on British rocks the Jurassic might well appear to have been a period of high diversity due to the abundance of plant remains in the deltaic deposits of Yorkshire and the rich literature which describes them.

Extinction and survival patterns through the Phanerozoic

1. *Background and methodology*

Originally geological boundaries were established where changes in the fossil record were perceived to occur. Biostratigraphy remains the single most important method of stratigraphical correlation. Variations in the number of taxa preserved in the fossil record may be due to biological extinction and evolutionary change, migrations relative to depositional environments, or periods of non-deposition throughout the habitat range occupied by the taxa in question. Whatever the cause of apparent taxonomic change, once a boundary is established it tends to become self sustaining and there is a serious risk that synchronous loss of taxa, an artificial 'mass extinction', may be noticed. This occurs due to the generally poor resolution and continuity of the sedimentary record and the extension of the stratigraphic range of taxa to the next convenient existing boundary irrespective of their actual occurrence. Such an extension of the range of taxa beyond the first and last actual occurrences may be deliberate or unintentional, but the result is an apparent mass biotic change at time-stratigraphic boundaries. In order to resolve the exact timing of biological change, and thereby detect true extinctions, pseudo-extinctions, and synchroneity of change across a broad biological spectrum, two important criteria should be sought. Firstly, detailed sedimentological investigations will help establish the continuity of sedimentation and the relative timing of biotic change. Secondly, geographically separated boundary sections are best correlated by evidence additional to the biological, otherwise circular reasoning will enhance the boundary and negate the value of the study. So the interpretations presented here are but a start and are offered to help other specialists assess their worth in comparison to their own work.

Several important periods of plant evolution occur during the Palaeozoic resulting in the appearance of all the major groups (of phyla) except the angiosperms. Thus, from the first records of land plants we can monitor changes in the composition of the genera or species making up the succeeding floras. Much effort has been put towards recognizing distinct plant assemblages that can be used as the basis for floristic zones and can be used as a means to correlating strata. This, of course, relies on a recognition of both the first appearance of plant taxa and their subsequent point of disappearance from the fossil record. Some species or genera persist longer than others so we have to deal with a series of overlapping stratigraphical distributions when trying to construct these floral zones. These disappearances of taxa are often taken as representing extinctions although, as we have already indicated, the 'extinction' of a plant organ rather than that of a whole plant instead might be the result of some evolutionary change within a species.

The fossil record of pollen and spores from vascular plants shares these features of obfuscation. As well as problems connected with taxonomy and preservation, pollen and spores also respond very sensitively to different catchment regimes. Different plants produce different quantities of pollen or spores which also confuses the record left by these fossils. Nevertheless, very large quantities of pollen and spores are preserved in many sediments and are used to deduce patterns of palaeoecology, plant migration, and biostratigraphy. The rich data from palynology is based on very large numbers of specimens, sometimes exceeding 100 000 in a single study, and may be of use to help interpret evolutionary processes. Not least of the difficulties to be aware of in such interpretations though is the different attitude to observation and taxonomy by different authors. Indeed, the palynological data currently available, voluminous though they may be (especially the results of pollen counts in many dissertations), has not been prepared with the aim of making a comparative synthesis of the history of plant evolution from the Silurian to the Pliocene. Rather, most palynological work aims to resolve problems of biostratigraphy and palaeoecology through relatively short periods of time, usually in a small geographical area.

Applied palynology in particular rigidly adheres to the principle of 'tops and bottoms'—the identification of the stratigraphic ranges of species and other fossil taxa. Little thought is given to the validity or evolutionary implications of this principle despite the clear influences of plant migration and evolution. In this article we are attempting to make interpretations about plant evolution from data that was largely devised to help the biostratigrapher. The data we have considered are from the work of different authors who follow a variety of different paradigms. Together with the innate limitations of the fossil plant record outlined above this makes the comparison of first and last appearances presented by the authors of dubious value. At least this analysis may encourage microscopists to make more appropriate observations for studies of evolution processes in the future.

We have taken what we believe to be reliable up-to-date charts of stratigraphic range, from each Period from the Silurian to the Neogene. Unfortunately, the charts vary in detail, particularly in the geographical region they represent and in their completeness of the palynoflora, but each does contain those taxa believed to start or stop in the particular region studied. Each set of data shows: (i) the time of first appearance (or 'bottom' of the applied palynologists, or 'origin' of some biologists — terms questioned here due to the lack of evidence of origin of other plant parts and of the geographical centre of origin); and (ii), the time of last appearance (or 'top' of the applied palynologist or 'extinction' of some biologists). The times of the first and last appearances are shown in our tables, at least to the level of stage and usually also to mid-stage. Tables 1.1–1.7 summarize the total first and last appearances at stage boundaries and at mid-stage. The tables also summarize the total number of species present during each stage.

In most of the data sets included here, there is a discrepancy between the number of taxa at the very top and bottom of each set; for example, at the top of the Palaeogene data set (Table 1.7) there are seven taxa and at the base of the Neogene data set there are 33 taxa. These differences are partly due to the different attitudes of the different authors and partly due to the number of taxa excluded due to uncertainty. Particularly from the Permian onwards many taxa are long ranging and have been omitted from the original author's charts altogether. Also, of course, some data sets come from quite different regions. As far as this present analysis is concerned interpretations from each data set should be considered separately and no reliable significance can be drawn about the boundaries of each Period.

We have also made subjective judgements from the literature of the megafossil plant record about the times of likely major events in plant evolution. These are also indicated on Tables 1.1–1.7 which thereby summarize the total palaeobotanical record as shown by our limited data and approach.

2. Late Silurian and Devonian

Megafossils Banks (1980) has revised the earlier attempts to recognize plant assemblages in the late Silurian and Devonian and has presented seven generic assemblage-zones based on the time of appearance of plant megafossils. These have been related to the palynological zones of

Richardson and McGregor (1986). Most genera that Banks included had very short geological time-spans although some last for considerably longer. At first sight, the plotting of stratigraphic ranges of these genera indicates a general and gradual series of overlapping ranges. However, there are recognizable periods of concurrent appearances of genera and others of apparent concurrent extinctions. Sometimes these concurrent extinctions coincide. This is most noticeable at the Emsian/Eifelian boundary where there is the first appearance of five genera and the disappearance of nine others.

Palynology Richardson and McGregor (1986) have recently published a detailed range chart of spore taxa as part of the zonation of the Old Red Sandstone Continent. There are 171 species included. Their zones are compared to the floral zones based on megafossils (Banks 1980) and with selected 'events' in the evolution of spore morphology.

There are two peaks in the total-species curve of these data in Table 1.1, the first in the Pridoli Stage being much less than that in the Givetian, about 25 Ma later. Of course, both are preceded by large numbers of first appearances and succeeded by more last appearances. As with similar patterns higher in the stratigraphic column this suggests two unequal periods with high rates of evolution of new taxa, followed by competition between them for survival in possibly new environments, followed by extinctions of those taxa that failed. The Givetian peak appears from this interpetation to be the most important, but that is perhaps because there were more plant taxa then.

When this palynological evidence is set against comparable data from the megafossil record (Banks 1980) the same peak of total taxa is seen in the Givetian Stage, and the patterns of first and last appearances are similar too. Such comparison is easiest in this part of the Palaeozoic because there were fewer plant species giving the palaeobotanical record less background noise from taxonomic and catchment effects. Table 1.1 shows the interpretation from Banks's range charts of 64 megafossil plant genera.

3. *Carboniferous*

Megafossils A great number of stratigraphic studies have been made on Carboniferous plant fossils, especially those of the Upper Carboniferous (Pennsylvanian). In the equatorial belt the climatic conditions led to widespread coastal shelf areas which favoured swamp and peat formation. These then became important sites for their preservation. For this reason most of our knowledge of Carboniferous plants is of those which grew in these paralic swamps. In contrast, we have only a limited knowledge of the

Table 1.1

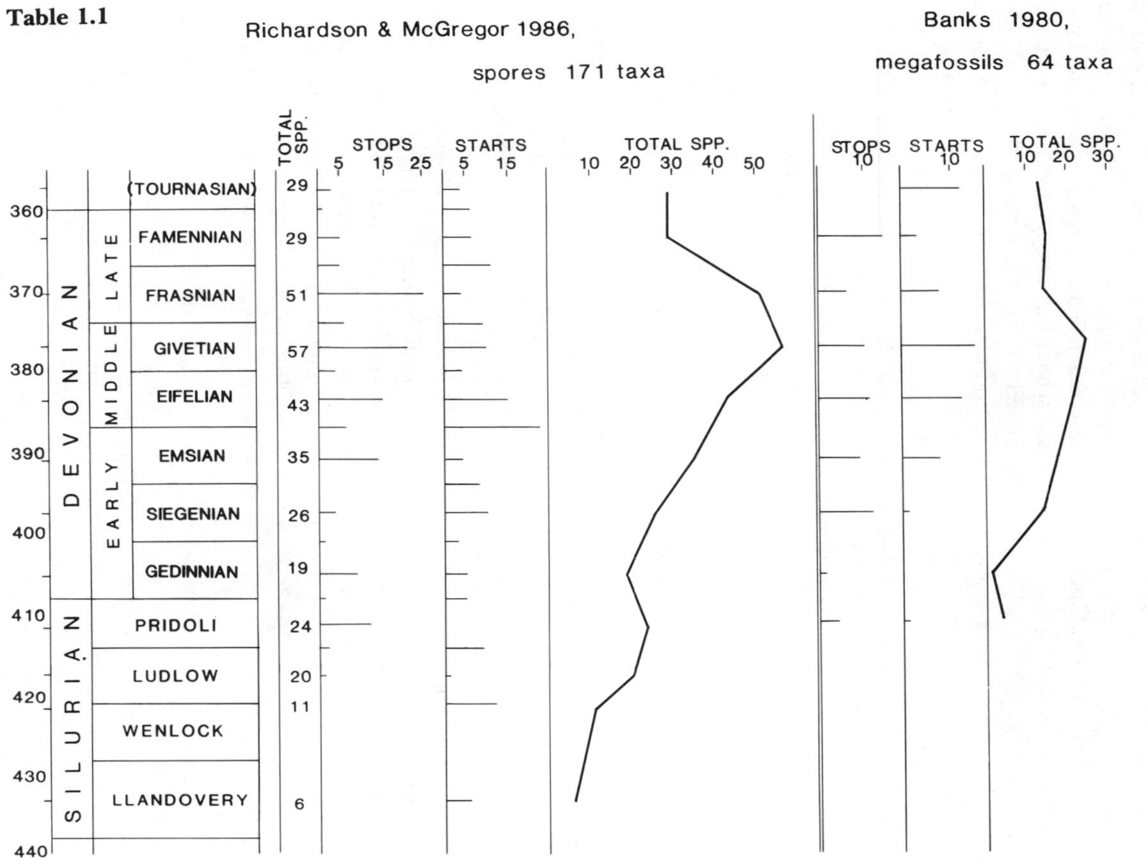
Richardson & McGregor 1986,
spores 171 taxa
Banks 1980,
megafossils 64 taxa
TOTAL SPP.
STOPS 5 15 25
STARTS 5 15
TOTAL SPP. 10 20 30 40 50
STOPS 10
STARTS 10
TOTAL SPP. 10 20 30
(TOURNASIAN)
FAMENNIAN
FRASNIAN
GIVETIAN
EIFELIAN
EMSIAN
SIEGENIAN
GEDINNIAN
PRIDOLI
LUDLOW
WENLOCK
LLANDOVERY
LATE
MIDDLE
EARLY
DEVONIAN
SILURIAN
29
29
51
57
43
35
26
19
24
20
11
6
360
370
380
390
400
410
420
430
440

'upland' floras which grew above the influence of the swamp water table (e.g. the flora of Rock Island, Illinois, described by Leary 1981).

Changes in the composition of Namurian megafossil assemblages have been the subject of much discussion since Kidston (1894, 1923) maintained that he recognized a floral break corresponding to a time gap in the strata of the British Isles. Gothan (1931) also described a sharp break in the floral sequence of the Upper Silesian basin which has been subsequently shown by Havlena (1982) to lie in an unbroken succession of strata.

The position and relevance of this mid-Carboniferous break is of course important in the recognition of the Mississippian–Pennsylvanian boundary. Bouroz *et al.* (1978) tentatively placed the boundary below the Kinderscoutian, attributing the event to the start of the Gondwana Ice Age. Wagner (1982), in his summary of floral changes near the Mississippian–Pennsylvanian boundary, suggested that this climatic event probably has a more immediate effect on land plants than on marine faunas. This implies that extinctions probably played an important part in this floral change. Jennings (1984) has also implied that the Mississippian–Pennsylvanian boundary in Illinois can be recognized by the simultaneous extinctions of several plant forms and the simultaneous appearance of several others.

In his division of the Carboniferous into 16 floral zones, Wagner (1984) stresses that the Mississippian–Pennsylvanian transition is fairly gradual, but that the paucity of records makes it uncertain whether it should be at the level of the Chokerian or at the base of the Kinderscoutian. An examination of Wagner's stratigraphic charts reveals a period of disappearances immediately before the Kinderscoutian followed by first appearances in the Kinderscoutian–Yorkian boundary. The most precise of Wagner's zones are in the Upper Carboniferous although even here some of the boundaries are often rather arbitrary and there is no absolute coincidence in all cases with the stage boundaries of the Westphalian and the Stephanian. Again it seems from the range charts that there is a gradual appearance of new taxa followed by a disappearance (maybe extinction) of others (Table 1.3).

Zoning of the Carboniferous in this way does not reveal the whole complex picture of plant survival, evolution, and extinction, for there is the further factor of migration. There is increasing evidence to suggest a northward migration of many plant groups throughout the generally ameliorating climate in the northern middle and high latitudes (Raymond 1985). Through such migrations some of the earlier types of pteridophytes of the Lower Carboniferous of Eurameria can be found in the Upper Carboniferous of Angara (Meyen 1966, 1982). Without the knowledge of these latter floras we would believe these forms to be extinct once they disappeared from the Euramerian fossil record.

Other work on Carboniferous plant fossils concentrates on palaeobotany rather than stratigraphy. Phillips *et al.* (1985) have analysed quantitatively Pennsylvanian coal swamp vegetation through their combined use of anatomically preserved plants in coal balls and dispersed spores. By this means they related changes in plant communities to climatic differences. Thus, the extinction of the cosmopolitan pteridosperm *Lyginopteris* and the rise in numbers of the cordaitalean gymnosperms just above the Westphalian A–B boundary was shown to be related to the beginning of what they recognized as a first drier interval. This corresponds to the junction between the *hoeningnausi/schlehani* and the *rugosa/urophylla* zones of Wagner (1984). Similarly, the onset of the second drier interval at the beginning of the Stephanian brought about the extinction of the arborescent lycophytes *Lepidodendron*, *Lepidophlois*, and *Paralycopodites*, and several species of the tree fern *Psaronius*, corresponding this time to the junction of Wagner's *cantabrica* and *lamuriana* zones. These changing conditions of the second drier interval were, however, quite complex because there was a short interval where other lycophytes and other plants dominated the swamps before the establishment of a *Psaronius* dominated community.

The major pteridophytes such as *Sigillaria* and *Psaronius* that survived the drying of the swamps must have been adapted to these new conditions for they survived in the limnic and paralic coal basins of the Stephanian palaeotropical belt. It was not until the final disappearance of these swamps that these also finally succumbed and became extinct. It was the plants of the surrounding areas just on and above the water table that then started to form the dominant plants of the subsequent fossil floras.

Palynology Recently, Clayton (1984) presented the ranges of 48 species from the Viséan and Tournasian of the British Isles, and made comparisons with the miospore zonations of Neves and others. The last appearances in his range chart are particularly associated with the Holkerian, though they also occur at the Asbian–Brigantian boundary. Each of these phases is preceded by an increase in the number of first appearances. Table 1.2 also shows a steady rise in the total number of species from the base of the Tournasian to the top of the Viséan, though the large number of last appearances during the Holkerian does cause a fall at that Stage.

Butterworth (1984) has presented the ranges of 349 species based on her extensive experience of British Westphalian and Namurian sections. Her range charts correlate the miospore occurrences with the goniatite zones of Ramsbottom and with the miospore zones of Clayton *et al.* Our summaries of total species and the first and last appearances of taxa are shown in Table 1.3.

There are maximum first and last appearances as well as total species

Table 1.2

Clayton 1984
48 SPP.

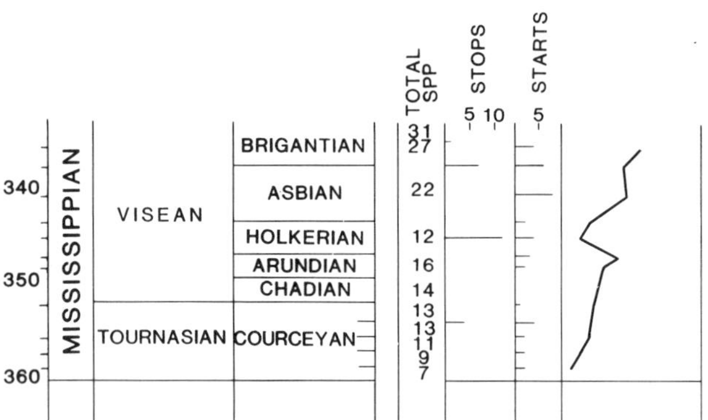

numbers at the Westphalian A, and all three reach maximum peaks again during the Arnsbergian. Once again, final appearances tend to be preceded by waves of first appearances.

4. *Triassic/Permian*

Megafossils Many genera of Triassic plant megafossils have very long ranges. Some such as *Todites, Phlebopteris, Clathropteris,* and *Williamsonia* are also known from the Jurassic, others such as *Neocalamites, Cladophlebis, Otozamites, Peloudea,* and *Brachyphyllum* extend to the Cretaceous, while the broadly based wood genus *Dadoxylon* and the equally broad conifer shoot genus *Pagiophyllum* even extend into the Tertiary (Ash 1972).

In contrast there are some genera which have extremely local distributions and ranges. On the basis of these, Ash (1980) has divided the Upper Triassic of North America into three floral zones. The oldest recognizable he termed the zone of *Eoginkgoites* (middle Carnian), the next is the zone of *Dinophyton* (late Carnian, ?Norian), with the un-named upper zone extending into the Lower Jurassic (?Rhaeto–Liassic). On the present evidence the diversity of the floras steadily increases throughout the Upper Triassic with disappearances (extinctions) and appearances marking the boundaries of the three zones.

Palynology G. Warrington (*pers. comm.*) has studied the ranges of about 600 taxa from this interval that have been described from well dated sections in Europe. Of these, he has justification to show 71 within a range chart (Warrington 1984) and these are mainly from alpine Triassic sequences — the remaining 500 or so range through the Triassic. As in

Table 1.3

Butterworth 1984, miospores 349 spp.
STOPS
10 20 30
STARTS
10 20 30 40
TOTAL SPP.
80 100 120 140 160 180
Wagner 1984, megafossils 70 spp.
STOPS
5 15
STARTS
5 15
TOTAL SPP.
10 20 30 40
STEPHANIAN
WESTPHALIAN
D
C
B
A
YEADONIAN
MARSDENIAN
KINDERSCOUTIAN
ALPORTIAN
CHOKIERIAN
ARNSBERGIAN
PENDLEIAN
BRIGANTIAN
290
300
310
320
330
340

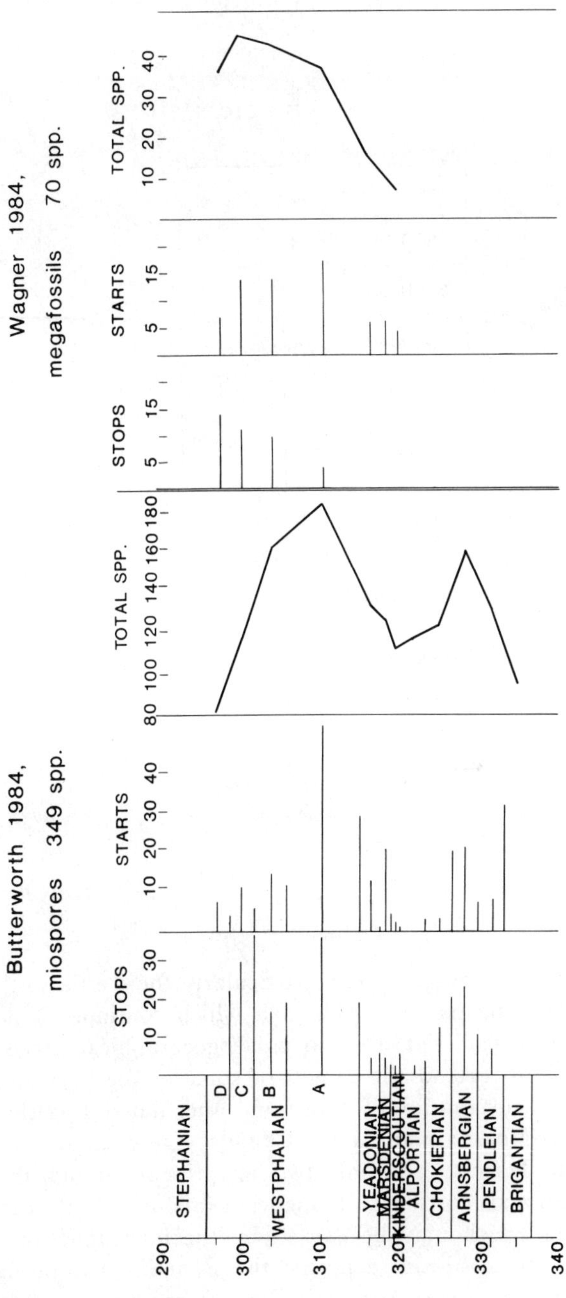

Table 1.4

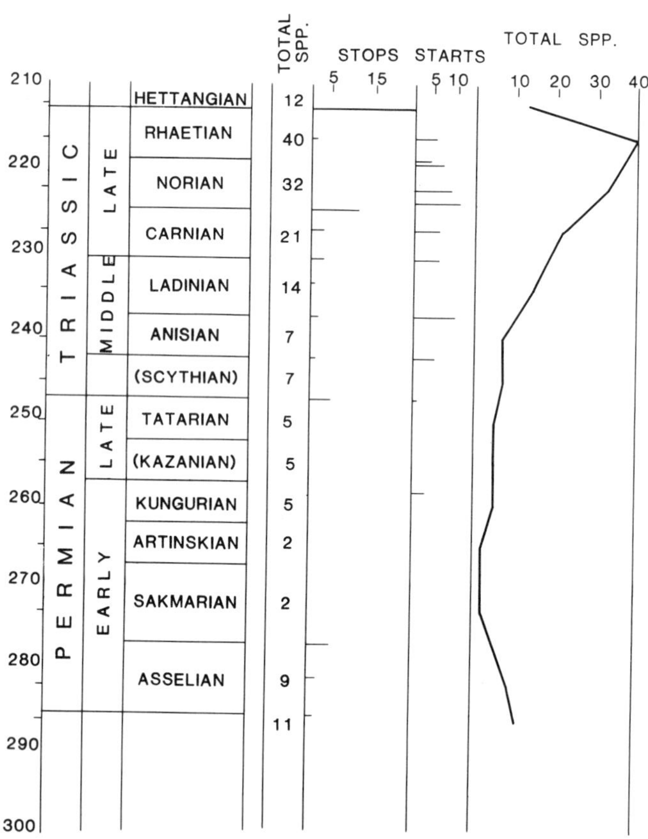

other sequences mentioned here, particularly the Tertiary, there are very many long ranging palynomorph taxa which continue right through the interval being considered. Due to palaeogeographical effects Warrington has been able to extend the ranges of some of the 71 taxa with limited range by the incorporation of data from other dated Triassic sequences in Greenland, Svalbard, and Arctic Canada. Only 12 of the species are present in the Permian and only two are present through the Sakmarian, the Artinskian and most of the Kungurian (about 20 Ma). Table 1.4 shows other features which may be caused by the inevitable concentration by palynologists on fossil-rich deposits: the Rhaetian has been very widely studied and is shown here to contain the maximum total number of species. The curve shows clearly that first appearances precede final ones.

5. *Jurassic*

Megafossils The Jurassic was a time of climatic equability and this is reflected to a large extent in the megafossil record. Global vegetation was dominated by gymnosperms and ferns with cycads, cycadeoides, conifers, ferns, pteriodosperms, and ginkgophytes undergoing gradual evolutionary change largely in response to developments in reproductive biology. Increasing levels of sophistication in biotic pollination were being developed by groups as diverse as cycadophytes, pteridosperms, and perhaps even in the coniferous Cheirolepidaceae. Strong selection in favour of ovule enclosure, rapid completion of the reproductive cycle, and insect pollination (whole characteristics were later most successfully developed by the angiosperms) brought about parallel and convergent evolution in such groups as the pteridosperms (e.g. *Caytonia*), cycadeoids (Thomas and Spicer 1987) and the Gnetales (G. R. Upchurch and P. R. Crane, *pers. comm.*).

The largest group to become extinct in the Jurassic was the southern hemisphere Glossopteridales, but this was the termination of a gradual decline that began in the Early Triassic.

There was little change in the physical environment and few extinctions above the generic level are evident in the megafossil record. The overall pattern is probably one of gradual turnover of taxa in response to competition from evolutionary novel forms.

Palynology In this present study we have considered two sets of data from Jurassic sediments, both long sections from borehole material covering the greater part of the Period. Guy-Ohlson's (1986) study is from the Vilhelmsfalt Bore No. 1, in Scania, southern Sweden, and extends from the Toarcian to the Aalenian. The stratigraphic ranges of 87 taxa are shown and each is described and illustrated. Filatoff (1975) has made a substantial study of Jurassic palynology of five boreholes from the Perth Basin in Western Australia. His range chart includes 85 taxa; a tentative comparison of the Australia sections to European Stages suggests that they may extend from the Pleinsbachian to the Kimmerridgian although this attempt by Filatoff to compare the Perth stratigraphy with that for Europe is offered by him very tentatively.

The inclusion of two such geographically widely separated locations of Jurassic data gives a useful opportunity to test the methodology being used here. The interpretations in Table 1.5 show that first appearances precede last appearances in both sets of data. However, the times of each peak are quite different, as are the times of maximum species through the Early Jurassic in both Sweden and Australia, as well as a decline in the Late Jurassic. The time differences in the Middle Jurassic may be real, due to patterns of plant geography, or they may be due to inaccurate correlation of the Australian sections with European Stages.

Table 1.5

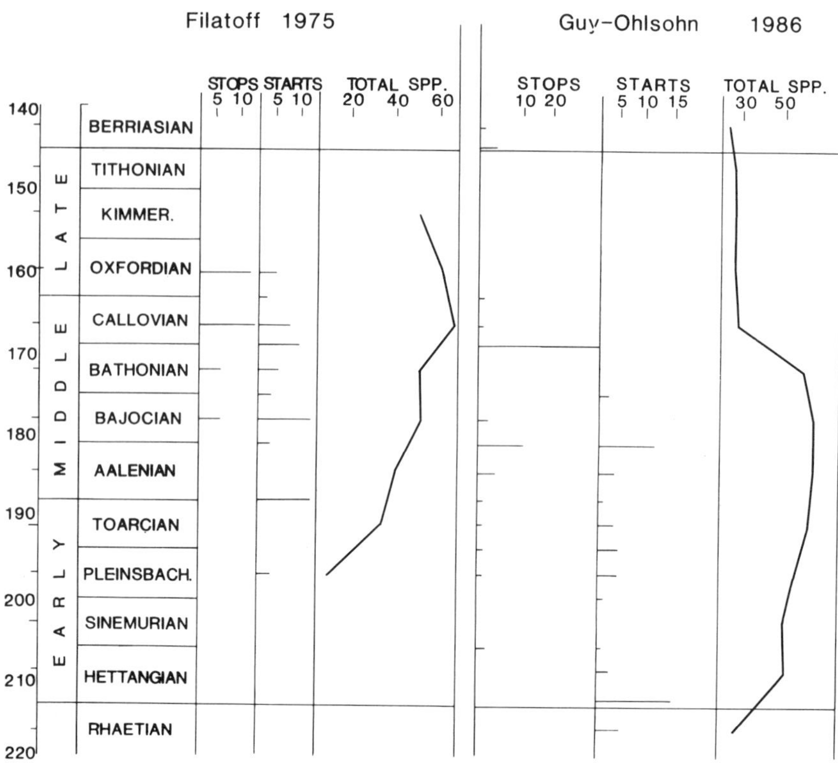

6. *Cretaceous*

Megafossils The advent of the angiosperms in the Early Cretaceous led to marked changes in vegetation, albeit slowly at first. The origin, or origins, of the angiosperms are still obscure but they are thought to have arisen at low latitudes (Axelrod 1959; Retallack and Dilcher 1981). They had many advantageous characters, both vegetative and reproductive, and passed a character combination threshold that enabled them to dramatically outcompete other plant groups. Beginning in the Aptian they underwent a global expansion and by the end of the Albian they had become established as a major component of vegetation in disturbed sites, even in polar environments (Spicer and Parish 1986). It is only following this radiation that we see a decline in other plant groups as the primitive angiosperms, adept in the role of early seral colonizers, continued to evolve and subsequently compete in mature forest ecosystems. Thus, the cycadeoids suffer a major decline in the Early Cretaceous to eventual extinction by Coniacian times while the pteridosperms disappear by the middle of the

Cretaceous. Even the conifers suffered a major loss with the gradual decline and disappearance of the Cheirolepidaceae: a family that prior to the evolution of the flowering plants had occupied a wide range of growth environments (Alvin *et al.* 1981; Upchurch and Doyle 1981; Alvin 1982). With a lower equator-to-pole temperature gradient than at present, and relatively minor global climatic changes throughout the Late Cretaceous, these attentuations and eventual extinctions of major groups must be attributed to inter-plant competition.

Palynology Srivastava (1981) summarized the stratigraphic ranges of 84 selected spores and pollen from the Fredericksburg Group (Albian) of the southern United States. Five species are omitted from the complete assemblage described by Srivastava in 1977 and 22 of these 84 species are restricted to the Albian and are excluded from this analysis to reduce the bias caused by the data coming from one location. The remaining 62 species do show a bias to an Albian maximum, partly no doubt, because they all occur in the same Fredericksburg Group.

The interpretations from these data (Table 1.6) shows very clearly that the first appearances of species precede the last appearances of other species, with the maximum number of species occurring at the Albian.

7. *The Cretaceous–Tertiary Boundary*

If the preceeding accounts of the plant fossil record have been somewhat negative the usefulness of palaeobotanical research in extinction and survival studies is demonstrated more successfully at the Cretaceous–Tertiary (K–T) boundary. The K–T boundary has attracted considerable attention in recent years largely because of the debate surrounding a possible bolide impact and its biotic consequences (Alvarez *et al.* 1980). Marine boundary sections have been examined in detail (e.g. Smit 1982; Surlyk and Johansen 1984; Alvarez *et al.* 1984) and the boundary itself is generally characterized by an iridium anomaly (and often shocked quartz grains: Bohor *et al.* 1983) in a boundary clay associated with a marked biotic change. The iridium anomaly appears to have a global occurrence, but much more work needs to be done to determine the presence or otherwise of iridium-rich horizons elsewhere in the geologic column. Iridium-rich boundary clays have also been identified in a number of non-marine sections, particularly in North America (Orth *et al.* 1981; Hotton 1984; Nichols *et al.* 1986; Pilmore *et al.* 1984; Smit and Van der Kaars 1984; Tschudy *et al.* 1984). Data from both palynological and megafossil assemblages demonstrate that terrestrial vegetation was profoundly affected by sudden environmental disturbance at the boundary, but the patterns of extinction are very different from those in animal groups.

Table 1.6

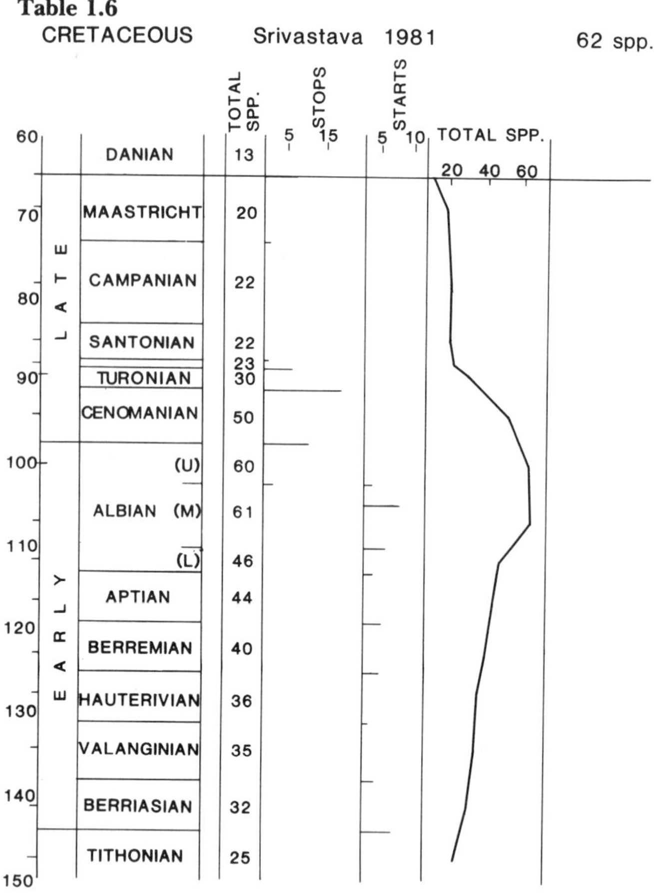

Megafossils A picture of terminal Cretaceous plant disturbance is beginning to emerge from the studies of plant megafossils. Although Hickey (1981) detected no mass plant extinctions at the boundary his work is questioned because he was not studying complete boundary sections. On the other hand, Wolfe and Upchurch (1986) have carried out detailed studies of leaf assemblages at a large number of apparently intact Western Interior boundary sections. Instead of concerning themselves with the taxonomic problems inherent in studies of leaves of this age they have studied the environmental and vegetational signals in the assemblages by taking into account foliar physiognomy and wood anatomy data. In addition to megafossils dispersed leaf cuticles were sampled at 1-cm intervals across the boundary.

Wolfe and Upchurch recognize five vegetational-floristic phases in the Raton Basin boundary sections that serve as a general comparison model for sections elsewhere (Fig. 1.1).

Phase 1 (Lancian age, latest Cretaceous) is characterized by broad-leaved evergreen vegetation with high diversity. Leaves tend to be small in size with thick hairy cuticles and very few have drip tips. This type of foliar physiognomy indicates a dry vegetation which persists up to the boundary clay. Evergreen conifers are also present in this megathermal (mean annual temperatures greater than 20°C) vegetation.

Phase 2 is immediately above the boundary and consists entirely of leaves and rhizomes of a fern species morphologically similar to those of extant *Stenochlana* and cuticles typical of herbs.

Phase 3 extends up to 2 m above the boundary and is typified by a depauperate flora of large leaves with drip tips and thick smooth cuticles. Overall, the leaves suggest early successional vegetation in an environment of high precipitation.

Phase 4 extends throughout the next 200 m of the section and is characterized by an increasing, but still low leaf diversity. Physiognomically, a warm and humid environment is indicated.

Phase 5 extends through the next 150 m of the section and indicates a low diversity megathermal rainforest.

The megafossil pattern of floristic change at the K–T boundary suggests, like the pollen and spores, an ecological trauma followed by a steady recovery that mimics normal seral succession, but over a timescale of perhaps 1.5 million years. The immediate post-boundary vegetation is fern dominated just as is modern vegetation that is devastated by volcanic eruption (Spicer *et al.* 1985), but that is not to say that the K–T boundary trauma was volcanic in origin. It is also evident that although the long-term thermal regime appears to have been little affected there was a significant increase in humidity, probably caused by precipitation, that lasted well into the Paleocene. This pattern of increased precipitation is not confined to the Raton Basin and is seen elsewhere in North America (Fig. 1.2), Egypt and India. Whatever the cause of the boundary event the global climate appears to have been shifted into a new metastable mode.

Leaf data from 66 collections at 8 localities ranging from the Mississippi Embayment to Alberta (Fig. 1.3) also show a major shift in vegetational patterns at the boundary. The latest Cretaceous vegetation at palaeolatitudes less than 65°N appears to have been rich in evergreens. However, the post-boundary recovery vegetation was essentially deciduous even at low palaeolatitudes. In the Raton Basin 75 per cent of leaf taxa become at least locally extinct at the boundary and extinction rates both in the Raton and Denver Basins are highest in the evergreen taxa. In central Alberta only 24 per cent of leaf taxa became extinct although the gymnosperms were

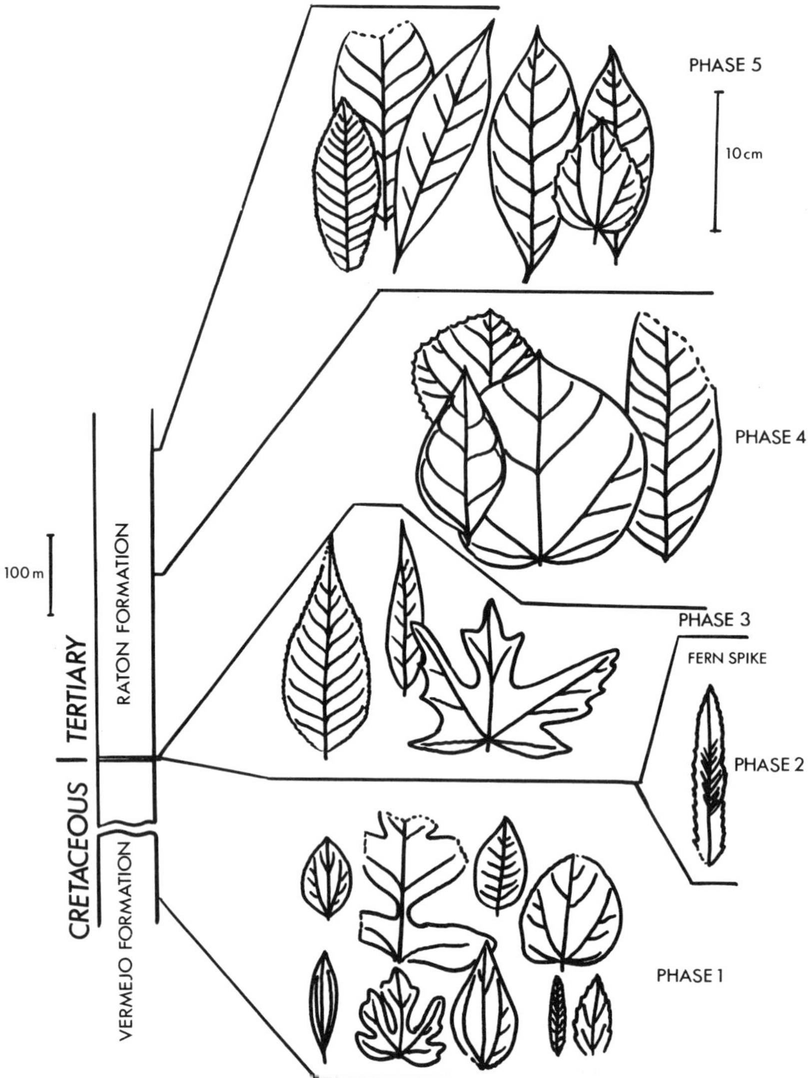

Fig. 1.1. Changes in leaf characteristics across the K–T boundary in the Vermejo–Raton sequence. Following the 'fern spike' there is an increase in leaf size and the depauperate early succession leaf forms of phase three gradually give way to a more diverse assemblage typical of a low diversity rain forest (J. A. Wolfe, *pers. comm.*).

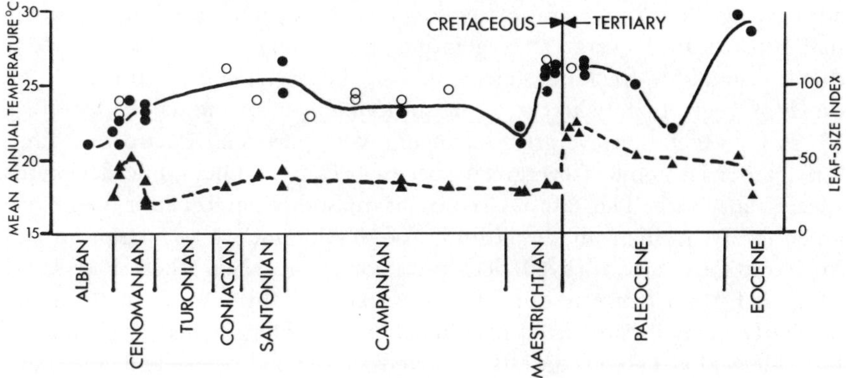

Fig. 1.2. Percentage of entire-margined leaf species and average leaf size for late Cretaceous and Tertiary lower–middle palaeolatitudes of North America. An increase in leaf size (dotted line) is seen at the K–T boundary, but following a Maastrichtian increase in mean annual temperature (solid line) no major temperature change can be resolved at the boundary. Leaf margin percentages have been standardized using inferred latitudinal temperature gradients to compensate for possible biases induced by selection in favour of deciduousness. Open circles: assemblages consisting of less than 30 species. Solid circles: assemblages consisting of more than 30 species. Redrawn from Upchurch and Wolfe (1987).

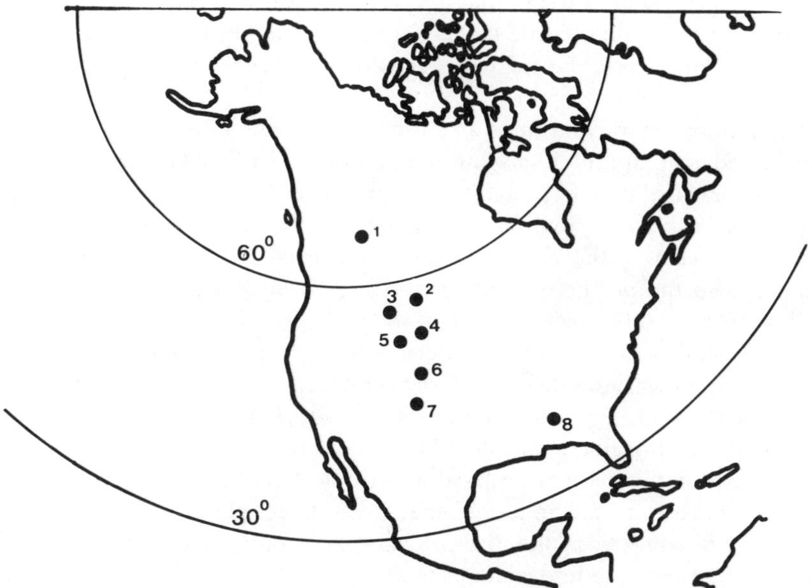

Fig. 1.3. Palaeocontinental map of the Paleocene showing positions of 8 leaf localities used by Wolfe and Upchurch for plotting late Cretaceous—early Tertiary vegetational changes. 1, Coalspur; 2, Hell Creek; 3, Bighorn; 4, Lances; 5, Medicine Bow; 6, Denver; 7, Raton; 8, Nabarton.

strongly affected. In general extinction was most pronounced in megathermal broad-leaved evergreen vegetation and lowest in broad-leaved deciduous vegetation. Deciduous elements had the lowest extinction rates in all types of vegetation, whereas in mesothermal (mean annual temperature 13–20°C) vegetation evergreen elements were particularly hard hit. Similarly, amongst conifers evergreen species became extinct while deciduous species survived. The latest Cretaceous mesothermal broad-leaved evergreen forests of high middle latitudes of North America were transformed to dominantly broad-leaved deciduous forests in the earliest Paleocene. Maastrichtian evergreen elements were replaced by taxa from the more northerly polar broad-leaved deciduous forests. The expansion of range of broad-leaved deciduous forests occurred in the Paleocene and also gave rise to an increase in genetic diversity. This is seen most strongly in the Junglandaceae, but overall a three-fold increase in dicotyledon families represented in the polar broad-leaved deciduous forest occurs during the Paleocene (J. A. Wolfe, *pers. comm.*). The increase in diversity occurred after the inoculation of mid-latitude vegetation by northern taxa as there is no evidence of major clades originating at high latitudes (Spicer *et al.* 1987).

The ability of some plants to enter dormancy appears to have conferred a major advantage at the boundary and strongly implies a short duration cold/dark excursion, during which evergreen taxa were adversely affected, before a new stable 'wet' climate regime was established. Such an interpretation clearly favours the possible existence of a 'post-impact winter'.

The reduced extinction rates at latitudes less than 45°N probably reflect the limited magnitude of the low temperatures. In tropical regimes such low temperatures may not have been experienced so that the megathermal vegetation survived. Southern hemisphere vegetation was much less severely affected at the K–T boundary and even today is evergreen-rich compared to the northern hemisphere. Evergreen refugia in mesothermal vegetation must have been because of the prolonged regional effect, and the successful and persistent influx of deciduous elements. A viable deciduous-rich ecosystem was established before megathermal evergreens could evolve into mesothermal vegetation. However, increased Paleocene precipitation in megathermal regimes may have been a critical factor in the origin of angiosperm-dominated paratropical and tropical rainforests.

The increase in precipitation may also have played a role in the extinction of some taxa, but it would not have brought about the overall pattern of an increase in deciduousness.

In spite of the apparent devastation of the vegetation 'standing crop' at the time of the boundary event, and attendant ecological and environmental trauma, most plant lineages including many evergreens were able to pass through the boundary and evolve in new directions in the changed post-boundary conditions. The key to plant success even under severe environmental stress undoubtedly lies in the ability of plants to enter dormancy either in a mature state (deciduousness or dying back to a perennating organ such as a rhizome) or in seed form. The exposure of buried seeds during increased post-boundary erosion of a denuded landscape would have quickly reestablished most such lineages. Most animals, on the other hand, are poorly equipped to survive prolonged adverse conditions.

Many boundary sections yield an abundance of fusain or other evidence of post-boundary fire (Tschudy and Tschudy 1986; Saito *et al.* 1986; Wolbach *et al.* 1985) and the suggestion has been made that wildfires were started as the result of the presumed meteoric impact (Wolbach *et al.* 1985). Such circumstances are difficult to envisage because at the time of such an impact most vegetation would have been living and with a high moisture content. Wet plant matter does not burn well. It is more likely that numerous wildfires started in post-event dead and dry vegetation as the result of frequent lightening strikes produced in a destabilized atmosphere. In a post-event world the recently killed and relatively desiccated forests would have provided ample fuel and would have been easily ignited.

These fires probably had little effect on land plant extinction because the distribution of fusain in the Late Cretaceous fossil record suggests that wild-fires were common and probably were an integral element in shaping Late Cretaceous vegetation.

Palynology The most widely recognized plant fossil boundary indicator is a sudden change in the palynological signal. Typically, Cretaceous palynomorphs suffer a severe attenuation in abundance and diversity at or near the boundary clay followed by a sudden rise in spore abundance in the immediate post-boundary sediments: the so-called 'fern-spike' (Fig. 1.4). Thereafter, the proportion of spores gradually declines up the section as pollen increases once more in abundance and diversity, but this time with characteristic Paleocene forms (Tschudy and Tschudy 1986, and references therein). This pattern of plant microfossil change was in large part recognized in the Western Interior of the United States long before the bolide theory was proposed (e.g. Leffingwell 1971; Tschudy 1971) and a formula for recognizing the boundary using plant remains (the lowest persistent lignite zone overlying the highest occurrence of dinosaur remains) was proposed by Brown in 1962. So consistent is the palynological signal that it is sometimes used to define the boundary in the absence of corroborating data (Tschudy 1973).

The most detailed palynological studies have been carried out in the Raton and Denver Basins, the Hell Creek, Lance Creek, and Fort Union Formations of the Western Interior, and represent a series of complete

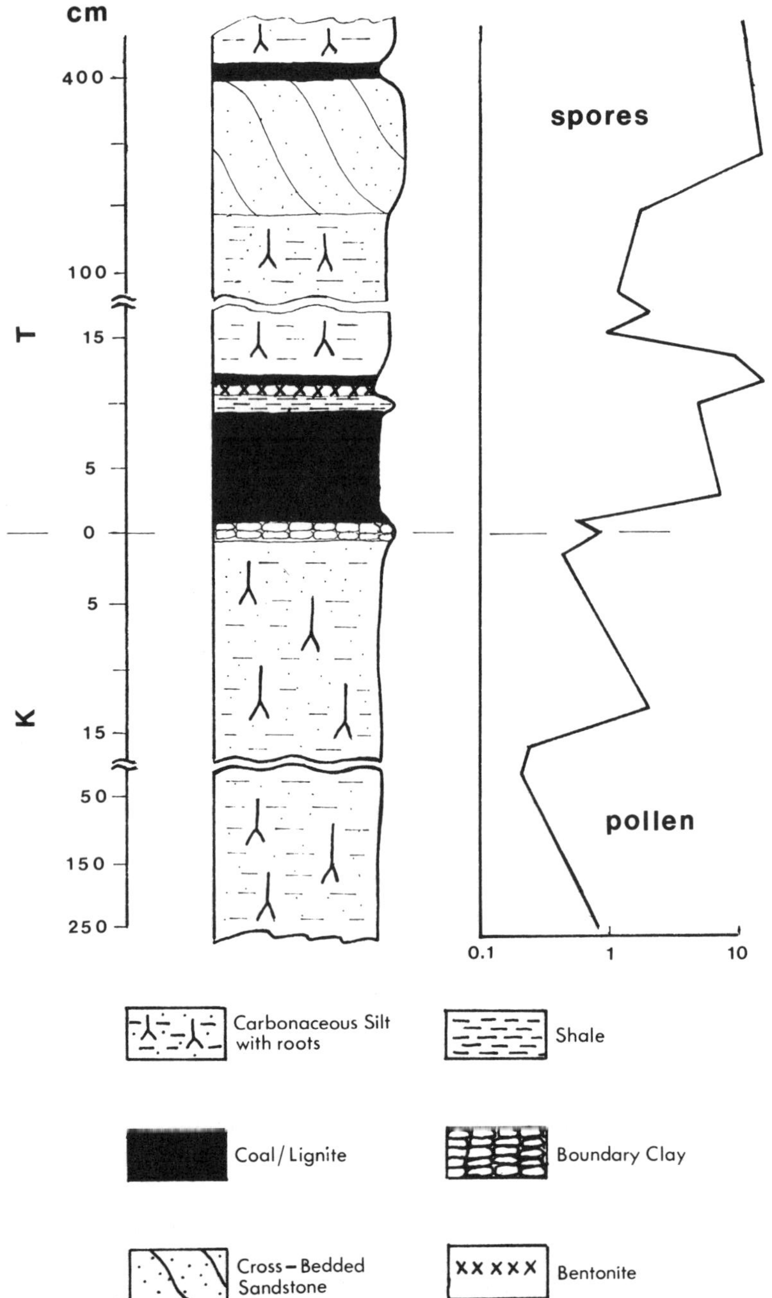

Fig. 1.4. Generalized lithologic column and spore pollen abundance across the Cretaceous–Tertiary boundary for the Western Interior of North America, based on Smit and Van der Kaars (1984).

non-marine boundary sections ranging from New Mexico to Saskatchewan. It is becoming evident that, although the general pattern of palynological change is consistent in all these areas, the pollen taxa that are affected varies geographically (Tschudy and Tschudy 1986). In the northern region of the Western Interior key pollen taxa such as *Proteacidites* and some species of *Aquilapollenites* are confined to the Cretaceous with other taxa (*Gunnera microreticulata, Lilacidites altimurus, Liliacidites complexus,* and *Cranwellia striata*) also disappearing at the boundary. Some taxa (*Wodehouseia spinata, Tricolpites parvistriatus,* and *Arecipites columellus*) suffer a dramatic decline at the boundary, while others (*Kurtzipites trispissatus, Triporopollenites plektosus, Ulmipollenites* spp., and *Alnipollenites* spp.) continued through relatively unaffected into the Tertiary (Leffingwell 1971). In the southern region, the Raton Basin, the only northern region pollen type to occur in Cretaceous rocks and then disappear at the boundary is *Proteacidites* spp. Typical southern region taxa to disappear are '*Tilia*' *wodehousei, Trisectoris,* and *Trichopeltinites* (probably a fungal thallus). Taxa common to both the uppermost Cretaceous and continuing into the basal Tertiary include: *Gunnera microreticulata, Fraxinopollenites variabilis, Liliacidites complexus, Thomsonipollis magnificus, Tricolpites anguloluminosus, Pandaniidites radicus, Arecipites columellus,* and *Salixipollenites* spp.

The loss of Cretaceous taxa occurs at the level of the boundary clay above which are a few millimetres of rock generally devoid of palynomorphs. Sapropel and fusain are often found in this layer suggesting rotting and/or burnt vegetation. This 'barren zone' is overlain by several centimetres of mudstone or more typically coal which yields abundant fern spores and very few pollen grains. This spore-rich horizon is usually interpreted as representing the first Tertiary vegetation, and indicates an abrupt and profound post-boundary change in the vegetation. Angiosperm pollen typically undergoes a slow recovery in abundance until about 10–15 cm above the boundary it again predominates (Tschudy and Tschudy 1986).

By concentrating on complete coal-bearing sections the likelihood of palynomorph reworking across the boundary has been minimized and therefore synchroneity of taxonomic change is demonstrated. In Saskatchewan approximately 30 per cent of angiosperm pollen taxa became extinct at the boundary (Nichols *et al.* 1986) and in the Hell Creek and Lance Creek areas 25 per cent (Hotton 1984) and 30 per cent (Leffingwell 1971) pollen extinctions were reported. While they give no figures Tschudy and Tschudy (1986) imply that the extinction rate was lower in the Raton Basin. However, they do note that all the extinctions are apparently only regional in that elsewhere in North America all the Cretaceous pollen types continue through into the Tertiary. What Tschudy and Tschudy fail to determine, however, is the extent of reworking there many be in other North American records. In the case of *Aquilapollenites* however there is a genuine continuation into the Tertiary of Japan (Saito *et al.* 1986).

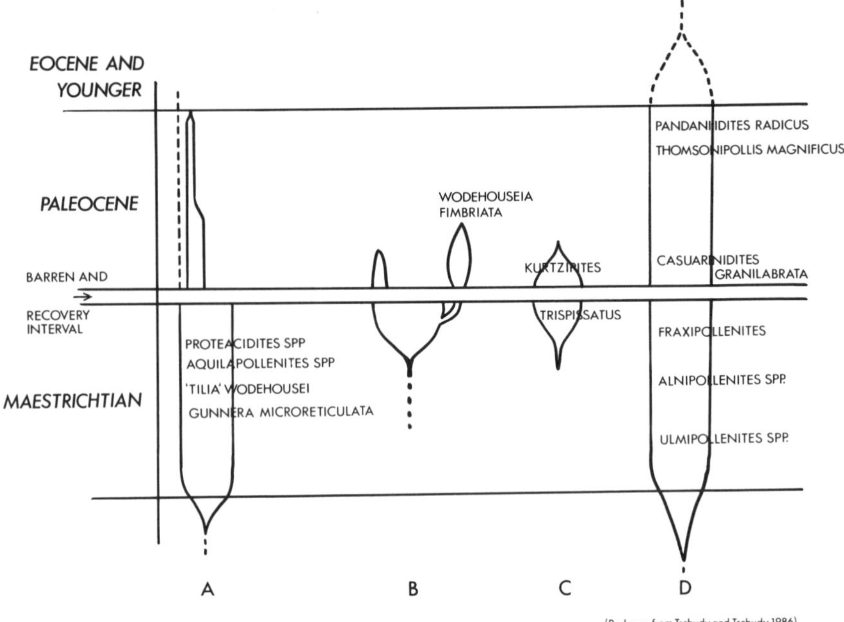

Fig. 1.5. Four types of changes in pollen abundances observed at the K–T boundary by Tschudy and Tschudy (1986). (A) Abrupt regional disappearance at the K–T boundary, but survival in other areas. (B) Pseudo-extinction of Cretaceous species followed by evolution of presumed derived species in the Tertiary. (C) Survival across the boundary followed by Tertiary extinction. (D) Cretaceous taxa unaffected by boundary events (after Tschudy and Tschudy 1986).

Whether or not the extinctions were regional or global, the plant microfossil record demonstrates four types of biotic change at the boundary (Tschudy and Tschudy 1986) as shown here in Fig. 1.5. In addition to the abrupt loss of certain taxa as typified by *Proteacidites* spp., some taxa, such as *Wodehouseia,* seem to undergo a pseudo-extinction at the 'species' level. In the northern region of the Western Interior the Cretaceous form *W. spinata* is replaced above the boundary by *W. fibrlata* and is assumed to be an example of evolution of one species to another. *Kurtzipites trispissatus* typifies a third type of boundary change in which a pollen type appears in the Maastrichtian, reaches its zenith at the boundary and passes into the Paleocene to become a significant element in the 'recovery' pollen flora. Other species of *Kurtzipites* exhibit this pattern, but all become extinct by the middle of the Paleocene. A large group of pollen types was little affected by the K–T boundary event. Many Cretaceous forms pass into the Paleocene and some survived to the Eocene or younger.

The palynological record suggests strongly that the Cretaceous vegetation was severely affected over a wide area by events at the boundary. The pattern of change mirrors that which occurs following profound ecological disturbance. The relationship between this disturbance and extinctions remains controversial, however, particularly as some suggest that most taxa survived in refugia (Tschudy *et al.* 1984) while others detect a pattern of increasing extinction poleward (Hickey 1981, 1984), a pattern incompatible with an 'impact winter'.

Evidence that the terrestrial Cretaceous non-marine ecological trauma may have been global in scale comes from a marine boundary sequence at Hokkaido, Japan (Saito *et al.* 1986). Here terrestrially-derived palynomorphs exhibit a pattern of change very similar to that seen in North America. There is an immediate post-boundary increase in the proportion of fern spores to pollen, but in Japan this is followed by an increase in pine pollen. Saito *et al.* (1986) interpret this as indicating that pine was an early colonizer of the devastated landscape. While pine can form a major component of modern successional vegetation it is also worth remembering that the bisaccate pine pollen grains are produced in large numbers and are particularly suited to long distance transport into marine environments (Boulter and Riddick 1986). It is likely, then, that the abundance of pine in the recovery vegetation, relative to other taxa, is exaggerated in this instance.

8. *Tertiary*

Megafossils By Tertiary times the diversity of land plants was very high and the number of taxa very large, many closely resembling plants living today. The background noise leading to the kinds of palaeobotanical confusion summarized in the section above was reaching its climax. The data available of this megafossil record are so vast that they tend to defeat review by textbook writers, and although Muller (1981, 1984) and others have made summaries of the times of first appearance of some plant groups it does not fit easily into the kind of methodology we are attempting here.

Palynology The IGCP 124 range chart of Boulter *et al.* (IGCP 124 Report 6, in press) includes 30 form taxa of angiosperm pollen (spores and conifer pollen are excluded) agreed to have restricted ranges through the NW European Palaeogene. As with the Neogene data, many other taxa with suspected limitations to their range through the Palaeogene are omitted because of the lack of reliable comparative occurrences and the confusing taxonomy.

The Palaeogene curves in Table 1.7 show the rise in new taxa that might be expected to occur after the K–T boundary. The number of last appearances in Europe increases through the Eocene and Oligocene due,

Table 1.7

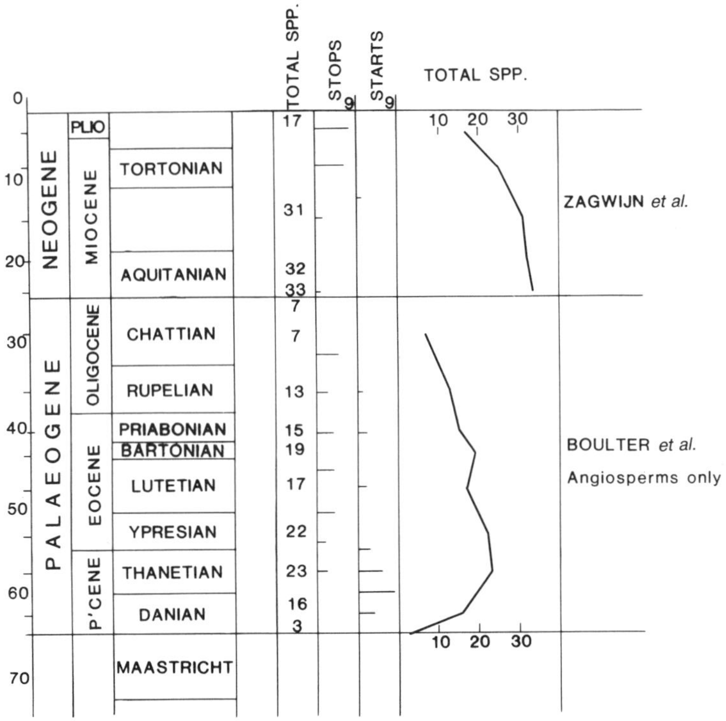

perhaps, as much to plant migration as due to falling temperatures. The profoundly different taxonomic approaches between palynologists working on Palaeogene material and those working in the Neogene makes the interpretation of changes across that boundary well-nigh impossible.

For the European Neogene the ranges of 35 taxa are included in the chart of Zagwijn *et al.* (IGCP Project 124, Report 6, in press) in the NW European Tertiary Basin. These are the angiosperm and conifer taxa (spores are omitted) agreed most likely to have restricted range.

The list of 35 taxa includes form-genera, form-speices, form-sub-species, modern genera, and modern families: a taxonomic mix familiar in the Neogene where there is the influence of Pleistocene and Recent plants. In the Neogene there are many thousands of megafossil plant taxa from Europe alone and the range of pollen types is equally bewildering. It is to be expected that the taxa included in the list of 35 have a wide range of morphology and each is likely to represent more than one original plant type. Many taxa with restricted range have been omitted from the chart because they are difficult to compare from one author's description to another's. Basically, the problem is that there were so many species of plants in the Neogene that palynological taxonomy cannot yet cope with identifying and naming them precisely. The situation will not improve until palynologists agree to use a standardized method of data storage and retrieval.

Conclusions on the palaeobotanical survey

Even under the most traumatic of environmental changes it seems that plant life shows a remarkable capacity for survival, and as far as can be ascertained, it is difficult to define true extinctions due to possibly catastrophic events. When extinctions have occurred in the fossil plant record, they appear to be the result of inter-plant competition or the loss of specific environments to which particular groups have become irreversibly specialized. Although the existing data are of poor quality as a basis for such arguments, we see no evidence for the kind of regular cyclical extinction proposed by Raup and Sepkoski (1984).

There are enormous problems involved in building up reliable sets of data to show overall patterns of plant extinctions. The existing literature has not been obtained for such purposes and should be replaced by more flexible systems of recording that may be more useful for such interpretations.

Nevertheless, in each set of data that is considered here, a distinct pattern emerges of the times of the sums of first and last appearances. The limitations of the data, particularly when attempting to compare values from different Periods which have been determined by different authors, make precise interpretations or useful statistical analysis impossible. What this crude attempt at surveying data of plant taxa ranges does show is a clear cyclical alternation between phases of first appearances and phases of last appearances. The timings of these phases vary, but this introduces the value judgement of magnitude for each such phase, for which some kind of statistical technique will be required to help resolve the extent of significant cycles, hopefully from a much larger data base.

Our findings are generally compatible with those of Niklas *et al.* (1980, 1986) who plotted 'diversity', 'turn-over', and 'net profit and loss' from 119 megafossil compression floras. We have found nothing from our survey to falsify their own cautious conclusions, though their publications suffer from a reluctance to reveal their data publicly.

Because the palynological data used here are so limited these deductions must be set against other evidence for plant evolution. The timing of peaks of maximum species, accurate to the half stage, does not fit events from any other evidence. Indeed, different plants respond differently to changes in environment and there is no theoretical reason to believe that these particular peaks reflect the times of any major physical events.

One pattern shown by this analysis may have a theoretical explanation that gives it some importance: that peaks of first appearances consistently precede peaks of last appearance by other taxa. In the intervals of time around the tops of these peaks the groups of new plant taxa may have been competing with older plant types which they replace as the peaks fall. To help test this observation we must now spend time considering the details of these groups to see if there is any biological and environmental evidence to explain the patterns that are shown here in Tables 1.1–1.7 from the pollen and spores and which are supported by some of the megafossil evidence.

Acknowledgements

We graciously thank Dr Jack Wolfe for preliminary discussions and permission to consider and comment on his unpublished results, Drs David Dilcher and Dianne Edwards for their critical readings of the first draft of the manuscript, and Howell Reynolds for drafting the figures and tables. Statoil, Norway, helped finance some of the work involved.

References

Alvarez, L. W., Alvarez, W., Asaro, F., and Michel, H. V. (1980). Extraterrestrial cause for the Cretaceous–Tertiary extinction. *Science* **208,** 1095–108.

Alvarez, W., Alvarez, L. W., Asaro, F., and Michel, H. V. (1984). The end of the Cretaceous: sharp boundary or Gradual transition? *Science* **223,** 1183–6.

Alvin, K. L. (1982). Cheirolepidiaceae: biology, structure and paleoecology, *Rev. Palaeobot. Palynol.* **37,** 71–98.

Alvin, K. L., Fraser, C. J., and Spicer, R. A. (1981). Anatomy and palaeoecology of *Pseudofrenelopsis* and associated conifers in the English Wealden, *Palaeontology* **24,** 759–78.

Ash, S. R. (1972). Plant megafossils of the Chinle Formation. In *Investigations in the Chinle Formation* (ed. S. Carol and W. J.), Breed Museum of Northern Arizona Bulletin, **47,** 23–44.

Ash, S. R. (1980). Upper Triassic floral zones of North America. In *Biostratigraphy of Fossil Plants* (ed. D. L. Dilcher and T. N. Taylor), pp. 153–70. Dowden, Hutchinson & Ross, Stroudsburg, Pa.

Axelrod, D. I. (1959). Poleward migration of early angiosperm flora. *Science* **130,** 203–7.

Banks, H. (1980). Floral assemblages in the siluro-Devonian. In *Biostratigraphy of Fossil Plants* (ed. D. L. Dilcher and T. N. Taylor), pp. 1–24. Dowden, Hutchinson & Ross, Stroudsburg, Pa.

Bohor, N. F., Foord, E. E. Modreski, P. J., and Triplehorn, D. M. (1983). Mineralogic evidence for an impact event at the Cretaceous–Tertiary boundary. *Science* **224,** 867–9.

Boulter, M. C. and Riddick, A. (1986). Classification and analysis of palynodebris from the Palaeocene sediments of the Forties Field. *Sedimentology* **33,** 871–86.

Boulter, M. C. *et al.* (in press). *Pollen and Spores of the North West European Tertiary Basin,* I.G.C.P. 124. Schweizerbart'sche Verlagbuchhandlung, Stuttgart.

Bouroz, A., Einor, O. L., Gordon, M., Meyen, S. V., and Wagner, R. H. (1978). Proposals for an international chronostratigraphic classification of the Carboniferous. *C.R. 8th Int. Congr. Strat. Geol.* (Moscow, 1975), **1,** (General problems of the Carboniferous Stratigraphy), 36–69.

Brown, R. W. (1962). Paleocene flora of the Rocky Mountains and Great Plains. *U.S. Geolog. Surv. Prof. Pap.* **375,** 119 pp.

Butterworth, M. (1984). Upper Carboniferous. *Pollen and Spore Biostratigraphy of the Phanerozoic in North-West Europe.* British Micropalaeontological Society, 3–11.

Clayton, G. (1984). Lower Carboniferous. *Pollen and Spore Biostratigraphy of the Phanerozoic in North-West Europe.* British Micropalaeontological Society, 2.

Ferguson, D. K. (1985). The origin of leaf assemblages — new light on an old problem. *Rev. Palaeobot. Palynol.* **46,** 117–88.

Filatoff, J. (1975). Jurassic palynology of the Perth basin, western Australia. *Palaeontographica B,* **154,** 1–113.

Gothan, W. (1931). Die Oberschleische Steinkohlenflora. 1. Farne und farnahnliche Gewachse (Cycadofilices bezw. Pteridospermen). *Kon. Preuss. Geol. L. A.,* 278 pp.

Guy-Ohlson, D. (1986). Jurassic palynology of the Vilhelmsfält bore no. 1, Scania, Sweden, Toarcian–Aalenian. Section of Palaeobotany, Swedish Museum of Natural History, Stockholm, 127 pp.

Havlena, V. (1982). The Namurian of Upper Silesia: floral breaks, lithological variations and the mid-Carboniferous boundary problem. In *Biostratigraphic Data for a Mid-Carboniferous Boundary* (*subcommission on Carboniferous stratigraphy*) (ed. W. H. C. Ramsbottom, W. B. Saunders and B. Owens), pp. 112–9. C.I.M.P., Leeds.

Hickey, L. J. (1981). Land plant evidence compatible with gradual not catastrophic change at the end of the Cretaceous. *Nature, Lond.* **292,** 529–31.

Hickey, L. J. (1984). Changes in the angiosperm flora across the Cretaceous–Tertiary boundary. In *Catastrophies in Earth History*: *The New Uniformitarianism* (ed. W. A. Bergren and J. A. Van Couvering), pp. 279–314. Princeton, Princeton University Press.

Hotton, C. (1984). Palynofloral changes across the Cretaceous–Tertiary Boundary in east-central Montana, U.S.A. *6th Int. Palynol. Conf. Calgary,* Abstracts, p. 66.

Jennings, J. R. (1984). Distribution of fossil plant taxa in the Upper Mississippian and Lower Pennsylvanian of the Illinois Basin. *C.R. IX Congr. Int. Strat. Géol. Carbonifière, Washington, Urbana, 1979* **2,** 301–12.

Kidston, R. (1894). On the various divisions of British Carboniferous rocks as determined by their fossil flora. *Proc. Roy. Phys. Soc. Edin.* **12,** 183–257.

Kidston, R. (1923). Fossil plants of the Carboniferous rocks of Great Britain (First Section). *Mem. Geol. Surv. G. B. Palaeontol.* **II** (1-b), 681 pp.

Leary, L. (1981). Early Pennsylvania geology and palaeobotany of the Rock Island County, Illinois area. *Illinois State Museum Report* **37,** 88p.

Leffingwell, H. A. (1971). Palynology of the Lance (Late Cretaceous) and Fort Union (Paleocene) Formations of the Lance area, Wyoming. In *Symposium on*

Palynology of the Late Cretaceous and Early Tertiary, Geological Society of America Special Paper No. 127 (ed. R. M. Kosanke and A. T. Cross), pp. 1–64.

Meyen, S. V. (1966). Carboniferous and Permian lycophytes of Angaraland. *Palaeontographica B* **157,** 112–57.

Meyen, S. V. (1982). The Carboniferous and Permian floras of Angaraland (a synthesis). *Biol. Mem.* **7,** 1–110.

Muller, J. (1981). Fossil pollen records of extant angiosperms. *Bot. Rev.* **47,** 1–142.

Muller, J. (1984). Significance of fossil pollen for angiosperm history. *Ann. Miss. Botan. Garden* **71,** 419–43.

Nichols, D. J., Jarzen, D. M., Orth, C. J., and Oliver, P. Q. (1986). Palynological and Iridium anomalies at the Cretaceous–Tertiary Boundary, Southern Saskatchewan. *Science* **231,** 714–7.

Niklas, K. J., Tiffney, B. H., and Knoll, A. H. (1980). Changes in diversity of fossil plants: a preliminary assessment. In *Evolutionary Biology* (ed. M. K. Hecht, W. C. Steere, and B. Wallace) Vol. 12, pp. 1–90. Plenum Press, New York.

Niklas, K. J., Tiffney, B. H., and Knoll, A. H. (1986). In *Phanerozoic Diversity Patterns. Profiles in Microevolution* (ed. J. W. Valentine), pp. 97–128. Princeton University Press.

Orth, C. J., Gilmore, J. S., Knight, J. D., Pillmore, C. L., Tschudy, R. H., and Fasset, J. E. (1981). An iridium abundance anomaly at the palynological Cretaceous–Tertiary boundary in northern New Mexico. *Science* **214,** 1341–3.

Parrish, J. T., Ziegler, A. M., and Scotese, C. R. (1982). Rainfall patterns and the distribution of coals and evaporites in the Mesozoic and Cenozoic. *Palaeogeogr. Palaeoclimatol. Palaeoecol.* **40,** 67–101.

Phillips, T. L., Peppers, R. A., and DiMichel, W. A. (1985). Stratigraphic and inter-regional changes in Pennsylvania coal-swamp vegetation: environmental inferences. In *Paleoclimatic Controls on Coal Resources of the Pennsylvanian System of North America,* International Journal of Coal Geology Publication No. 5 (ed. T. L. Phillips and C. B. Cecil), pp. 43–109.

Pillmore, C. L., Tschudy, R. H., Orth, C. J., Gilmore, J. S., and Knight, J. D. (1984). Geologic framework of non-marine Cretaceous–Tertiary boundary sites, Raton Basin, New Mexico and Colorado. *Science* **223,** 1180–3.

Raymond, A. (1985). Floral diversity, phytogeography, and climatic amelioration during the Early Carboniferous (Dinantian). *Paleobiology* **11,** 293–309.

Raup, D. M. and Sepkoski, J. J. (1984). Periodicity of extinctions in the geologic past. *Proc. nat. Acad. Sci.* **81,** 801–5.

Retallack, G. J. and Dilcher, D. L. (1981). A coastal hypothesis for the dispersal and rise to dominance of flowering plants. In *Paleobotany, Paleoecology and Evolution* Vol. 2 (ed. K. J. Niklas), pp. 27–77. Praeger Publishers, New York.

Richardson, J. B. and McGregor, D. C. (1986). Silurian and Devonian spore zones of the Old Red Sandstone continent and adjacent regions. *Geol. Surv. Canada Bull.* **364,** 1–79.

Saito, T., Yamanoi, T., and Kaiho, K. (1986). Devastation of the terrestrial flora at the end of the Cretaceous in the Boreal Far East. *Nature, Lond.* **323,** 253–6.

Smit, J. (1982). Extinction and evolution of planktonic foraminifera at the Cretaceous–Tertiary boundary after a major impact. In *Geological Implications of Impacts of Large Asteroids and Comets on the Earth,* Geological Society of America Special Paper 190 (ed. L. T. Silver and R. H. Shulz), pp. 329–52.

Smit, J. and Van der Kaars, S. (1984). Terminal Cretaceous extinctions in the Hell Creek Area Montana compatible with a catastrophic extinction. *Science* **223,** 1177–79.

Spicer, R. A. (1986). Comparative leaf architectural analysis of Cretaceous radiating angiosperms. In *Systematic and Taxonomic Approaches in Palaeobotany,* Systematics Association Special Volume No. 31 (ed. R. A. Spicer and B. A. Thomas), pp. 223–34.

Spicer, R. A. and Greer, A. G. (1986). Plant taphonomy in fluvial and lacustrine systems. In *Land Plants* (ed. T. Broadhead), pp. 10–26. University of Tennessee Department of Geological Sciences Studies in Geology 15.

Spicer, R. A. and Parrish, J. T. (1986). Paleobotancial evidence for cool North Polar climates in the mid-Cretaceous (Albian–Cenomanian). *Geology* **14,** 703–6.

Spicer, R. A., Burnham, R. J., Grant, P., and Glicken, H. (1985). *Pityrogramma calomelanos,* the primary post-eruption colonizer of Volcan Chichonal, Chiapas, Mexico. *Am. Fern J.* **75,** 1–5.

Spicer, R. A., Wolfe, J. A., and Nichols, D. J. (1987). The arctic origin hypothesis: evidence from Cretaceous and Tertiary floras of Alaska. *Paleobiology* **13,** 73–83.

Srivastava, S. K. (1977). Microspores from the Fredericksburg Group (Albian) of the Southern United States. *Paleobiol. Continent.* 6, **2,** 1–119.

Srivastava, S. K. (1981). Stratigraphic ranges of selected spores and pollen from the Fredericksburg Group. *Palynology* **5,** 1–28.

Steward, W. N. (1983). *Paleobotany and the Evolution of Plants.* Cambridge University Press, Cambridge. 405 p.

Surlyk, F. and Johansen M. B. (1984). End Cretaceous brachiopod extinctions in the chalk of Denmoole. *Science* **223,** 1174–7.

Sylvester-Bradley, P. C. (1977). Biostratigraphical tests of evolutionary theory. In *Concepts and Methods of Biostratigraphy* (ed. E. Kauffman and J. Hazel), pp. 41–63. Dowden, Hutchinson & Ross, Stroudsburg, Pa.

Thomas, B. A. and Spicer R. A. (1987). *The Evolution and Palaeobiology of Land Plants.* Croom Helm, London. 309 pp.

Tiffney, B. H. (1981). Diversity and major events in the evolution of land plants. In *Paleobotany, Paleoecology and Evolution* (ed. K. J. Niklas), pp. 193–230. Praeger, New York.

Tschudy, R. H. (1971). Palynology of the Cretaceous–Tertiary boundary in the northern Rocky Mountains and Mississippi Embayment regions. In *Symposium on Palynology of the Late Cretaceous and Early Tertiary,* Geological Society of America Special Paper 127 (ed. R. M. Kosanke and A. T. Cross), pp. 65–111.

Tschudy, R. H. (1973). The Gasbuggy Core — a palynological appraisal in Cretaceous and Tertiary rocks of the southern Colorado Plateau. In *Memoir: Durango, Colorado* (ed. J. E. Fassett), pp. 131–43. Four Corners Geological Society, Colorado.

Tschudy, R. H. and Tschudy, B. D. (1986). Extinction and Survival of plant life following the Cretaceous–Tertiary boundary event, Western Interior, North America. *Geology* **14,** 667–70.

Tschudy, R. H., Pillmore, C. L., Orth, C. J., Gilmore, J. S., and Knight, J. D. (1984). Disruption of the terrestrial plant ecosystem at the Cretaceous–Tertiary boundary: Western Interior. *Science* **225,** 1030–2.

Upchurch, G. R. and Doyle, J. A. (1981). Paleoecology of the conifers *Frenelopsis* and *Pseudofrenelopsis*. In *Geobotany II* (ed. R. C. Romans), pp. 167–202. Plenum Press, New York.

Upchurch, G. R. and Wolfe, J. A. (1987). Mid-Cretaceous to Early Tertiary vegetation and climate: evidence from fossil leaves and woods In *Origins of the Angiosperms and their Biological Consequences* (ed. E. M. Friis, W. G. Chaloner, and P. R. Crane). Cambridge University Press.

Wagner, R. H. (1982). Floral changes near the Mississippian-Pennsylvanian boundary: an appraisal. In *Biostratigraphic Data for a Mid-carboniferous Boundary* pp. 120–7. Sub-commission on Carboniferous Stratigraphy. Leeds.

Wagner, R. H. (1984). Megafloral zones of the Carboniferous. *C.R. IX Congr. Int. Strat. Géol. Carbonifère, Washington, Urbana, 1979* **2,** 109–34.

Warrington, G. (1984). Permian/Triassic. *Pollen and Spore Biostratigraphy of the Phanerozoic in North-west Europe.* British Micropalaeontological Society, 12.

Wolbach, W. S., Lewis, R. S., and Anders, E. (1985). Cretaceous extinctions: evidence for wildfires and search for meteoritic material. *Science* **230,** 167–70.

Wolfe, J. A. and Upchurch, G. R. (1986). Vegetation, climatic and floral changes at the Cretaceous–Tertiary boundary. *Nature, Lond.* **324,** 148–52.

Zagwijn, W. H. *et al.* (in press). Pollen and spores from the Neogene of the North West Europe Tertiary Basin. I.G.C.P. 124. Schweizerbart'sche Verlagsbuchhandlung, Stuttgart.

2. Foraminiferid extinction and ecological collapse during global biological events

M. D. BRASIER

Department of Geology, University of Hull, Hull, UK

Abstract

The Cambrian to Recent fossil record of foraminifera provides a reasonable data base for evolutionary and extinction studies. Numerous iterative trends can be traced from simple generalized to complex specialized tests, the latter including forms with endosymbionts. Non-septate to simple septate forms appear to have wide ecological tolerance and long stratigraphic ranges; these comprise the bulk of survivors. Larger benthic foraminifera provide an evolutionary 'thermometer' from mid-Devonian times onwards, as do planktonic foraminifera from the mid-Cretaceous onwards. Extinction events within mainly transgressive periods punctuated adaptive radiations involving depth stratification; they were associated with anoxia and black shales, notably in the late Frasnian, Norian, early Toarcian, and late Cenomanian. Anoxia and 'drowning' were possible causes of extinction here. Extinction events within more regressive, cooling periods terminated longer patterns of decline, as in the late Permian and late Eocene. Falling sea level was associated with changes in substrate, temperature, salinity, and chemistry of the milieu. The K–T boundary marked abrupt impact-associated extinction of a fauna suffering decline during a period of cooling and falling sea level. A model is envisaged in which adaptive raditions led to niche partitioning involving depth stratification among symbiont-bearing forms, especially during periods of rising sea level. These adaptive tiers were vulnerable to sudden changes in water depth, either through transgression or regression, or to climatic and

Extinction and Survival in the Fossil Record (ed. G. P. Larwood), Systematics Association Special Volume No. 34, pp. 37–64. Clarendon Press, Oxford, 1988.

impact-associated perturbations. Collapse of the adaptive tiers left mainly the 'base' of shallow water forms.

Introduction

Does the foraminiferid fossil record provide support for the hypothesis of periodicity in extinction of marine families and genera, advanced by Raup and Sepkoski (1984, 1986), Sepkoski and Raup (1986)? These authors find eight major episodes of extinction above background level over the last 250 Ma, with a strong indication of a 26 Ma periodicity. Extinctions at the end of the better-dated events in the Cenomanian, Maastrichtian, upper Eocene (Priabonian), and middle Miocene (Serravallian) show a significant periodicity at the generic level.

What information does the foraminiferid fossil record provide on the rate of change, and on the ecological effects of such extinction events? Does this evidence provide any clue to a single ultimate cause, as implied by the periodicity data of Sepkoski and Raup (1986)?. These authors reviewed possible extraterrestrial causes, such as meteorite impacts, galactic oscillations, the tenth planet, and unobserved binary companions, owing to the apparent absence of Earth processes operating on suitable time scales.

Does the foraminiferid data lend support to sea level forcing of adaptive radiations and extinctions? If rising sea level is associated with the adaptive radiation of depth-stratified communities (e.g. Brasier 1981; Hart 1980) is it not logical that falling sea level must have caused their partial or complete collapse?

The foraminiferid fossil record holds a key to the answers. The group spans the Cambrian to Recent, is often abundant enough to reach rock-forming proportions (e.g. *Globigerina* oozes and tropical limestones), ecologically varied (abyssal to fresh water benthics and marine planktics) and taxonomically diverse (over 30 000 species; e.g. Haynes 1981). These characteristics have given them an eminent place in biostratigraphy, providing a standard biostratigraphic scale for much of the geological column, especially in the Tournaisian to end Permian and Albian to Holocene (e.g. Jenkins and Murray 1981). Added to this is the virtue of providing stable isotopes of oxygen and carbon for palaeoenvironmental analyses (e.g. Shackleton *et al.* 1985).

Ideally, a well-funded multinational team working full-time should computer-catalogue all recorded stratigraphic occurrences of foraminifera and assess current information on originations and extinctions. Such a project would provide a very valuable data base for biostratigraphy and evolutionary studies, but there are some obvious scientific (as well as financial and logistical) difficulties. Firstly, there is the problem of systematics: variability, plus the large and diverse literature, has generated many dubious species. In several major groups (e.g. the Nodosariina), the genus may be a grade rather than a clade. Higher taxonomic categories are also dubious and in a state of flux. Secondly, there is the problem of sampling bias: biostratigraphically useful forms, such as planktonics and larger benthics are well-documented, but agglutinated taxa, lower Palaeozoic taxa, and deep sea taxa are less well known. Thirdly, there is the problem of palaeobiological interpretation: it is obvious that origination and extinction curves using combined data on all benthic foraminifera must obscure the palaeobiological aspects of change. This study, therefore, concentrates on the better known larger benthic and planktonic foraminiferid record, utilizing some of the numerous monographs and compilations available, following the timescale and chronostratigraphy of Harland *et al.* (1982).

Foraminiferid evolutionary palaeobiology

Aspects of the functional morphology and broad pattern of evolution in foraminiferid tests have been examined recently by the writer (e.g. Brasier 1982*a,b*, 1984*a,b*). Their tests reveal a tendency towards shorter relative lines of communication through time, arguably explained as a trend towards greater cytoplasmic efficiency in the single, often uninucleate, and relatively large cell. Adaptive aspects of growth strategy, surface area, test density, and buoyancy have also been modelled (Brasier 1986*a,b*). More work is needed on the adaptive implications of reproductive strategies, which are varied and relatively complex in Foraminifera (e.g. Haynes 1981). The group does, however, provide clear examples of r- and K-selection strategies. Examples of the former include the rotaliid *Ammonia beccarii,* living in the physically unstable conditions of estuaries and lagoons (e.g. Bradshaw 1961) with small sized individuals, shortlife cycles, and reproduction that produces numerous small daughter cells. Larger foraminifera such as living *Heterostegina* provide examples of K-strategy, with large individuals having complex cytoplasmic differentiation, long life cycles and reproduction producing a smaller number of large, well-provided daughter cells (e.g. Rottger *et al.* 1986). Gigantism of up to *c.* 150 mm is found at the apogee of certain extinct larger foraminiferid lineages (e.g. *Lepidocyclina*). This trend towards larger individuals is also seen within smaller benthic lineages, and even in small planktonics such as the *Orbulina* lineage.

Some general palaeobiological differences between primitive and advanced architecture should also be mooted. Primitive unilocular tests may be considered to have only simple strategies for biomass increase, e.g. by colonial organization of globular forms, or by continuous growth (Brasier 1986*a,b*). Since many simple foraminifera now graze on diatoms and other small organisms in shallow seas, or may be omnivorous in deeper water (e.g. Lipps 1983), an omnivorous or bacteriophage diet is arguable for primitive foraminifera prior to emergence of diatoms in the Cretaceous. It is suggested that simple forms living in shallow water and marginal habitats are r-strategists, although bathyal and abyssal forms of the Cainozoic have also evolved towards larger size.

Episodic constrictions of the tubular test resulted in septate growth. Septa were initially for added protection, but were subsequently exploited for their potential to allow changes in chamber shape, expansion rate, and apertural form (Brasier 1986*a,b*). The palaeoecological setting for these important transformations deserves to be better known, but is here suspected to have taken place in shallow waters. Evolutionary radiations frequently followed septation, culminating in end members regarded by Brasier (1986*a,b*) as adaptive-constructional bauplans. Uncoiling, for example, has developed by gradual or abrupt changes in chamber shape, leading to elongate tests (*ibid.*) that are possibly adaptive for bacterial or detritus feeding in muds (e.g. Haynes 1981; Lipps 1983). Here again, r-strategies seem to prevail, but larger (*c.* 10 mm) specimens are known from deeper shelf and slope habitats.

Benthic species that rely greatly on endosymbiotic algae may have 'larger' complex tests whose form varies with light intensity (e.g. Hottinger 1982) though not all of the 'larger' type have endosymbionts. These larger foraminifera show tests with lines of communication that are nearly minimal for each geometrical form, though the latter is largely dictated by surface area requirements relating to light intensity (Brasier 1986*a,b*). Many of these larger foraminifera are K-strategists whose dependence on symbionts varies.

Modern shallow planktonic forms may have globular, spinose chambers and bubble-like tests, and tend to be carnivorous; they often contain endosymbiotic algae and show a degree of dependence upon them (e.g. Bé 1977). Modern deeper planktonics have keels and discoidal to conical tests and tend to be herbivorous (e.g. Bé 1977; Hemleben and Spindler 1983). Modelling indicates that planktonic foraminifera modulate their test surface area through ontogeny and evolution, leading to changes in test density and frictional resistance that affect their vertical position in the water column (Brasier 1986*a,b*). Changes in test porosity may also be used to modulate buoyancy (Parsons and Brasier 1987).

It seems that the remarkable variety of test form has no simple, single explanation, but rather reflects the interplay between four main factors: the demands of the ecological niche, the architecture of the ancestor, the physical and constructional constraints of test morphogenesis, and the biological limitations of a unicellular organization.

Cycles of evolution-extinction

An intriguing aspect of foraminiferid evolution is the evidence for cycles of evolution from primitive to advanced test morphology. A general impression of this has been given by Brasier (1982*b*, fig. 1) based on the British data set of Jenkins and Murray (1981). Three rather unequal cycles from primitive- to advanced-dominant assemblages are recorded: Wenlockian to Frasnian; Fammenian to Upper Namurian (no Westphalian to mid-Permian data); and Zechstein to Quaternary. Although these patterns seem clear enough and demand explanation, further studies suggest they compound a variety of palaeoecological, palaeogeographical, and evolutionary phenomena, tending to obscure some of the qualitatively significant turnover points at the end-Permian, Cenomanian–Turonian boundary and end-Cretaceous. It is better to focus on groups whose ecology and evolution is reasonably well understood, i.e. the larger benthic and the planktonic foraminifera.

Cycles in the evolution and extinction of larger benthic foraminifera can be traced in at least a dozen lineages (e.g. Hottinger 1982) for which the earlier stages are often obscure but evidence from the fossil record, ontogeny of extant *Discospirina* and geometrical modelling (Brasier 1982*a*,*b*, 1984*a*,*b*, 1985*a*,*b*) suggests the following evolutionary steps.

1. Globular forms acquire the basic advantages of a test.
2. Tubular forms achieve growth of the individual.
3. Longithalamous forms (with chambers longer than wide) develop from tubes through gnomonic growth, with the protective barriers of septa.
4. Brevithalamous forms (with chambers wider than long) exploit the adaptive potential of varied chamber shape and apertural configuration, and save time and energy through shortened lines of communication.
5. Simple chambers develop 'partitions' and multiple apertures; these give extra strength and encourage progressively more-differentiated radial and concentric cytoplasmic streaming in the central zone, and quieter areas towards the epiderm.
6. Progressive changes in chamber shape transform the test shape from the ancestral type towards the ideal types or bauplans of discoidal, fusiform or conical shapes; each of these is adaptive for a symbiotic lifestyle in different ecological settings or from different ancestries, and has often been accompanied by a progressive increase in size.
7. (A) sexual and (B) asexual generations become more differentiated, with the macrospheric proloculus of the A-generation tending to enlarge, usually at the expense of recapitulated nepionic growth stages (i.e. accelerating-out of the nepion) and the adult portion of the B-generation tending towards gigantism. Some of the larger forms seem to have lived in deeper water and an evolutionary progression from shallow to deeper photic waters is implied.

8. Extinction affects the lineage, especially in the more -advanced forms occupying deeper waters. Notable extinctions took place near the Frasnian–Fammenian boundary, through the late Permian, near the Cenomanian–Turonian boundary, in the late Maastrichtian, through the Late Eocene, and through the Miocene.

Planktic foraminifera evolved from benthic ancestors in the Jurassic and underwent several evolutionary cycles. Here the pattern is arguably from small, generalized, globular-chambered forms living in the upper layers of the water column, to enlarging forms more specially adapted to life in shallow, intermediate or deeper layers (e.g. orbuline, clavate, keeled discoidal, and conical forms). Increasing K-strategy and niche partitioning through depth stratification are implied. Environmental perturbations periodically caused the collapse of adaptive tiers, with resulting extinctions. Three such cycles seem to have occurred during the Cretaceous and two during the Cainozoic, with notable extinctions at the end of the Cenomanian, Santonian, Maastrichtian, and late Eocene (e.g. Hart 1980; Hart and Ball 1986).

The taxonomic and palaeobiological character of faunal changes over these events is outlined below, with some reference to their general geological setting.

The Fransnian–Fammenian boundary event

The first major foraminiferid radiation occurred in the Middle Devonian (Eifelian to Givetian) and culminated in the Upper Devonian (Frasnian) at a time of widespread reefal carbonates. This Tethyan microfauna (Fig. 2.1) is characterized by Semitextulariina such as *Nanicella, Paratextularia, Paratikhinella, Pseudopalmula,* and *Semitextularia*; plus nodosarioids such as *Eogeinitzina, Eonodosaria, Frondilina,* and *Multiseptida.* The upper Frasnian also had the first rare representatives of the Tournayellidae: *Tournayella* (= *Eotournayella*), *Glomospiranella* and possible *Septaglomospiranella* (Kalvoda 1986). The most advanced forms were *Semitextularia* (with multiple apertures and partitions) and *Multiseptida* with multiple partitions. By analogy with recent partitioned forms, these may have contained endosymbiotic algae. Agglutinated faunas also include *Oxinoxis* by the Givetian and uniserial septate *Reophax* by the Upper Devonian (Conkin *et al.* 1981).

Extinctions of reefal faunas at the end of the Frasnian appear to have been associated with major transgressive pulses and anoxic black shales, especially of the 'Kelwasser Event' (House 1975, 1985). The Upper Kellwasser Limestone is connected with a $\delta^{13}C$ spike indicating phytoplankton blooms but lacks signs of lowered temperature or impact (McGhee *et al.* 1986). The distinctive Semitextulariina–Nodosarioid assemblage became extinct at about this time in North America (Toomey and

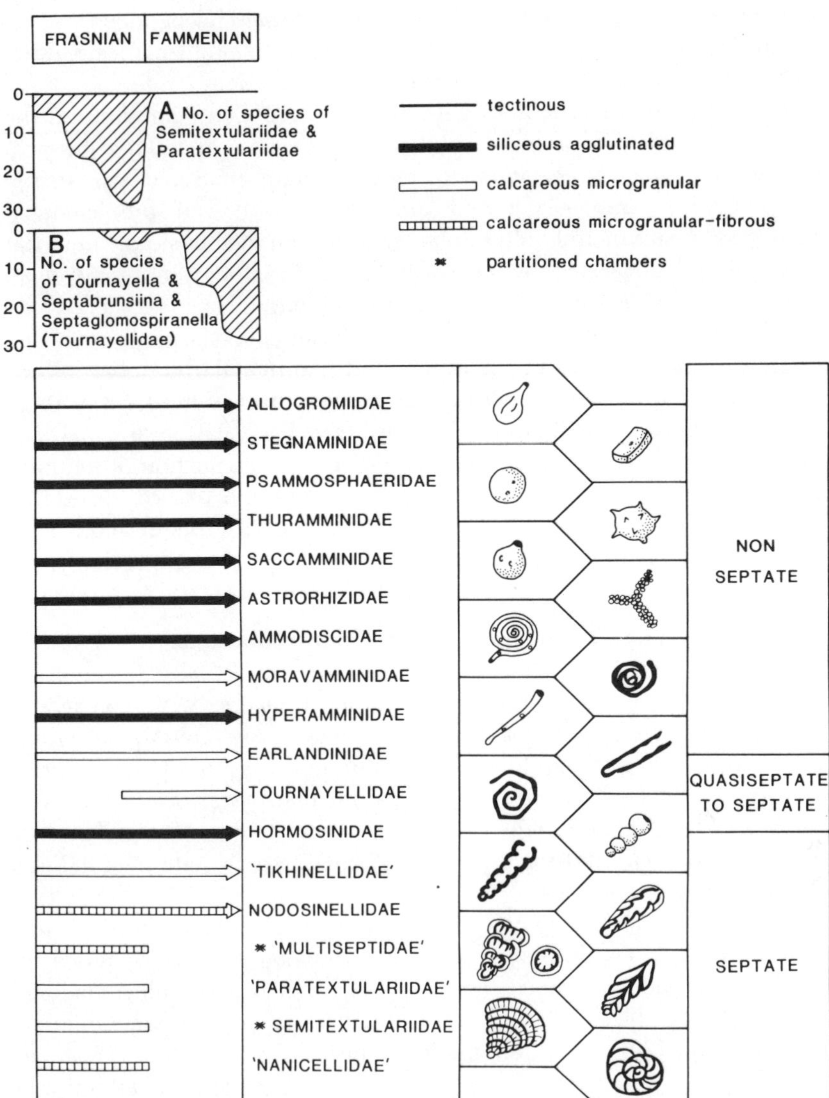

Fig. 2.1. Distribution of foraminiferid families across the Frasnian–Fammenian boundary event, showing also the extinctions of species of semitextulariids and paratextulariids (A) and tournayellids (B). Data mainly from Poyarkov (1979).

Mamet 1979) and USSR (Poyarkov 1979). According to Kalvoda (1986) the foraminiferid decline was neither abrupt nor connected only with the late Frasnian–early Fammenian (uppermost *gigas*–lower *triangularis* Zones) Kellwasser Event in Eastern Europe. There, the major rate of decline seems to have been at the time of the middle *triangularis* Zone Crickites

Event, in the early Fammenian. Succeeding *crepida* Zone deposits are characterized mainly by unulocular foraminifera and problematical calcareous spheres, while siliceous radiolaria became abundant (*ibid.*).

Figure 2.1 plots the general distribution of foraminiferid families plus the number of species in two main groups across the boundary interval. Survivors included non-septate to pseudoseptate tournayellids, simple agglutinated saccaminaceans and ammodiscaceans, and problematical parathuramminacean and volvocacean spheres. Survival of simple uniserial nodosarioids is suggested by *Lunacammina* (= *Geinitzina*), *Nodosinella* (= *Eonodosaria*) and *Frondina* in the Carboniferous. Thus the Frasnian-Fammenian extinction event eliminated forms of relatively advanced architecture (notably semitextulariids and paratextulariids; Fig. 2.1A) while those of primitive to intermediate architecture survived. Even these suffered a setback in the number of species (e.g. Fig. 2.1B).

It therefore seems that the Frasnian–Fammenian extinction of foraminifera resulted from widespread anoxia over shelves and phased 'drowning' of the reef-associated fauna. This seems to have been brought about by a rapid transgressive pulse.

The Fammenian–Tournaisian radiations

A second major cycle developed from the survivors, with non-septate *Tournayella* and *Brunsiina* transforming to pseudoseptate *Septatournayella* and *Septabrunsiina* during the middle Fammenian and even to septate endothyrids such as *Quasiendothyra* by middle Fammenian *marginifera* Zone times (Kalvoda 1986). Their further evolution seems to have coincided with major transgressions and the expension of suitable carbonate shelf habitats in upper Fammenian times.

A return to muddy, anoxic, and brackish conditions in the latest Fammenian-early Tournaisian resulted in extinction of relatively advanced *Quasiendothyra*, though simpler Tournayellidae and Chernyshinellidae survived (Kalvoda 1986) and underwent further major adaptive radiations (Conil and Lys 1977). Agglutinated forms also thrived over the Devonian–Carboniferous boundary interval with evolution of species of uniserial pseudoseptate *Hyperammina*, uniserial septate *Reophax*, and the first septate lituolacean, planispiral-uncoiled *Ammobaculites* in the Kinderkookian (Conkin *et al.* 1981). It is possible that these were able to flourish at a time of transgression with local carbonate depletion.

Late Permian extinctions and the P–TR boundary event

The trend from simple endothyrids towards more advanced types developed gradually through the Tournaisian and Visean (e.g. Conil and

Lys 1977) to Pennsylvanian and Permian, culminating in Endothyridae such as *Bradyina* and *Janischewskina* (with open sutural apertures and mural pores), Ozawainellidae (with discoidal involute forms), and Fusulinidae, Schwagerinidae, and Verbeekinidae (with a fusiform shape, chamberlets, and mural pores).

Progressive extinction then affected this fauna (top of Figure 2.2, mainly from Ross and Ross 1979). The schwagerinid lineage, which culminated in large, very elongate *Polydiexodina*, died out at the end of the Kazanian. The neoschwagerinids, which culminated in the advanced form of *Lepidolina*, died out a little later, at the end of the *Lepidolina* Zone.

The reduced and relatively diminutive Dzhulfian fusulinacean fauna was derived from long-ranging *Eoschubertella* and included uncoiled, long-ranging *Codonofusiella*. A considerable diversification of smaller benthic Foraminifera took place from the later part of the *Neoschwagerina* Zone (i.e. prior to extinction of the neoschwagerinids) and thrived up to the end of the *Palaeofusulina* Zone in late Dzhulfian times (Lys 1984). This fauna included the advanced nodosarioid *Pachyphloia* and the partitioned miliolid *Shanita*. Late Dzhulfian faunas notably contain the partitioned nodosarioid *Colaniella* and biseriamminid *Paradagmarita*, fragmentary remains of which are common near the Permo-Triassic boundary (Lys 1984; Altiner 1984).

Fusulinacean larger foraminiferids were therefore in relative decline by late Permian times. Nodosarioids such as *Lunucammina, Frondina* and *Ichthyolaria* had begun to dominate in early Dzhulfian carbonates, contemporaneous with scarcer *Codonofusiella* (e.g. Johnson 1981) while nodosarioids, biseriamminids, and hemigordiopsid miliolids thrived in late Dzhulfian times, contemporaneous with *Palaeofusulina* (Altiner 1984) and duostominaceans had already appeared by latest Permian times in the Himalayas (Dr P. Khalia, *pers. comm.* 1986). Nodosarioids, miliolids, and agglutinated forms also predominated in Boreal Zechstein facies of late Permian age (e.g. Pattison, in Jenkins and Murray 1981). Thus, late Permian faunas of Tethyan regions had a transitional aspect, while Boreal ones were essentially 'Mesozoic'.

Xu Dao-Yi *et al.* (1985) report that latest Permian faunas in south China were diverse in fusulinacean and other foraminiferid species (*c.* 133 and 131, respectively). All but a few tens of species disappeared during a short time interval represented by *c.* 30–50 cm of strata at the base of the lower Triassic. This mass extinction in the boundary layer coincides with a negative $\delta^{13}C$ anomaly in Chinese sections, marking a major shift seen also in western USA, the Zechstein Basin, and the Tethys from the Alps to China (Holser *et al.* 1986). The latter suspect a rapid fall in sea level of *c.* 250 m during the last stage of the Permian.

The late Permian extinctions most conspicuously affected the architecturally more-advanced forms (e.g. fusulinaceans, partitioned nodosarioids, and miliolids, uncoiled palaeotextulariids) of tropical shelves. Survivors

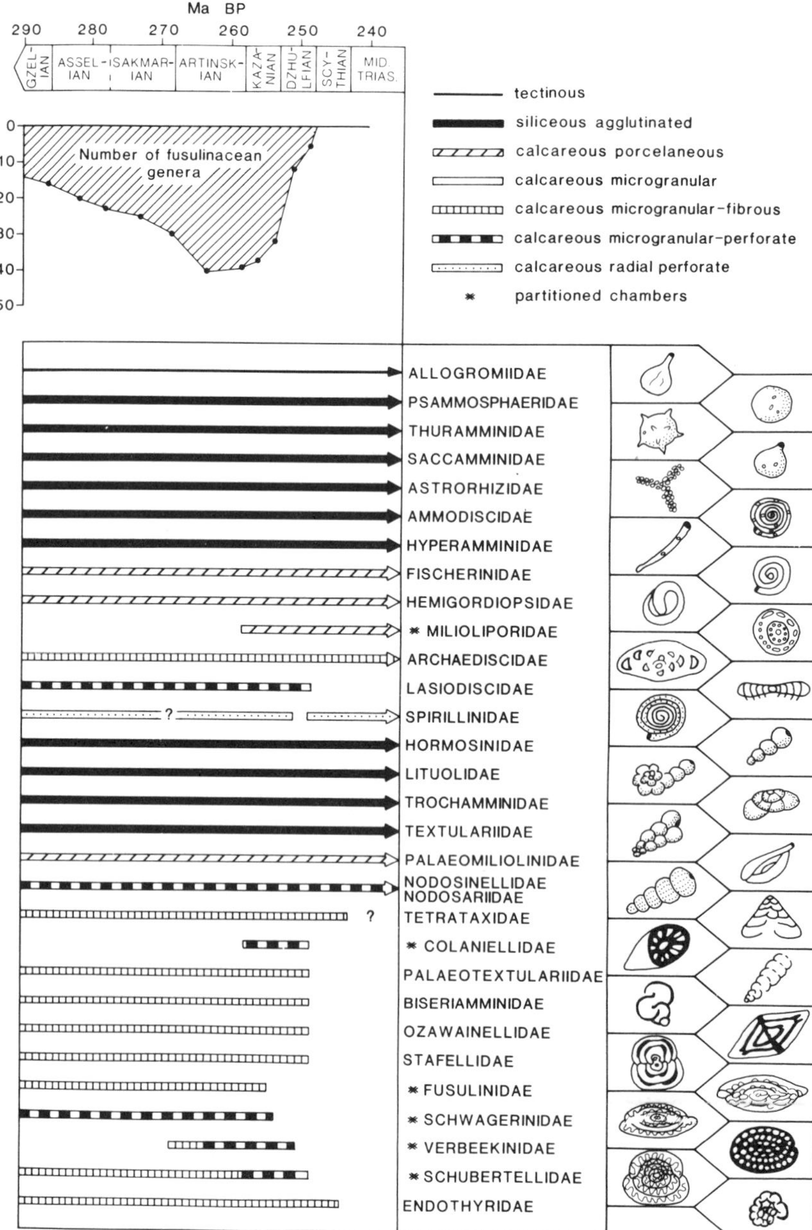

Fig. 2.2. Distribution of foraminiferid families through the Permian to mid-Triassic, showing also the rise and fall in the number of fusulinacean genera. Data from Ross and Ross (1979), and other sources in the text.

included agglutinated ammodiscaceans and lituolaceans (e.g. *Haplophragmoides, Textularia, Bigenerina, Verneuilina*) and simple Miliolina (e.g. *Agathammina*) and nodosarioids (e.g. *Geinitzina, Nodosinella*), which show no marked change from late Palaeozoic to early Mesozoic seas. The details of survival are severely obscured by adverse facies, the 'Lazarus effect', and the probability that several Triassic stocks appear to have developed a radial hyaline wall structure from Permian microgranular-fibrous ancestors over this interval, tending to obscure their relationships (i.e. Nodosariina from Palaeozoic nodosarioids; Involutinina from Palaeozoic archaediscids; Spirillinina from Palaeozoic archaediscids; see Hohenegger and Piller 1975). It is from this simpler stock, rather than the more-advanced Fusulinina, that the Mesozoic foraminiferid fauna was largely derived. A possible exception is *Duostomina,* a late Permian–Triassic trochospiral form with a complex aperture and tooth-plates, which was possibly derived from the fusuline-walled *Tetrataxis* of late Palaeozoic to Triassic seas (Hohenegger and Piller 1975). *Neoendothyra reicheli* was a rare survivor into the early Triassic (Lys 1984).

Thus, the end-Permian extinction event compares with the Frasnian-Fammenian event in the preferential extinction of specialized partitioned forms and the survival of simpler, more tolerant ones. It differs, however, in being part of a prolonged decline, involving low origination rates of fusulinaceans in the Dzhulfian (cf. Tappan 1982). Falling sea level, with salinity, temperature, and substrate effects, may have led to squeezing and elimination of specialized niches.

Triassic faunas and the Norian extinction event

Despire the apparently slow recovery of foraminiferid faunas in the Triassic, several new aragonitic-walled forms made their appearance in shallow Tethyan facies. Their aragonitic mineralogy is interesting in the light of other evidence for carbonate mineralogy and biomineralization at this time (e.g. Sandberg 1983; Brasier 1986*a*). The duostominids are relatively advanced trochospiral forms with extraumbilical apertures and tooth plates that were possibly ancestral to Jurassic epistominids and thence to Cretaceous rotaliids (e.g. Loeblich and Tappan 1974). Non-septate, involute, aragonitic-walled Involutinina also developed at this time (Piller 1978) and may be considered to have reached a climax in the Norian (including Rhaetian) with the partitioned form of *Triasina.* Like many molluscs, marine reptiles and conodonts (e.g. Hallam 1981) these reefal forms did not survive the end of the Triassic, though simpler, non-septate *Involutina* and *Trocholina* survived. Late Triassic foraminifera of boreal regions show close affinities with the Hettangian assemblages (e.g. P. Copestake, in Jenkins and Murray 1981).

The Toarcian event

A major transgressive pulse and an associated oceanic anoxic event (OAE 1) are recorded widely in the early Toarcian *Harpoceras falciferum* Zone (e.g. Hallam 1975; Jenkyns 1980, 1985). Over this interval occurred high rates of organic carbon burial and regional positive δ^{13}C isotope anomalies (Jenkyns 1985), blooms of prasinophyceans, plus a drop in the proportion of dinoflagellates, acritarchs, pollen, and spores (Loh *et al.* 1986), and anoxic seafloor conditions preserving mainly pelagic ammonites and epibyssate bivalves (Hallam 1975; Kauffman 1981; Seilacher 1982). Major extinctions and diversifications in ammonites actually took place during the preceding *spinatum* and *tenuicostatum* Zones across the Pliensbachian–Toarcian boundary, at a time of lowered δ^{13}C (Jenkyns 1985) and regression to early transgression (Hallam 1986) whereas the benthos was more drastically reduced at the base of the *falciferum* Zone (Hallam 1986) at the time of raised δ^{13}C. Lowered salinities of surface waters may account for both the drastic change in phytoplankton composition over the Jet Rock—Posidonia Shale interval and the establishment of a pycnocline in the water column leading to anoxia at the basin floor (Loh *et al.* 1986).

The effects of this early Toarcian *falciferum* event on foraminiferid faunas have been examined in Britain by Mr R. Young of Hull University. Regressive Marlstone to peak transgressive Jet Rock conditions were accompanied by a succesive change in the dominant forms: calcitic nodosariids—aragonitic *Rheinholdella* or '*Brizalina*'—agglutinated *Textularia.* This is associated with a decrease in size and diversity, while the climax paper shale facies themselves are usually barren. In the bituminous Alum Shales above the anoxic event, agglutinated and nodosariid faunas reappeared. Lowered oxygen, pH, and Ca^{++} therefore seem to be implicated in the Jet Rock event.

No major foraminiferid group suffered extinction over this interval but the important Liassic *Lingulina tenera, Frondicularia terquemi,* and *Marginulina prima* plexi were eliminated in boreal regions (Copestake and Johnson 1984); these seem to have been relatively shallow–mid-shelf forms. There was also a marked change in the character of younger nodosariid assemblages: the place of uniserial *Nodosaria, Frondicularia,* and *Lingulina* was largely taken up by coiled *Lenticulina.* The first larger lituolacean foraminiferids *Orbitopsella* and *Cyclorbitopsella* developed in the Tethyan region during the Pliensbachian (Cherchi *et al.* 1984). Their extinction needs further study, but may relate to events at this boundary interval.

Thus, benthic foraminifera were clearly affected by rising anoxia during the early Toarcian of western Europe. After the event, moderately specialized shallow-mid shelf nodosariids seem to have been displaced by very tolerant *Lenticulina.* Minor faunal turnovers occurred, as did anoxic events, through the Callovian to Lower Cretaceous, but there is little available evidence for foraminiferid mass extinction events in this interval.

The Cenomanian–Turonian boundary event

Organic-rich pelagic shales are widely developed in shelf to oceanic environments over the late Barremian to Albian, the Cenomanian–Turonian boundary interval, and to a lesser extent, the Coniacian–Santonian (Jenkyns 1980). Of these anoxic events, that at the Cenomanian–Turonian boundary, known in England as the Black Band or Plenus Marl, is most sharply marked, extending across the Atlantic Ocean into the western Tethys (de Graciansky *et al.* 1984). This is associated with a major transgression (Hancock and Kauffman 1979; Jenkyns 1980), a positive δ^{13}C spike in the late Cenomanian *archaeocretacea* Zone (Jenkyns 1985), an expansion of the oxygen minimum zone onto the shelf (Hart 1985), blooms of siliceous radiolaria (de Graciansky *et al.* 1984), a minor mass extinction of bivalves (Kauffman cited *in* Flessa and Jablonski 1983), abrupt fall in the diversity of planktonic forminifera plus major ecological changes in benthics (Hart and Bigg 1981; Hart 1985; Hart and Ball 1986). Although there is no significant iridium anomaly, the smectitic clays suggest a volcanic origin (Jenkyns 1980). The trace metal accumulations suggest that complete stagnation of the ocean basins, rather than dynamic upwelling, was responsible for anoxia with very low sedimentation rates, over a period of less than half a million years (Brumsack 1986; Jenkyns 1985).

This Cenomanian–Turonian boundary event clearly shows the effects of physical perturbation on early planktonic foraminifera (Fig. 2.3B,C). During the Cenomanian their morphology and distribution suggests an adaptive radiation from shallow into deeper water layers up to the *Rotalipora cushmani* Zone (Hart and Bailey 1979; Hart and Ball 1986). This initial radiation of hedbergellids and rotaliporids was selectively extinguished by the Black Band event, which mostly spans the *R. cushmani* to *Whiteinella archaeocretacea* Zones of the upper Cenomanian, locally extending into the *Praeglobotruncana helvetica* Zone. In SW England, extinction of the *Rotalipora* lineage was phased, with more heavily-calcified and thicker-shelled *R. greenhornensis* disappearing before *R. cushmani,* suggesting that the former taxon lived deeper and was first affected by a rising oxygen minimum zone (Hart 1985). Simple agglutinated ammodiscaceans (Hart and Bigg 1981), *Reophax* (Hart 1985) or small, deposit-feeding *Brizalina* (Brasier, unpublished) bloomed over the anoxic, black, *Chondrites*-burrowed interval, suggesting that low oxygen levels extended down to the seafloor. Planktonic foraminiferid diversity and dinoflagellate abundance fell with the rise in ^{13}C, indicating ecological restriction over the isotope event (Hilbrecht *et al.* 1986).

Survivors over the δ^{13}C spike included globular chambered hedbergellids *Hedbergella* and *Whiteinella,* suggested to have occupied the upper 50 m of the water column, and *Praeglobotruncana* and *Dicarinella* whose discoidal keeled form suggests intermediate depths of 50–100 m (Hart 1985). These shallower forms provided the root stock for a second major adaptive

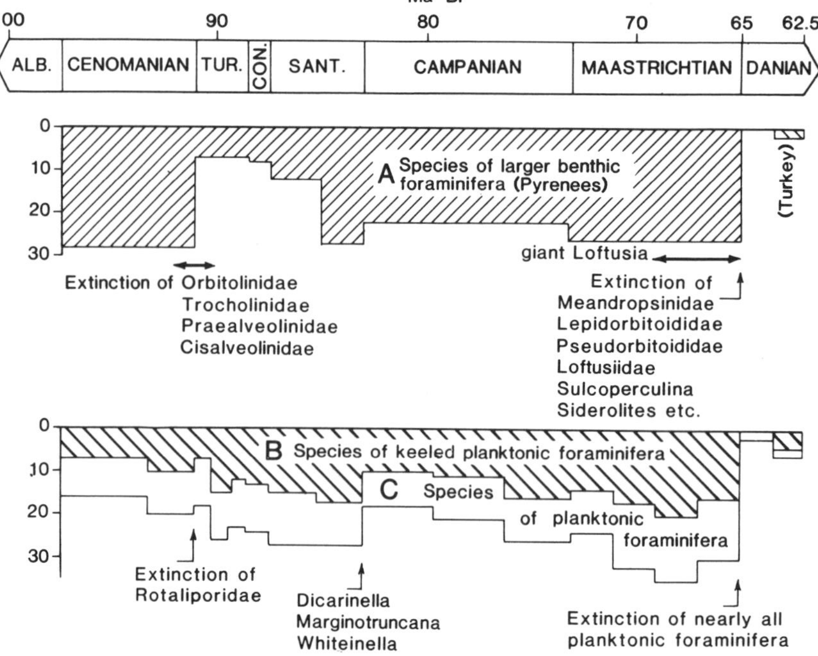

Fig. 2.3. Diversity of late Cretaceous larger benthic and planktonic foraminifera. (A) Species diversity of larger benthic foraminifera from the Pyrenees (from data in Bilotte 1984) with some additional data from Sirel *et al.* (1986). (B) Species of keeled planktonic foraminifera. (C) Total species of planktonic foraminifera (from data in Caron 1985).

radiation, rapidly assuming architectural forms suggestive of the exploitation of deeper water masses (Hart 1980; Hart and Ball 1986). The appearance of the fully twin-keeled *Marginotrunca* indicates middle to deep water assemblages that were absent during the carbon isotope anomaly (Hilbrecht *et al.* 1986).

Thus, the Black Band event may be argued to have affected planktonics more than smaller benthics and deeper planktics more than shallow ones. The postulated rise in the oxygen minimum layer not only accompanied the eustatic rise upwards, but was also expanding within the water column, approaching perhaps to within 100–150 m of the surface (Hart 1985). Small benthic foraminifera adapted to organic rich muds (such as ammodiscaceans and buliminaceans) were predictably less affected by oxygen minimum conditions.

Contemporaneous effects on larger foraminiferid faunas are also suspected. Mid- and late Cenomanian assemblages include diverse larger lituolid and miliolid faunas, few of which survived into the latest Cenomanian–Turonian. Figure 2.3A shows a significant decline in the number of larger foraminifera in the Pyrenees at this time (from data in Bilotte 1984). The extinction of some boreal deeper shelf *Arenobulimina* at this level is also notable, since these had a relatively sophisticated architecture.

Both larger benthic and planktonic foraminiferid data sets (Fig. 2.3) show an increase in the number of species through the Turonian, Coniacian, and Santonian, with extinction at the end of the Santonian, and further adaptive radiation through the Campanian and Maastrichtian. The late Santonian extinction of planktonic *Whiteinella, Dicarinella,* and *Marginotruncana,* thought to have occupied successively deeper layers, also coincided with the final phase of anoxic event OAE 3 (Hart and Ball 1986). Evolution of deep-dwelling *Globotruncana calcarata* in the late Campanian coincided with probable maximum transgression while its extinction took place during the end Campanian regression (Hart and Ball 1986, fig. 5).

The Cretaceous–Tertiary boundary event

By late Maastrichtian times, both planktonic and benthic foraminiferid faunas were well-developed. Planktonic faunas contain a wide variety of architectural types, including relatively large deeper-water conical forms (e.g. *Rosita contusa*) and morphological analysis suggests considerable depth stratification (Parsons and Brasier 1987). Although there was only a slight drop in diversity in the late Maastrichtian *mayaroensis* Zone (e.g. Fig. 2.3B, C; also Hart and Ball, fig. 1) the rate of origination of new taxa was declining markedly, well in advance of the terminal Cretaceous event.

Larger foraminifera were also highly developed in the late Maastrichtian, with lituolaceans reaching the giant fusiform proportions of *Loftusia persica,* and rotaliids achieving canaliculate nummulitid forms (e.g. *Sulcoperculina*) and multilayered orbitoids (e.g. *Lepidorbitoides*). Thus, Maastrichtian foraminifera present a picture of benthic and planktonic faunas near, or just passing their evolutionary apogee.

Events at the Cretaceous–Tertiary boundary are fairly well documented, with evidence for mass extinctions of invertebrates and vertebrates, an iridium anomaly in the boundary clay (Alvarez *et al.* 1980) and locally just below it (Rast and Graup 1985), associated with strained quartz and microtektites (e.g. Graup 1985), a global negative $\delta^{13}C$ anomaly prior to the iridium anomaly (Hansen *et al.* 1986), a eustatic fall in sea level (Vail *et al.* 1977) and massive volcanism associated with Laramide movements (e.g. Ekdale and Bromley 1984). The evidence for a meteorite-impact forcing agent is considerable (e.g. Silver and Schulz 1982).

In most pelagic chalk sequences, the boundary is marked by up to about 1 m of carbonate-depleted clay. Changes in planktonic foraminifera over this interval provide evidence for rapid ecological breakdown and catastrophic extinction. Smit (1982) has reviewed much of the data, summarized in Fig. 2.4. Chalks of the late Maastrichtian *mayaroensis* Zone contain a rich planktic assemblage including large, sculptured deep water forms with slow rates of origination and extinction. Hence, the abrupt extermination at the end of the zone involved many long established taxa. The boundary clay is often disturbed, condensed, and incomplete. A supposedly more complete section occurs at Gredero in SE Spain, where anomalies at the sharp base of the clay are associated with an impoverished relict fauna of *Guembelitria cretacea*, prior to the widespread appearance of *Globigerina eugubina* and its relatives (Fig. 2.4; Smit 1982). In Israel, this Danian form appeared within the negative $\delta^{13}C$ anomaly and before the terminal extinctions of *mayaroensis* Zone planktonic foraminifera (Margaritz *et al.* 1985). The $\delta^{13}C$ anomaly at Nye Kløv in Denmark also precedes extinction of the *mayaroensis* Zone plankton and the iridium spike (Hansen *et al.* 1986). Several iridium anomalies occur in Bavaria, including one beneath the boundary clay extinctions (Rast and Graup 1985). Clearly, the plankton extinctions and/or cessation of chalk deposition was not instantaneous.

The pattern of survival among planktonic foraminifera is curious: all deep water forms became extinct, while some surficial forms survived and others did not. Minute forms of *Guembelitria cretacea, Globotruncanella monmouthensis,* and *G. caravacensis* survived in the Gredero section (Fig. 2.4). Margaritz *et al.* (1985) have also shown that small *Guembelitria*, '*Globigerina*' *eugubina,* and small pustulose *Hedbergella* survived into the Danian. These were not accompanied, however, by morphologically and hydrodynamically similar, but exclusively Cretaceous *Globigerinelloides* and *Heterohelix.*

Benthic foraminifera across continuous boundary sediments in southern Israel have been studied in detail by Mr A. Chepstow-Lusty of Hull University. These reveal no major change in composition, with a Midway-type shelf microfauna of Danian aspect appearing before extinction of the *mayaroensis* Zone biota. At the boundary is seen a marked rise in the abundance of *Angulogavelinella,* a deeper water form which may indicate cooling. There is general continuity of other deep sea microfaunas across the boundary (Berggren 1984).

Sections showing events across the boundary in shallow tropical carbonates often lack the critical interval because of tectonic-eustatic events, though this may be preserved in central Turkey (Sirel *et al.* 1986). Here, a late Maastrichtian larger foraminiferid fauna of diverse orbitoids, *Siderolites, Sulcoperculina* and *Loftusia* occurs in sandy limestones with algae. Overlying algal limestones of Danian age contain a reduced microfauna including *Planorbulina, Mississipina,* Ataxophragmiidae and miliolids. *Quinqueloculina* and *Peneroplis* and rotaliid *Rotalia* seem to have provided the root-stock for subsequent development of larger foraminifera, in many cases convergent with Cretaceous precursors. These three taxa occur as diminutive forms in relatively barren carbonate shoal and beach sands today.

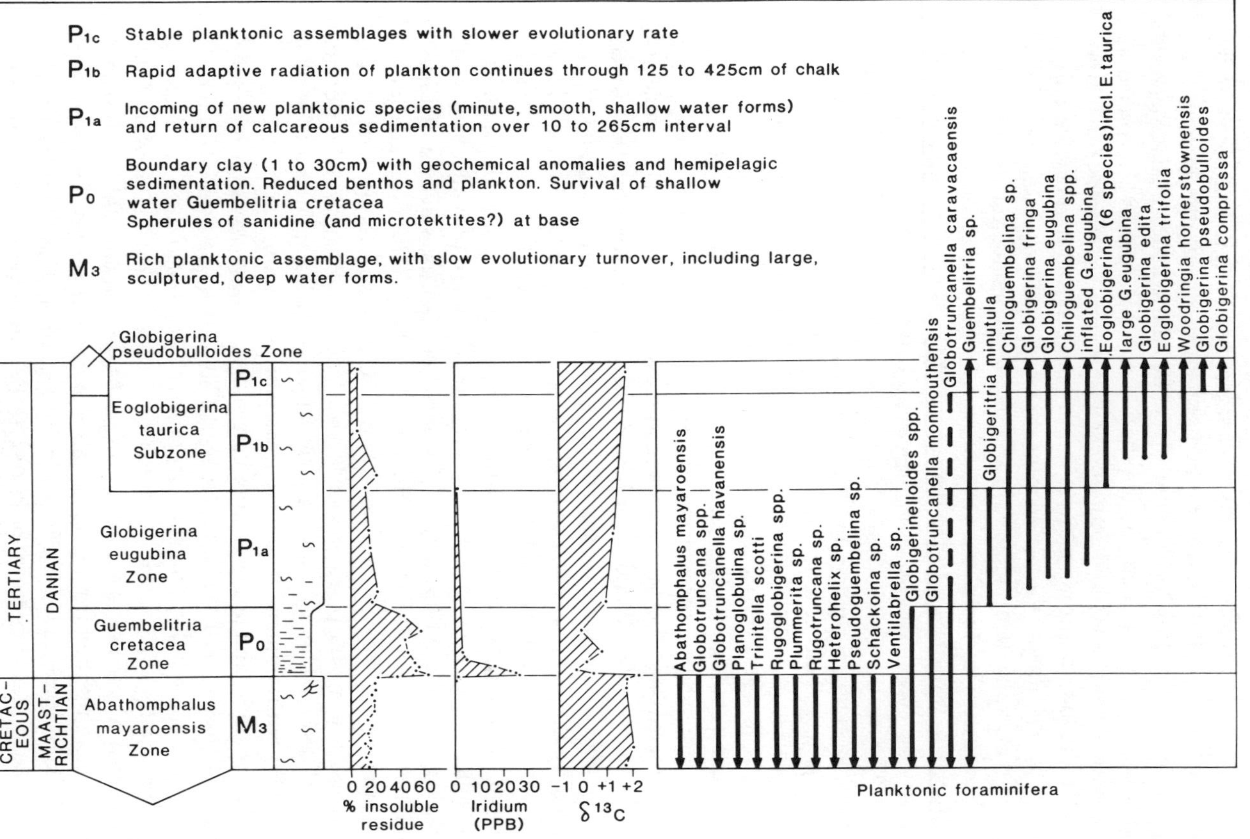

Fig. 2.4. Planktonic foraminiferid distributions, sedimentary changes and geochemical anomalies across the Cretaceous–Tertiary boundary clay, based largely on the Gredero section (from data in Smit 1982).

Palaeogene radiations and the Eocene–Oligocene boundary event

Gradual adaptive radiations of both planktonic and larger benthic foraminifera took place in the Palaeocene and expanded during the Eocene (Fig. 2.5). Planktonic foraminifera again diversified in a manner that suggests progressive depth stratification, culminating in conical forms interpreted as deeper (e.g. *Turborotalia*) and bubble-like forms (e.g. *Orbulinoides*) that were probably shallow to intermediate in depth (cf. Hart 1980). They were little affected by events at the Palaeocene/Eocene boundary, signalled by a negative $\delta^{13}C$ spike (Corfield 1987) and major taxonomic turnover in deep water benthics, with the elimination of many relict Cretaceous species (K. G. Miller, in Broadhead 1982; Berggren 1984). Planktonic taxa suffered more in the late Middle Eocene–early Late Eocene crisis (P14–15 boundary) with the elimination of long ranging *Acarinina, Morozovella,* and *Planorotalites,* and of more specialized forms such as *Clavigerinella* and *Orbulinoides* (e.g. Toumarkine and Luterbacher 1985).

The Palaeogene radiations of larger benthic foraminifera are exemplified by patterns in the diversity and size of miliolid alveolines (Figs. 2.5A–C) and rotaliid nummulites (Figs. 2.5D–F). Both originated in late Palaeocene times and diversified through the lower to middle Eocene, with trends towards increasing size, greater elongation and architectural complexity (e.g. Hottinger 1960; Blondeau 1972). Orbitoid discocyclines also thrived at this time. Nummulites reached their zenith in late Lutetian times, followed by elimination of all the large flat species in the late Eocene. Thus, late Eocene nummulites tend to be smaller, simpler and more variable. Tectonic movements in the Tethyan region may have been partly responsible for this, with evidence for late Lutetian provincialism contemporaneous with the formation of submarine nappes; stacking of thrust sheets near the Lutetian–Bartonian boundary led to the progradation of deltas and decline of nummulitids (Martin-Closas and Sierra-Kiel 1986). There was also a sharp fall in sea level at the end of the middle Eocene (Vail *et al.* 1977). The diversity pattern of planktonic foraminifera is similar (Fig. 2.5H) with an acme in the late Lutetian and decline thereafter.

Both planktonic and larger foraminifera suffered mass extinction across the Eocene–Oligocene (P17–18) boundary. The spread of events (anomaly 15 — extinction of disc-shaped *Discoaster, Globigerinatheka* — *Turborotalia cerroazulensis* s.l. — *Hantkenina* — *Cryptoprora ornata–Theocyrtis tuberosa* Zone radiolarians) does not suggest an impact-type catastrophe and seems rather to have resulted from a major fall in oceanic temperatures (Saunders 1985; Jenkins 1986). This is suggested by a gradual replacement of warm by cool water planktics from early to late Eocene (Boersma and Premoli-Silva 1985), by dwarfing of *T. cerrozazulensis* before extinction (Nocchi *et al.* 1985), by a sudden enrichment in ^{18}O in benthic *Oridorsalis* in the basal Oligocene (Saunders 1985), by a drop in the CCD at the boundary (Van

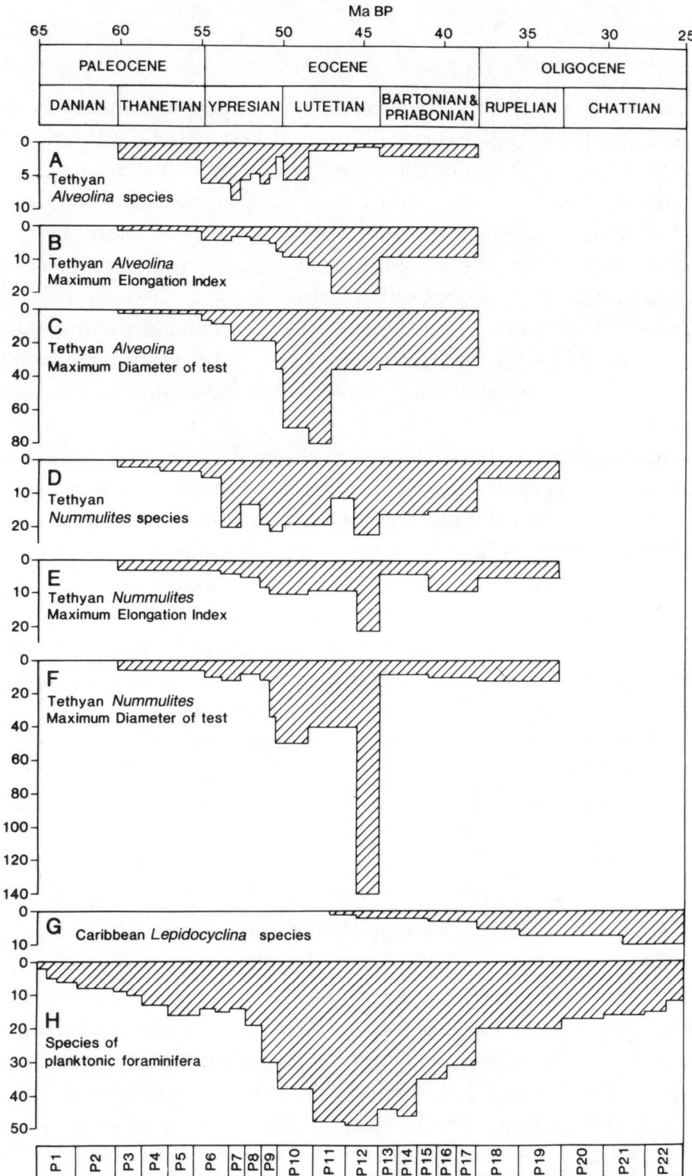

Fig. 2.5. Species diversity and size variation in selected larger benthic foraminifera and species diversity in planktonic foraminifera through the Palaeogene. A, B, and C show the diversity, maximum elongation and maximum diameter of Tethyan *Alveolina* spp. (from data in Hottinger 1960). D, E, and F show the diversity, maximum elongation (i.e. diameter divided by thickness) and diameter of Tethyan *Nummulites* spp. (from data in Blondeau 1972). G shows the independent development of Caribbean *Lepidocyclina* spp. (mainly from Butterlin 1984). H shows the rise and fall in the diversity of planktonic foraminiferid species (from data in Toumarkine and Luterbacher 1985).

Couvering *et al.* 1981) and by latitudinal patterns of extinction (Jenkins 1986). Sea level also fell over this interval (e.g. K. G. Miller, p. 122 in Broadhead 1982). The major global drop of 5–6°C in bottom and surface water temperatures actually occurred after the extinction event (Jenkins 1986). Extinctions of larger foraminifera were also significant in this interval, including 15 out of 29 Late Eocene genera cited by Wagner (1964), including true alveolinids, *Orbitolites* and discocylinids. Further extinction of small gibbous and flat *Nummulites* reduced the group to a few simpler forms. Among other survivors were small, globular, alveoline-like *Borelis* in the Tethyan area and orbitoid lepidocyclines in the Caribbean (Fig. 2.5G).

Planktonic species that continued across the Eocene–Oligocene boundary were those which lived at deepest levels and/or at coolest (i.e. high latitude) temperatures during the late Eocene according to Boersma and Primoli-Silva (1985). Oligocene assemblages were therefore largely reduced to simple, shallow *Globigerina* and deeper *Turborotalia*, with diversity remaining low throughout the Oligocene.

The Eocene–Oligocene extinction event appears to have been part of a progressive phenomenon, initiated in the middle Eocene, increasing during the late Eocene, and reaching its peak in latest Eocene by the combined action of oceanographic, tectonic, volcanic, and cosmic accelerations (Pomerol, in Jenkins 1984). The decline and extinction of shallow, low latitude foraminiferids clearly suggests regression and cooling as major factors.

Oligocene to Holocene events

Simple *Globigerina* survived through the relatively regressive Oligocene and underwent adaptive radiations from the end of this epoch into the middle Miocene (Fig. 2.6B). The succeeding decline in planktonic species diversity recorded by Tappan and Loeblich (1973) is not supported by more recent data (Fig. 2.6B). Larger foraminiferid faunas certainly suffered conspicuous decline after late Oligocene-early middle Miocene times (Fig. 2.6A) when giant *Lepidocyclina* (*Eulepidina*) thrived in the Caribbean and Mediterranean. The lepidocycline stock then died out diachronously during zones N6–7 in the Caribbean and Mediterranean, and through the late Miocene to early Pliocene in the Indo-Pacific (e.g. Berggren 1972; Adams 1983). Miogypsinids were another group that developed rapidly in the Early Miocene, only to suffer diachronous extinction (*c.* N10–11 in the Mediterranean and Caribbean, but N14 in Far East; Berggren 1972). The phased decline of Caribbean and Mediterranean larger foraminifera (e.g. Fig. 2.6A) is probably related to falling surface temperatures and the effects of successive closure of the eastern Mediterranean, western Mediterranean, and Panama

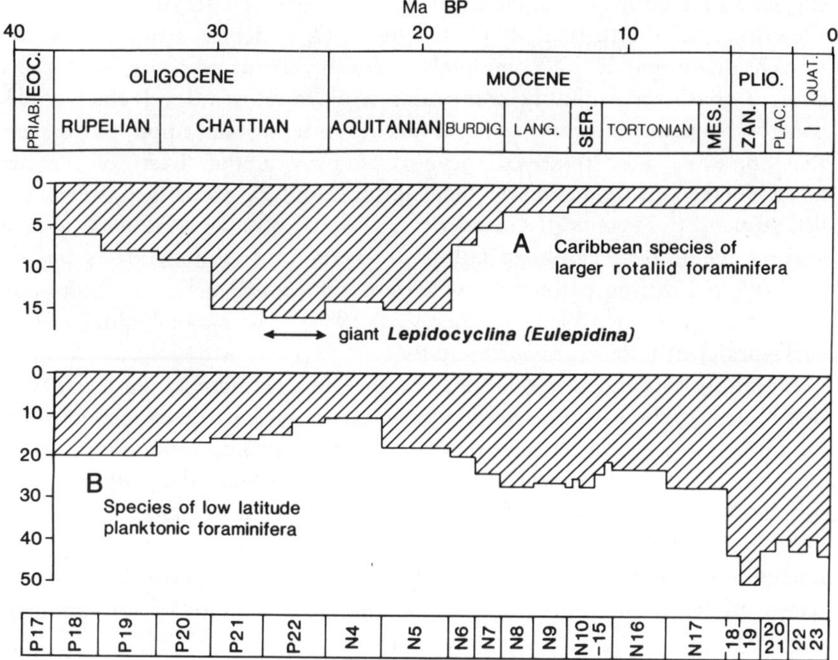

Fig. 2.6. Species diversity of selected Oligocene to Quaternary assemblages of Oligocene larger benthic and planktonic foraminifera. (A) Caribbean species of larger rotaliid foraminifera (unpublished compilation). (B) Species of low latitude planktonic foraminifera (from data in Bolli and Saunders 1985).

isthmus over the Aquitanian to Middle Pliocene interval (e.g. Adams 1983). Geographical isolation of the Caribbean also prevented the migration of new forms into an area with rather low origination rates. Surviving larger foraminifera in these two areas are known to be rather tolerant forms such as are found in seagrass communities and coastal lagoons, contrasting with the survival of more specialized forms in the Indo-Pacific.

The mid-Miocene event of Raup and Sepkoski (1986) does not emerge clearly from this data set.

Conclusions

This review has mainly considered the pattern of evolution and extinction in larger benthic and planktonic foraminifera. Both are marine groups known to thrive in the photic zone for all or part of their life history, and both include forms with dependence on endosymbionts and/or a tendency towards K-strategy. Preliminary studies on evolutionary palaeobiology suggests that periods of their diversification involved stratification of taxa according to depth-related parameters. This depth stratification was broken down partly or completely, presumably as a result of physical stresses that were notably coincident with rapid sea level change. The palaeobiological pattern of collapse has yet to be worked upon, but it seems that shallow water relatives survived to provide the 'base' of the next adaptive tier. More primitive benthic foraminifera seem to have suffered little during the events discussed here, perhaps because of deposit feeding, scavenging or bacteriophage habits and general r-strategy. It is possible they suffered during extinction events not highlighted here; this is certainly true for deep sea benthics (e.g. Berggren 1984). The palaeobiology of these survivors clearly deserves closer study.

Emerging evidence for relationships between extinction, sea level, water depth, and carbon isotope anomalies is certainly difficult to reconcile with an impact origin, except at the Cretaceous–Tertiary boundary. Of the periodic events noted by Raup and Sepkoski (1986), the Tithonian and middle Miocene events do not emerge clearly from available data. Thus, only the Cenomanian, Maastrichtian, and upper Eocene events are distinct among the better-dated events. A simple causal link between even these seems questionable, however, because four distinct kinds of situation were involved.

1. Anoxic pulse, e.g. the Cenomanian–Turonian boundary event. A positive $\delta^{13}C$ anomaly was associated with a major transgression and expansion of the midwater oxygen minimum zone forming black shales. Events caused a pause in the radiation of deeper planktonic foraminifera (perhaps through a rise in the oxygen minimum zone) and possibly coeval extinction of specialized larger benthic foraminifera of 'reefal' aspect (perhaps through drowning). The Toarcian anoxic event had a roughly similar character while the Frasnian–Fammenian event provides a Palaeozoic example. The Norian–Rhaetian extinctions are also close to an anoxic event.

2. Cooling pulse, e.g. the Eocene–Oligocene boundary event. This terminated a period of longer decline planktonic foraminifera and larger benthic foraminifera of 'reefal' aspect, associated with changes in sea level and temperature that were falling from an earlier high. The origination rate was low, following mass extinction at the end of the middle Eocene, with a series of continuing extinctions up to and after the boundary event.

3. Regressive pulse; e.g. the end Permian event. This terminated a period of longer decline in larger benthic foraminifera of 'reefal aspect' associated with falling sea level, and a negative $\delta^{13}C$ anomaly at the boundary. The origination rate was low and extinction culled the fauna in stages through the Kazanian and Dzhulfian.

4. Impact-associated pulse, i.e. the end-Cretaceous event. This falls within a regressive setting, with falling temperature and taxonomic

diversity in slight decline. Terminal extinction was associated with a negative $\delta^{13}C$ anomaly, iridium anomalies, and a sharp fall in temperature across the boundary. It seems that a mature depth-stratified fauna, already stressed by regression and falling temperatures, may have been pushed to extinction by impact-associated changes in sea level and climate.

Thus, the writer is generally in accord with Kauffman (1979), Hallam (1984), and many others in concluding that rising sea levels were closely associated with adaptive radiations, and either anoxic events or falling sea levels with mass extinctions. Evidence for impact-forced extinctions is only persuasive at the Cretaceous–Tertiary boundary, and even those were not instantaneous. More detailed studies on the pattern of collapse are needed to help us understand the palaeobiological character of these events. These data would obviously provide for a more fully reasoned discussion of their causes.

References

Adams, C. G. (1983). Speciation, phylogenesis, tectonism, climate and eustasy; factors in the evolution of Cenozoic larger foraminiferid bioprovinces. In *Evolution, Time and Space: the Emergence of the Biosphere* (ed. R. W. Sims, J. H. Price and P. E. S. Whalley), Systematics Association Special Volume No. 23, pp. 255–89. Academic Press, London.

Alvarez, L. W., Alvarez, W., Asaro, F., and Michel, H. V. (1980). Extraterrestrial cause for the Cretaceous-Tertiary extinction. *Science* **208,** 1095–8.

Altiner, D. (1984). The genus *Paradagmarita* and its biostratigraphic significance in south and southeast Turkey. In *Benthos '83; 2nd Int. Symp. Benthic Formaminifera* (ed. H. J. Oertli), p. 15. EPF Aquitaine, REP, and Total CFP, Pau.

Bé, A. W. H. (1977). An ecological, zoogeographic and taxonomic review of recent planktonic foraminifera. In *Oceanic Micropalaeontology,* Vol. 1 (ed. A. T. S. Ramsay), pp. 1–100. Academic Press, London.

Berggren, W. (1972). A Cenozoic time-scale — some implications for regional geology and paleobiology. *Lethaia* **5,** 195–215.

Berggren, W. (1984). Cenozoic deep water benthic foraminifera: a review of major developments since Benthonics '75. In *Benthos '83; 2nd Int. Symp. Benthic Foraminifera* (ed. H. J. Oertli), pp. 41–3. EPF Aquitaine. REP, and Total CFP, Pau.

Bilotte, M. (1984). Les grandes foraminiferes benthiques du Cretace superieur Pyreneen. Biostratigraphie. Reflexion sur les correlations Mesogeennes. In *Benthos '83; 2nd Int. Symp. Benthic Foraminifera* (ed. H. J. Oertli), pp. 61–7. EPF Aquitaine, REP, and Total CFP, Pau.

Blondeau, A. (1972. *Les Nummulites*. Vuibert, Paris.

Boersma, A. and Premoli-Silva, I. (1985). Eocene to Oligocene planktonic foraminiferal biogeography reflecting major climatic changes in the Atlantic. *Terra Cognita* **5,** 117.

Bolli, H. M. and Saunders, J. B. (1985). Oligocene to Holocene low latitude planktic foraminifera. In *Plankton Stratigraphy* (ed. H. M. Bolli, J. B. Saunders, and K. Perch-Nielsen), pp. 155–262. Cambridge University Press, Cambridge.

Bradshaw, J. S. (1961). Laboratory experiments on the ecology of foraminifera. *Contrib. Cushman Found. Foramin. Res.* **10,** 25–64.

Brasier, M. D. (1981). Sea level changes, facies changes and the late Precambrian–early Cambrian evolutionary explosion. *Precamb. Res.*, **17,** 105–23.

Brasier, M. D. (1982*a*). Architecture and evolution of the foraminiferid test—a theoretical approach. In *Aspects of Micropalaeontology* (ed. F. T. Banner and A. R. Lord), pp. 1–41. George Allen & Unwin, London.

Brasier, M. D. (1982*b*). Foraminiferid architectural history; a review using the MinLOC and PI methods. *J. Micropalaeont.* **1,** 95–105.

Brasier, M. D. (1984*a*). *Discospirina* and the pattern of evolution in foraminiferid architecture. In *Benthos '83; 2nd Int. Symp. Benthic Foraminifera* (ed. H. J. Oertli), pp. 87–90. EPF Aquitaine, REP, and Total CFP, Pau.

Brasier, M. D. (1984*b*), Some geometrical aspects of fusiform planispiral shape in larger Foraminifera. *J. Micropalaeont.* **3,** 11–5.

Brasier, M. D. (1986*a*). Form, function, and evolution in benthic and planktic foraminiferid test architecture. In *Biomineralization in Lower Plants and Animals* (ed. B. S. C. Leadbeater and R. Riding), Systematics Association Special Volume No. 30, pp. 251–68. Clarendon Press, Oxford.

Brasier, M. D. (1986*b*). Why do lower plants and animals biomineralize? *Paleobiology* **12,** 241–50.

Broadhead, T. W. (1982). *Foraminifera. Notes for a short course.* University of Tennessee, Knoxville.

Brumsack, H. J. (1986). Trace metal accumulation in black shales from the Cenomanian–Turonian boundary event. *Summaries of the First International Workshop of IGCP Project* 216, *Göttingen*, 13–9.

Butterlin, J. (1984). Remarques sur des espèces de grands foraminifères du Tertiare des Petites Antilles Françaises et sur la phylogenie des espèces americaines du genre *Lepidocyclina*. In *Benthos '83; 2nd Int. Symp. Benthic Foraminifera* (ed. H. J. Oertli), pp. 105–15. EPF Aquitaine, REP, and Total CFP, Pau.

Caron, M. (1985). Cretaceous Planktic foraminifera. In *Plankton Stratigraphy* (ed. H. M. Bolli, J. B. Saunders, and K. Perch-Nielsen), pp. 17–86. Cambridge University Press, Cambridge.

Cherchi, A., Schroeder, R., and Zhang, B.-G. (1984). *Cyclorbitopsella tibetica* n. gen., n. sp., a lituolacean foraminifer from the Lias of southern Tibet. In *Benthos '83; 2nd Int. Symp. Benthic Foraminifera* (ed. H. J. Oertli), pp. 159–65, EPF Aquitaine, REP, and Total CFP, Pau.

Conil, R. and Lys, M. (1977). Les transgressions Dinantiennes et leur influence sur la dispersion et l'evolution des foraminifères. *Mem. Inst. géol. Univ. Louvain.* **29,** 0–55.

Conkin, J. E., Conkin, B. M., Walton, M. M., and Neff, E. D. (1981). Devonian and early Mississippian smaller foraminiferans of southern Indiana and northwestern Kentucky. *GSA '81–Field Trip No. 2.*, 87–112.

Copestake, P. and Johnson, B. (1984). Lower Jurassic (Hettangian–Toarcian) foraminifera from the Mochras borehole, North Wales (UK) and their application to worldwise biozonation. In *Benthos '83; 2nd Int. Symp. Benthic*

Foraminifera (ed. H. J. Oertli), pp. 183–4. EPF Aquitaine, REP, and Total CFP, Pau.

Corfield, R. M. (1987). Patterns of evolution in Palaeocene and Eocene planktonic foraminifera. In *Micropalaeontology of Carbonate Environments* (ed. M. B. Hart), pp. 93–110, Ellis Horwood, Chichester.

Ekdale, A. A. and Bromley, R. G. (1984). Sedimentology and ichnology of the Cretaceous–Tertiary boundary in Denmark: implications for the causes of the terminal Cretaceous extinction. *J. sed. Pet.* **54,** 681–703.

Flessa, K. W. and Jablonski, D. (1983). Extinction is here to stay. *Paleobiology* **9,** 315–21.

Graciansky, P. C. de, *et al.* (1984). Ocean-wide stagnation in the late Cretaceous. *Nature, Lond.* **308,** 346–9.

Graup, G. (1985). Impact glasses (microtektites) from the C/T boundary, Lattengebirge, Bavarian Alps. *Terra Cognita* **5,** 246.

Hallam, A. (1975). *Jurassic Environments.* Cambridge University Press, Cambridge.

Hallam, A. (1981). The end-Triassic bivalve extinction event. *Palaeogeogr. Palaeoclimatol. Palaeoecol.* **23,** 1–32.

Hallam, A. (1984). The causes of mass extinctions. *Nature, Lond.* **308,** 686–7.

Hallam, A. (1986). The Pliensbachian and Tithonian extinction events. *Nature, Lond.* **319,** 765–8.

Hancock, J. M. and Kauffman, E. G. (1979). The great transgressions of the Late Cretaceous. *J. geol. Soc. Lond.* **136,** 175–86.

Hansen, H. J., Gwozdz, R., Hansen, J. M., Bromley, R. G., and Rasmussen, K. L. (1986). The diachronous C/T plankton extinction in the Danish Basin. *First International Workshop of IGCP Project 216, Göttingen,* 40–5.

Harland, W. B., Cox, A. V., Llewellyn, P. G., Pickten, C. A. G., Smith, A. G., and Walters, R. (1982) *A Geologic Time Scale.* Cambridge University Press, Cambridge.

Hart, M. B. (1980). A water depth model for the evolution of the planktonic Foraminiferida. *Nature, Lond.* **286,** 252–4.

Hart, M. B. (1985). Oceanic anoxic event 2, on shore and off-shore SW England. *Proc. Ussher Soc. 1985*, 183–90.

Hart, M. B. and Bailey, H. W. (1979). The distribution of planktonic Foraminiferida in the mid Cretaceous of N. W. Europe. *Aspekte der Kreide Europas, IUGS Ser A* **6,** 527–42.

Hart, M. B. and Ball, K. C. (1986). Late Cretaceous anoxic events, sea-level changes and the evolution of the planktonic foraminifera. In *North Atlantic Palaeoceanography* (ed. C. P. Summerhayes and N. J. Shackleton), Geological Society Special Publication No. 21, pp. 67–78. Blackwell Scientific Publications, Oxford.

Hart, M. B. and Bigg, P. J. (1981). Anoxic events in the late Cretaceous shelf sea of north-west Europe. In *Microfossils from Recent and Fossil Shelf Seas* (ed. J. W. Neale and M. D. Brasier), pp. 177–85. Ellis Horwood, Chichester.

Haynes, J. (1981). *Foraminifera.* Macmillan, London.

Hemleben, C. and Spindler, M. (1983). Recent advances in research on living planktonic Foraminifera. *Utrecht micropalaeont. Bull.* **30,** 141–70.

Hilbrecht, H., Arthur, M. A., and Schlanger, S. O. (1986). The Cenomanian-Turonian boundary event: sedimentary, faunal and geochemical criteria developed from stratigraphic studies in NW Germany. *First International Workshop of the IGCP Project 216, Göttingen,* 46–50.

Hohenegger, J. and Piller, W. (1975). Wandstrukturen und Grossgliederung der Foraminiferen. *Oesterr. Akad. Wiss. Sitzber. mathem.-Naturwiss. K*l. Abt. I, **184,** 67–96.

Holser, W. T., Margaritz, M., and Wright, J. (1986). Chemical and isotopic variations in the World Ocean during Phanerozoic time. *First International Workshop of the IGCP Project 216, Göttingen,* 55–67.

Hottinger, L. (1960). Recherches sur les Alveolines du Paleocene et de l'Eocene. *Schweiz. Paleont. Abh. Mem. Suisse Paleont.* **75/76,** 1–243.

Hottinger, L. (1982). Larger foraminifera, giant cells with a historical background. *Naturwissenschaften* **69,** 361–71.

House, M. R. (1975). Facies and time in the marine Devonian. *Proc. Yorks. geol. Soc.* **40,** 233–88.

House, M. R. (1985). Correlation of mid-Palaeozoic ammonoid evolutionary events with global sedimentary perturbations. *Nature, Lond.* **313,** 17–22.

Jenkins, D. G. (1984). Geological events at the Eocene–Oligocene boundary. *UK Contrib. IGCP, 1984 Rept,* pp. 32–6, The Royal Society, London.

Jenkins, D. G. (1986). The Eocene/Oligocene boundary in deep sea deposits. In *Terminal Eocene Events* (ed. Ch. Pomerol and I. Premoli-Silva), pp. 203–7. Elsevier, Amsterdam.

Jenkins, D. G. and Murray, J. W. (1981). *Stratigraphical Atlas of Fossil Foraminifera.* Ellis Horwood, Chichester.

Jenkyns, H. C. (1980). Cretaceous anoxic events: from continents to oceans. *J. geol. Soc. Lond.* **137,** 171–88.

Jenkyns, H. C. (1985). The early Toarcian and Cenomanian-Turonian anoxic events in Europe: comparisons and contrasts. *Geol. Rundschau* **74/3,** 505–18.

Johnson, B. (1981) Microfaunal biostratigraphy of the Dalan Formation (Permian), Zagros Basin, South-West Iran. In *Microfossils from Recent and Fossil Shelf Seas* (ed. J. W. Neale and M. D. Brasier), pp. 52–60. Ellis Horwood, Chichester.

Kalvoda, J. (1986). Upper Frasnian-Lower Tournaisian events and evolution of calcareous foraminifera — close links to climatic changes. *First International Workshop of the IGCP Project 216, Göttingen,* 68–76.

Kauffman, E. G. (1979). Cretaceous. In *Treatise on Invertebrate Paleontology. Part A, Introduction* (ed. R. A. Robinson and C. Teichert), pp. A418–87. Geological Society of America and University of Kansas Press, Boulder, Colorado, and Lawrence, Kansas.

Kauffman, E. G. (1981). Ecological reappraisal of the German Posidonienschiefer (Toarcian) and the stagnant basin model. In *Communities of the Past* (ed. J. Gray, A. J. Boucot, and W. B. N. Berry), pp. 311–81. Hutchinson & Ross, Stroudsberg, Pennsylvania.

Lipps, J. H. (1983). Biotic intercations in benthic Foraminifera. In *Biotic Interactions in Recent and Fossil Benthic Communities* (ed. M. J. S. Tevesz and P. and L. McCall), pp. 331–76. Plenum, New York.

Loeblich, A. R., Jr, and Tappan, H. (1974). Recent advances in the classification of the Foraminiferida. In *Foraminifera,* 1 (ed. R. H. Hedley and C. G. Adams), pp. 1–53. Academic Press, London.

Loh, H., Maul, B., Prauss, M., and Riegel, W. (1986). Effects and causes in a black shale event — the Toarcian Posidonia Shale of NW Germany. *First International Workshop of the IGCP Project 216, Göttingen,* 80–5.

Lys, M. (1984). Foraminiferes benthiques (non fusulinida): critères mineurs de biozones dans la biostratigraphie du Carbonifère et du Permien. In *Benthos '83; 2nd Int. Symp. Benthic Foraminifera* (ed. H. J. Oertli), pp. 393–400. EPF Aquitaine, REP, and Total CFP, Pau.

Margaritz, M., Moshovitz, S., Benjamini, C., Hansen, J. H., Hakansson, E. & Rasmussen, K.-L. (1985). Carbon isotope-, bio- and magnetostratigraphy across the Cretaceous–Tertiary boundary in the Zin Valley, Negev, Israel. *Newsl. Stratigr.* **15,** 100–13.

Martin-Closas,, C. and Sierra-Kiel, J. (1986). Two examples of evolution controlled by large scale abiotic processes: Eocene Nummulitids of the South Pyrenean Basin and Cretaceous Charophyta of Western Europe. *First International Workshop of the IGCP Project 216, Göttingen,* 86–90.

McGhee, G. R., Orth, C. J., Quintana, L. R., Gilmore, J. S., and Olsen, E. J. (1986). Geochemical analyses of the Late Devonian "Kellwasser Event" stratigraphic horizon at Steinbruch Schmidt (Federal Republic of Germany). *First International Workshop of the IGCP Project 216, Göttingen,* 91–6.

Nocchi, M. *et al.* (1985). The Eocene–Oligocene boundary in the Umbrian pelagic sequence in Italy. *Terra Cognita* **5,** 116.

Parsons, D. G. and Brasier, M. D. (1987). Changes in planktonic and benthic Foraminifera through Campanian–Maastrichtian phosphogenic cycles, southwest Atlas. In *Micropalaeontology of Carbonate Environments* (ed. M. B. Hart), pp. 111–20. Ellis Horwood, Chichester.

Piller, W. (1978). Involutinacea (Foraminifera) der Trias und des Lias. *Beitr. Palaont. Oesterreich* **5,** 1–164.

Poyarkov, B. V. (1979). *Evolution and Distribution of Devonian Foraminifera.* Izdat. Nauka, Moscow (In Russian).

Rast, U. and Graup, G. (1985). Iridium anomaly at the Cretaceous–Tertiary boundary, Lattengebirge, Bavarian Alps. *Terra Cognita* **5,** 246.

Raup, D. M. and Sepkoski, J. J., Jr (1984). Periodicity of extinctions in the geologic past. *Proc. nat. Acad. Sci. USA* **81,** 801–5.

Raup, D. M. and Sepkoski, J. J., Jr (1986). Periodic extinction of families and genera. *Science* **231,** 833–6.

Ross, C. A. and Ross, J. R. P. (1979). Permian. In *Treatise on Invertebrate Paleontology. Part A, Introduction* (ed. R. A. Robinson and C. Teichert), pp. A291–350. Geological Society of American and University of Kansas Press, Boulder, Colorado, and Lawrence, Kansas.

Rottger, R., Fladung, M., Schmaljohann, R., Spindler, M., and Zacharias, H. (1986). A new hypothesis: the so-called megalospheric schizont of the larger foraminifer *Heterostegina depressa* d'Orbigny, 1826, is a separate species. *J. foramin. Res.* **16,** 141–9.

Sandberg, P. A. (1983). An oscillating trend in Phanerozoic non-skeletal carbonate mineralogy. *Nature, Lond.* **305,** 19–22.

Saunders, J. B. (1985). Events around the Eocene–Oligocene boundary in the Bath Cliff section, Barbados, West Indies. *Terra Cognita* **5,** 115–6.

Seilacher, A. (1982). Ammonite shells as habitats in the Posidonia shales of Holzmaden — floats or benthic islands? *Neues Jb. Geol. Palaontol. Mh.* **1982,** 98–114.

Sepkoski, J. J. and Raup, D. M. (1986). Periodicity in marine extinction events. In *Dynamics of Extinction* (ed. D. K. Elliot), pp. 3–336. John Wiley & Sons, Chichester.

Silver, L. T. and Schulz, P. H. (eds.) (1982). Geological implications of impacts of large asteroids and comets on Earth. *Spec. Pap. Geol. Soc. Am.* **190.**

Sirel, E., Dager, Z., and Sozeri, B. (1986). Biostratigraphy of the Upper Cretaceous (Maastrichtian) — Paleocene (Danian) boundary in the Haymana-Polatli region, Central Turkey. *First International Workshop of the IGCP Project 216, Göttingen,* 122–3.

Smit, J. (1982). Extinction and evolution of planktonic foraminifera at the Cretaceous–Tertiary boundary after a major impact. *Spec. Pap. Geol. Soc. Am.* **190,** 329–52.

Tappan, H. (1982). Extinction or survival: selectivity and causes of Phanerozoic crises. *Spec. Pap. Geol. Soc. Am.* **190,** 265–76.

Tappan, H. and Loeblich, A. R., Jr. (1973). Evolution of the ocean plankton. *Earth Sci. Rev.* **9,** 207–40.

Toomey, D. and Mamet, B. L. (1979). Devonian Protozoa. *Spec. Pap. Palaeontol.* **23,** 189–92.

Toumarkine, M. and Luterbacher, H. (1985). Paleocene and Eocene planktic foraminifera. In *Plankton Stratigraphy* (eds. H. M. Bolli, J. B. Saunders, and K. Perch-Nielsen), pp. 87–154. Cambridge University Press, Cambridge.

Vail, P. R., Mitchum, R. M., and Thompson, S., III. (1977). Seismic stratigraphy and global changes of sea level. Pt 4: Global cycles of relative changes of sea level. *Mem. Am. Ass. Petrol. Geol.* **26,** 83–97.

Van Couvering, J. A. *et al.* (1981). The terminal Eocene event and the Polish connection. *Palaeogeogr. Palaeoclimatol. Palaeoecol.* **36,** 321–62.

Wagner, C. W. (1964). *Manual of Larger Foraminifera.* Bataafse Int. Petrol., Maatschappij NV, The Hague.

Xu Dao-Yi, Zhang Qin-Wen , Sun Yi-Ying, and Yan Zheng (1985). Three main mass extinctions—significant indicators of major natural divisions of geological history in the Phanerozoic. *Modern Geology* **9,** 1–11.

3. Patterns of extinction and survival in Palaeozoic corals

COLIN T. SCRUTTON

Department of Geology, The University, Newcastle upon Tyne, UK

Abstract

The two major subclasses of Palaeozoic corals, the Tabulata and the Rugosa, appear respectively in the Lower and Middle Ordovician. Analysed at the generic level, both show similar patterns of peaks of extinction close to successive system boundaries accompanied or followed by enhanced rates of diversification. Prior to the final disappearance of both groups at the end of the Permian, the most important extinction event occurs in the late Givetian, followed by the loss of most remaining genera about the Frasnian–Famennian boundary. Both subclasses survive into the Carboniferous through a very small number of long ranging genera, together with the continued evolution of distinctive Rugosa, mainly adapted to deeper water fine clastic environments. Although the Rugosa eventually rediversify vigorously in the Lower Carboniferous, the Tabulata never regain their early-mid Palaeozoic prominence. Coloniality and colonial integration appear to have little value overall in enhanced survivorship among Rugosa. Tabulate faunas are dominated successively by massive coenenchymal, massive perforate cerioid, and communicate fasciculate colony forms. A detailed analysis of generic ranges from the late Frasnian to the early Tournaisian reveals three extinction events among the rugose corals which can be matched to those in other groups. Environmental perturbations of eustatic and/or climatic origin are favoured as the principal causal factors.

Extinction and Survival in the Fossil Record (ed. G. P. Larwood), Systematics Association Special Volume No. 34, pp. 65–88. Clarendon Press, Oxford, 1988.

Introduction

Rugose and tabulate corals were an important and sometimes dominant element in shallow water carbonate and fine calcareous clastic shelf environments in the Palaeozoic. In addition, some rugose corals, although few tabulate corals, were specifically adapated for muddy sediments, usually considered to be of poorly oxygenated deeper water origin and in which goniatites are often a significant faunal element. The Rugosa and Tabulata are interpreted here in the sense of Hill (1981) and Scrutton (1979, 1984), respectively. The small group of Heterocorallia, also assigned to the Anthozoa and possibly related to the Rugosa (Hill 1981), are not included in this review.

Analysis of broad evolutionary trends in these two subclasses is facilitated by the recent documentation of generic and higher taxonomic level ranges by Hill (1981) (Figs 3.3 and 3.4). Although this data base is now somewhat dated and its stratigraphical resolution crude, its quality is otherwise excellent and has the advantage of internal consistency in taxonomic interpretation and selectivity. The first part of this review, concerned with overall patterns of extinction and survival in Palaeozoic corals, is based almost entirely on this source. Those few generic ranges shown as originating or terminating within the time units used (mainly epochs) by Hill (1981, tables 1, 3) are extended to the nearest boundary. Ranges indicated with a query are ignored; otherwise, all ranges are taken as continuous.

Apart from the end Permian extinction event, which resulted in the loss of both subclasses, the most significant period in the evolutionary history of the Palaeozoic corals was the late Devonian. In the second part of this review, the generic ranges of the Rugosa and Tabulata are analysed in detail from the late Frasnian, through the Famennian, and into the Tournaisian, based on data from various recent sources.

Characteristics of Palaeozoic corals

The general assumption is made that the corallites of all Palaeozoic corals were the products of skeletal deposition by polyps by analogy with the scleractinian corals. The presence of coenenchyme, deposited by interpolypal common colonial tissue, coenosarc, is not clearly differentiated in colonial rugose corals, but is a feature of some tabulate corals such as the heliolitids. Excellent general accounts of coloniality and skeletal structure in Palaeozoic corals may be found in Coates and Oliver (1973), and Hill (1981). Two doubtful solitary Tabulata (Hill 1981) are discounted here.

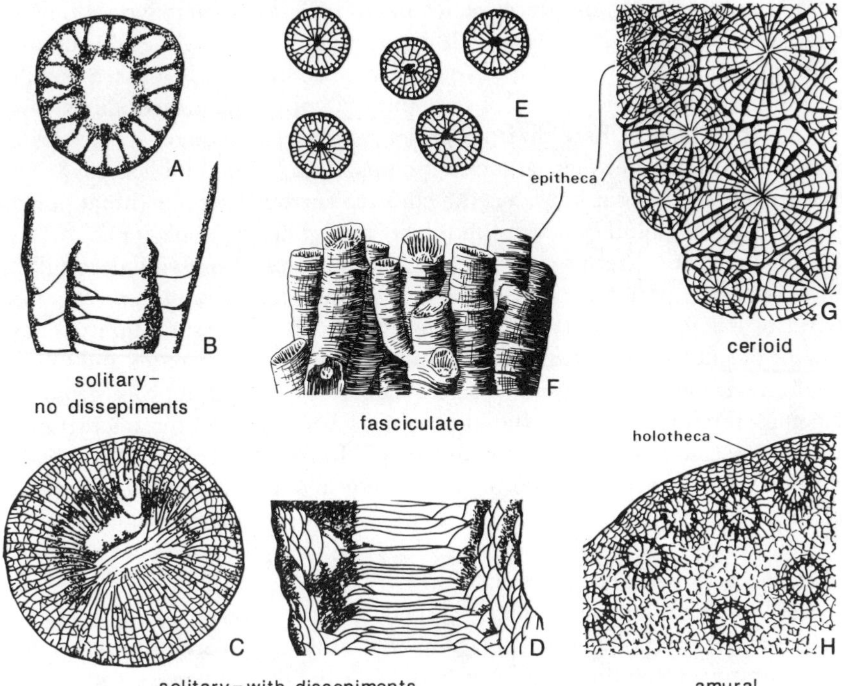

Fig. 3.1. Structural form in rugose coral coralla. A–B, solitary, non-dissepimentate: A, cross-section; B, longitudinal section; C–D, solitary dissepimentate; C, cross-section; D, longitudinal section. E–F, fasciculate; E, cross-section: F, exterior view. G, cerioid, cross-section. H, amural, cross-section. Based on various sources; not to scale.

1. *Rugosa*

Rugose corals occur as both solitary and colonial skeletons (coralla) composed of corallites varying from about 4 to 110 mm diameter (Fig. 3.1). For general purposes here, solitary corals are treated as a single group, but for more detailed analysis, they are split into two groups on the presence or absence of dissepiments. This division broadly coincides with the two ecological groups of shallow water forms with dissepiments and small, deeper water coralla with thickened skeletal elements, and no dissepiments. However, some poorly dissepimentate solitary Rugosa may also tolerate deeper, muddy environments.

Colonial Rugosa are divided into fasciculate, cerioid and more 'advanced' colonial forms (Fig. 3.1). Fasciculate coralla among Rugosa were clones of discrete polyps whose spatial relationships were determined by a continuous basal skeleton of cylindrical corallites (Fig. 3.1E,F). In cerioid coralla, individuals are in close contact, forming a massive colony of polygonal corallites. However, each is surrounded by an epithecal wall, and the external wall of the colony is not a single continuous holotheca but the sum of the external segments of the epithecae of peripheral corallites (Fig. 3.1G). It is inferred here that the presence of an epitheca reflects a lack of tissue continuity between adjacent polyps (Coates and Oliver 1973). In some massive Rugosa, however, the epitheca surrounding individual polyps is lost and a holotheca surrounds the colony as a whole (Fig. 3.1H). Various structural terms describe the relationships of individual corallites to each other, pseudocerioid, astraeoid, thamnasterioid, aphroid, etc. (see Scrutton 1968, Hill 1981), but they are united in the assumption of some degree of soft tissue continuity across the colony surface (Coates and Oliver 1973). Previously, the term plocoid has been applied to this group of colonies (Hill 1956, 1981), but this term is best retained for scleractinian corals possessing a clear coenenchyme (see Rosen 1986). No clear division into polyp and coenosarc can be inferred for at least the bulk of this group of integrated rugosan colonies by analogy with the scleractinian skeleton. Therefore, I introduce the term *amural* for this group in reference to the loss of epitheca (Fig. 3.1H).

The trend solitary–fasciculate–cerioid–amural is taken to represent an increase in interpolypal cooperation and ultimately continuity, together with improved stability resulting from an enlarged and more solid basal skeleton. These various structural states and resulting growth forms are of considerable ecological significance, and are analysed separately to test their importance in the evolution of the group as a whole. Where genera include species of more than one structural form, they are scored in the graphs (Fig. 3.3) proportionately.

2. *Tabulata*

Tabulate corals have smaller corallites than rugose corals, in the range 1–20 mm diameter. They are wholly colonial and include a wide variety of structural organization (Fig. 3.2). At the moment, not all are easily analysed in terms of the degree of soft tissue continuity and integration they represent. Many of the early cerioid imperforate group (Fig. 3.2J,K) distinguished here have obscure wall structure and all may not have possessed an epitheca surrounding individual corallites. Similarly, the fasciculate incommunicate group (Fig. 3.2B–D) may not all possess internal partitions isolating the terminal polyps and in some, external soft tissue continuity between polyps, investing the skeletal surface, is possible. Otherwise, these are thought to represent the lowest levels of colonial integration amongst tabulates. The cateniform group (Fig. 3.2P,Q) includes those both with and without mesocorallites. The latter are thought definitely to be coenenchymal, but all halysitids probably had polyps with

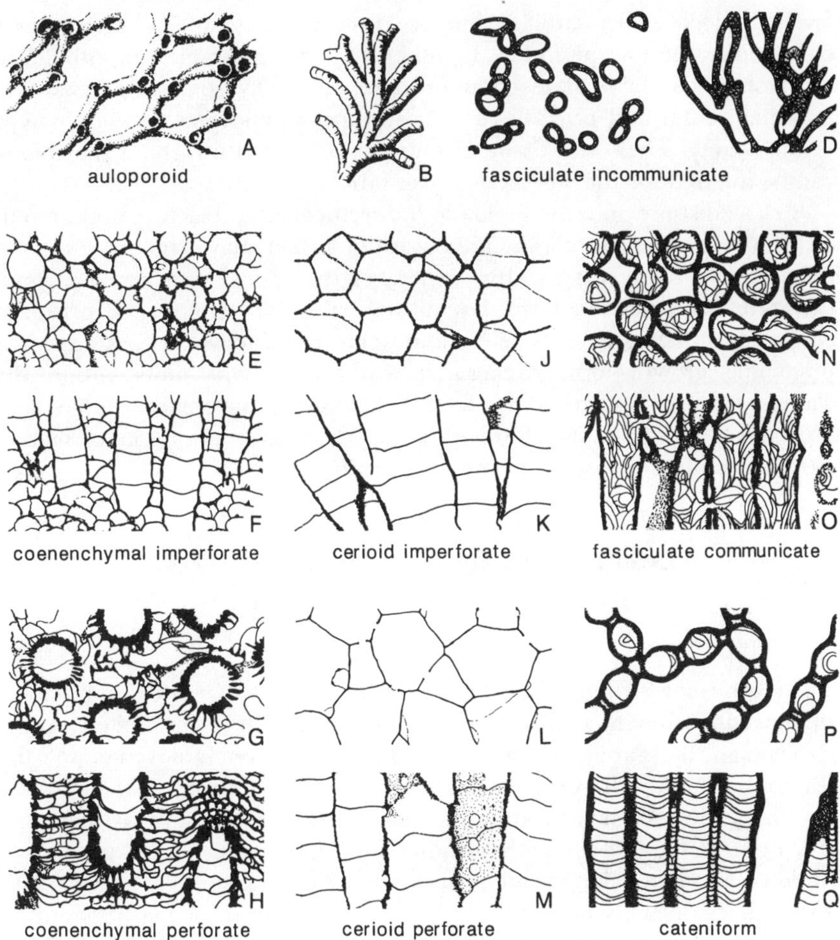

Fig. 3.2. Structural form in tabulate coral colonies. A, auloporoid, exterior view of adnate colony. B–D, fasciculate incommunicate: B, exterior view of erect colony; C, cross-section; D, longitudinal section. E–F, coenenchymal imperforate, G–H, coenenchymal perforate. J–K, cerioid incommunicate. L–M, cerioid communicate. N–O, fasciculate communicate. P–Q, cateniform. For E–Q in each case the first of the pair of figures is a cross-section, the second a longitudinal section. Based on various sources, not to scale.

continuous tissue laterally along the ranks of corallites. Auloporoid tabulate corals mostly, if not all, had internal continuity between polyps (Laub, in Coates and Oliver 1973, p. 13; Fig. 3.2A). The heliolitids (coenenchymal imperforate, Fig. 3.2E,F) and the mixed group comprising some sarcinulids, some syringoporids and specialized favositids (coenenchymal perforate, Fig 3.2G,H) would appear to represent comparable integration levels to plocoid scleractinians (but see discussion below). The mural pores of favositids (cerioid perforate, Fig. 3.2L,M) and the connecting tubules of syringoporoids (fasciculate communicate, Fig. 3.2N,O) suggest channels allowing neural and probably also gastric communication between polyps in the colony. Thus, soft tissue linkage of some sort throughout the colony can be inferred for the vast majority of tabulate corals.

A clear distinction must be made between colonial structure and growth form in some of these corals, although the former may strongly constrain the latter in many cases. Most genera of the cerioid and coenenchymal categories have massive form, but substantial numbers of cerioid perforate corals (the thamnoporoids, etc.) and some coenenchymal corals have a branching growth form. Generally, major taxonomic units within the Tabulata are characterised by corals of a specific colonial structure and often also a limited range of growth form. The relative evolutionary success of these groups is analysed in the next section.

Evolutionary patterns in Palaeozoic corals

1. General features

There are some striking similarities in the overall patterns of evolution of rugose and tabulate corals (Figs. 3.3 and 3.4). Both originated in the Ordovician, the Tabulata slightly earlier, in the Lower Ordovician and the Rugosa in the Middle Ordovician. It is considered unlikely that there was any direct evolutionary relationship between the two groups at that time (Scrutton 1979, 1984). Both steadily increased in numbers up to the Middle Devonian. The Tabulata show a more convex curve (Fig. 3.4), with a subsidiary diversity peak in the Upper Ordovician, reflecting pre-Silurian numerical generic dominance and a significant end Ordovician extinction event. Thereafter, the Rugosa always outstripped the Tabulata in generic diversity. Both subclasses suffered massive extinctions in the late Devonian. In the Lower Carboniferous, the Rugosa rediversified almost to their previous maxima before declining steadily in numbers of genera until their extinction at the end of the Permian. The Tabulata rediversify much more weakly, however, before showing a similar decline.

The graphs for rate of appearance of new genera (Figs 3.3D and 3.4D) clearly show the rapid initial stocking and post-Devonian restocking of coralline ecosystems and the effect of progressive saturation of the available ecospace. The graphs relative to continuing and all immediately preceding genera respectively relate to replacive and displacive evolution. Lower Carboniferous diversification was clearly replacive (see later section) and early Silurian diversification probably largely so. At other times the pattern is less clear. Enhanced rates of turnover near system boundaries are

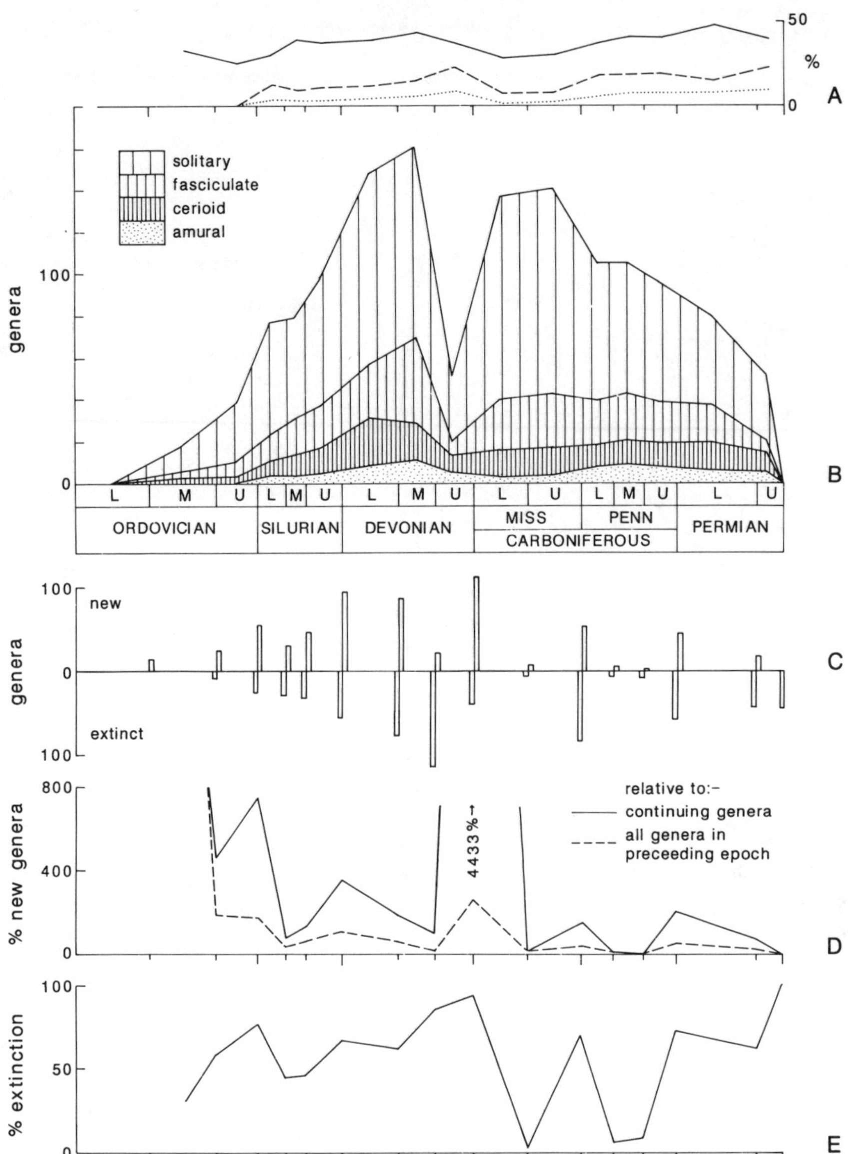

Fig. 3.3. Evolutionary patterns in Rugosa. (A) Colonial genera and amural genera both as a percentage of all contemporary Rugosa, and amural genera as a percentage of contemporary colonial Rugosa. (B) Cumulative standing diversity in numbers of genera of different structural categories of Rugosa. Categories discussed in text. (C) Number of genera becoming extinct during preceding unit, and new genera appearing in succeeding unit. (D) New genera appearing during succeeding unit, as a percentage of all genera surviving across unit boundary, and as a percentage of all genera recorded in preceding unit. (E) Extinctions as a percentage of all genera in preceding unit. All graphs relate to time scale in (B); units (mainly epochs) and data from Hill (1981); scaling after Harland *et al.* (1982).

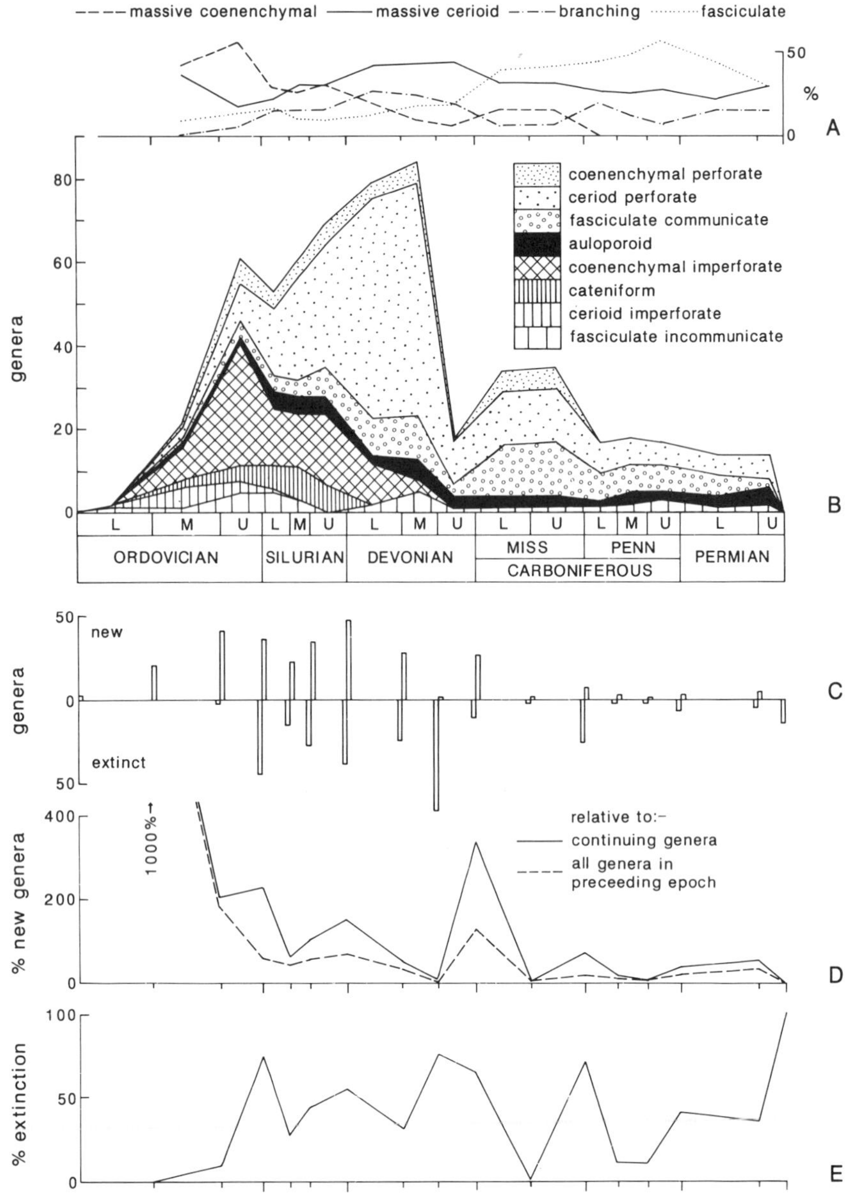

Fig. 3.4. Evolutionary patterns in Tabulata. (A) Massive coenenchymal, massive cerioid, branching, and fasiculate genera, each as a percentage of all contemporary Tabulata. (B) Cumulative standing diversity in numbers of genera of different structural categories of Tabulata: coenenchymal perforate (mainly Sarcinulida, some Favositida, and some Syringoporida), cerioid perforate (most Favositida), fasciculate communicate (most Syringoporida, some erect Auloporida), auloporoid

generally well marked (Figs 3.3C–E and 3.4C–E), but analysis at the series level obscures the full picture and artificially concentrates extinctions at stratigraphic boundaries (Newell 1982). Some Ordovician genera persist into the early Llandovery, which was a period of very low coral diversity. Most members of the characteristic Silurian fauna appeared in the middle and upper Llandovery (Scrutton, in press). There was a gradual change-over in faunas in the late Silurian and Devonian with a peak of diversification in the Pragian. The Lochkovian also was a period of relatively reduced diversity (Oliver and Pedder 1979; Scrutton, in press). The events at the Devonian–Carboniferous boundary are analysed in detail below, but briefly, the Famennian and early Tournaisian was an extended period of very low diversity. Permo-Carboniferous faunas show much less differentiation and are usually taken as a single evolutionary cycle (Federowski 1981; Hill 1981).

The bar charts of generic turnover (Figs 3.3C and 3.4C) emphasize the contrast in evolutionary activity in both groups pre- and post- the Devonian–Carboniferous boundary.

2. *Rugosa*

A striking feature of the Rugosa is the relative uniformity of the solitary to colonial ratio throughout the bulk of the Palaeozoic. However, there is a weak, but clear trend towards both a higher percentage of colonial genera and, amongst them, a higher percentage of amural genera (Fig. 3.3A). Both trends were reset by the end Devonian extinctions. As Bambach (1985, p. 205, fig. 8) emphasizes, colonial forms are always less diverse in numbers of genera post-Devonian than during their early to mid-Devonian acme (Fig. 3.3B). In percentage terms, however, the Rugosa are slightly more than equally successful in their trend to coloniality in the later Palaeozoic than in the early to mid-Palaeozoic.

It is interesting that towards the end of both the Devonian and the Permian, percentages of colonial corals generally drop whilst those of amural colonies continue to rise, suggesting enhanced survivorship among the latter as the shelf ecosystems were disrupted. However, when the shelf

---

(reptant Auloporida), coenenchymal imperforate (mainly Heliolitida), cateniform (mainly Halysitida), cerioid imperforate (mainly some Lichenariida), and fasciculate incommunicate (some Lichenariida, some erect Auloporida): see text for discussion. (C) Number of genera becoming extinct during preceding unit and new genera appearing in succeding unit. (D) New genera appearing during succeding unit, as a percentage of genera surviving across unit boundary and as a percentage of all genera recorded in preceding unit. (E) Extinctions as a percentage of all genera in preceding unit. All graphs relate to time scale in (E); units (mainly epochs) and data from Hill (1981); scaling after Harland *et al.* (1982).

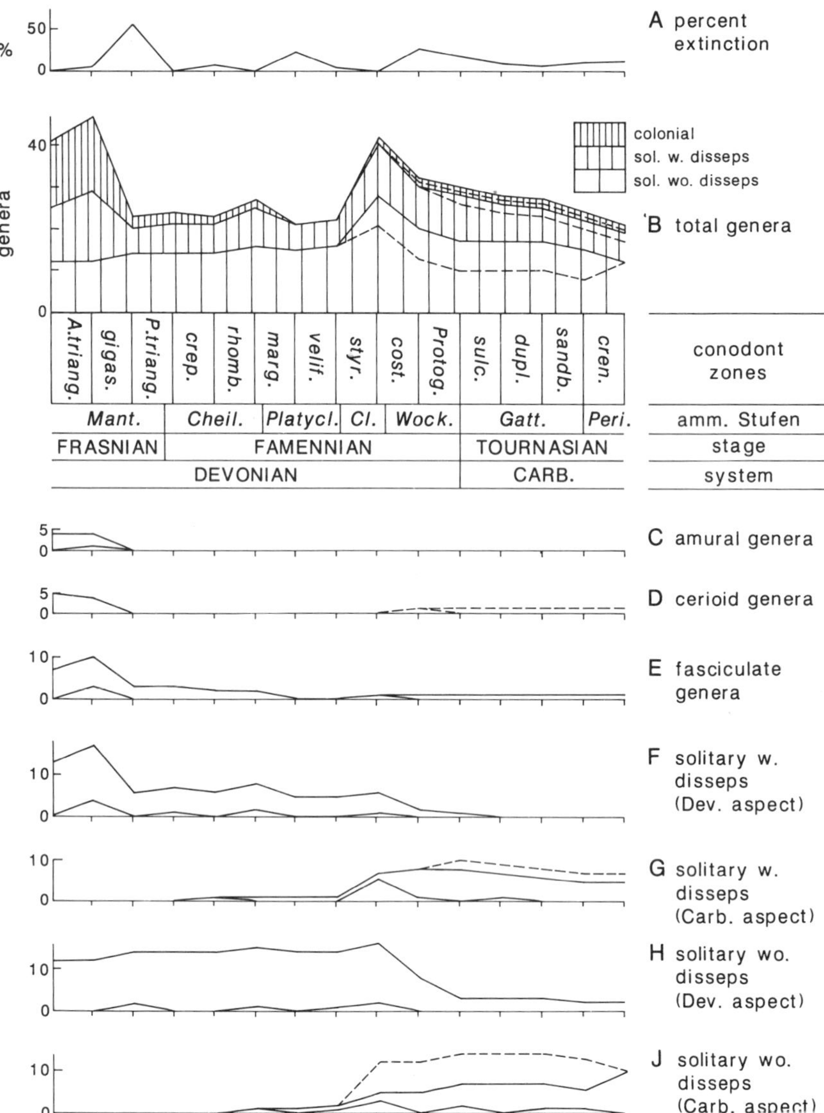

Fig. 3.5. Evolutionary history of Rugosa in the late Devonian and early Carboniferous. (A) Extinctions as a percentage of all genera in preceding biozone. (B) Cumulative standing diversity in numbers of genera of different structural categories of Rugosa. (C–J) Standing diversity in numbers of genera of: (C) amural genera; (D) cerioid genera; (E) fasciculate genera; (F) dissepimentate solitary genera of Devonian aspect; (G) dissepimentate solitary genera of Carboniferous aspect; (H) non-dissepimentate solitary genera of Devonian aspect; (J) non-dissepimentate solitary genera of Carboniferous aspect. Lower solid line indicates proportion of genera first appearing in succeding biozone. Dashed line indicates

coral faunas are finally almost eliminated at the end of the Frasnian, the amural forms disappear whilst some fasciculate and a few solitary genera survive for a while (Fig. 3.5). Rugosa persist, but in steadily dwindling numbers of genera, to the very end of the Permian (Flügel 1970: note that his 'Trias' is now regarded as Djulfian). An earliest Triassic fauna had been claimed (Ilina 1965), but its age disputed (Waterhouse 1976). The Scleractinia are regarded not as descendants of rugosans, but as representing the acquisition of a skeleton by a previously skeletonless group of anemones (Oliver 1980).

Mean generic survivorship from the Hill (1981) data is 28 Ma and shows no significant difference between the four structural categories recognised and little difference in their survivorship curves. As ranges are extended to series boundaries, however, this figure could be an overestimate by as much as 12 Ma.

3. Tabulata

By the Middle Ordovician, the full range of tabulate colonial organization had evolved and it is striking how it is progressively reduced through the history of the group (Fig. 3.4B). Different colonial structures dominate the group in turn, massive coenenchymal imperforate (heliolitids) up to the early Silurian, massive cerioid perforate (favositids) in the Devonian and fasciculate communicate (syringoporoids) post-Devonian (Fig. 3.4A). Branching coenenchymal and cerioid perforate colonies, those structurally closest to the successful modern scleractinian genus *Acropora*, are a significant, but never a dominant group among the tabulates. However, during the history of the group, there was a fairly steady trend from massive form towards less dense errect growth (branching/fasciculate/cateniform strategies). Excluding auloporids, massive growth forms made up 68 per cent on average of Ordovician tabulate corals by genus, then successively 58 per cent (Silurian), 57 per cent (Devonian), 49 per cent (Mississippian), 30 per cent (Pennsylvanian), and 44 per cent (Permian). The low level of evolutionary activity of the Tabulata in the Upper Palaeozoic is very apparent (Fig. 3.4C).

Mean generic survivorship calculated from the Hill (1981) data is 31 Ma for the Tabulata as a whole, but is subject to overestimate in the same way as for the Rugosa.

---

newly appearing genera of uncertain biozonal age. Genera of Devonian aspect are those of wholly or predominantly Devonian age, or which range throughout the scale of the diagram. Genera of Carboniferous aspect are those of wholly or predominantly Carboniferous age or which first appear in the Strunian fauna. All graphs relate to time scale in (B); conodont biozones arbitrarily of equal length. Data from various sources listed in text.

Events at the Devonian–Carboniferous boundary

1. *Introduction*

An analysis of generic ranges to the series level can demonstrate evolutionary events only in their crudest form and obscures the number, nature, and precise timing of extinction events. The most important period for the Palaeozoic corals was undoubtedly the late Devonian. Sufficient well dated faunas are now available, mostly assembled in a valuable review by Poty (1986), to document generic ranges against the conodont chronology for the late Frasnian to the early Tournaisian and to demonstrate the complex changes occurring over that interval. Additional data from Hill and Jell (1970), Fedorowski (1981), Friakova *et al.* (1985), Hill (1981), McLean (1984), McLean and Pedder (1984), Poty (1985), Sando and Bamber (1984), Shilo *et al.* (1984), and Weyer (1981, 1984) have been incorporated here with those of Poty (1986) with due regard to limitations in the accuracy of their dating. Questionable data have been ignored, including single Famennian records of colonial Rugosa which may be based on derived specimens.

2. *Rugosa*

The Rugosa had already been massively reduced in diversity in the late Givetian (*varcus*-Zone) when the Streptelasmatina, Arachnophyllina, Lycophyllina and the bulk of the Columnariina, Cyathophyllina, Ketophyllina, Ptenophyllina and Cystiphyllida became extinct. A more modestly diverse but successful, cosmopolitan and distinctive shallow water fauna evolved through the Frasnian to be largely eliminated by substantial extinctions in the *gigas*-Zone (Fig. 3.5). All remaining massive Rugosa apparently disappeared then (unless doubtful Famennian records are substantiated), whilst fasciculate and solitary dissepimentate genera were much reduced. It is striking, however, that solitary non-dissepimentate Rugosa were unaffected at this time, at least at the generic level. In fact, Pedder's (1982) documentation of rugose coral species extinction immediately across the Frasnian–Famennian boundary shows that only 6 of 150 shallow water and 4 of 10 deeper water species survive.

The assumed continuity of generic range in Fig. 3.5 obscures the fact that virtually no early Famennian corals are known. Small numbers of new solitary genera appear in the *rhomboidea*- and *marginifera*-Zones forming a distinctive fauna of corals with thin skeletal elements best known from Poland where they occur in muddy sediments interpreted as deposited in a calm, turbid, poorly aerated environment (Rozkowska 1969, 1981). The first Rugosa of the distinctive shallow water Strunian (late Famennian) fauna (*Campophyllum*) and of the Carboniferous (*Cyathaxonia*), also appear at this time. A minor extinction event in the *marginifera*-Zone eliminated surviving fasciculate genera among others.

Solitary dissepimentate corals of Devonian aspect persisted into the beginning of the Tournaisian (Fig. 3.5F). The longest surviving representative of this group, *Tabulophyllum,* seems to have played an important role in the origin of the Carboniferous shallow water fauna. Species of this genus are accepted as likely ancestors for several Strunian corals arising in the *costatus*-Zone, principally '*Dibunophyllum*' *praecursor, Siphonophyllia,* and possibly also *Molophyllum* (Poty 1986). Some typical Strunian corals disappeared very early in the Tournaisian, leaving a relict and highly impoverished shallow water Tournaisian fauna. The great rediversification of the Rugosa, stocking up the extensive carbonate shelves of the Lower Carboniferous, did not take place until late Tournaisian and early Visean times. *Palaeosmilia* and solitary dissepimentate corals with axial structures, present in the latest Devonian, become uncommon or absent until that time (Poty 1986).

Evolutionary activity in the *costatus*-Zone also saw the development of a second distinctive suite recorded from Poland, this time of largely non-dissepimentate solitary corals with thick skeletal elements (Rozkowska 1969). This fauna appears to have inhabited well aerated, shallower waters, but elsewhere, in deeper water, goniatite-bearing muds, more new genera of non-dissepimentate Rugosa appeared, less precisely dated, over the period of the *Clymenia* to early *Gattendorfia* Stufen. At the same time, similar corals of Devonian aspect were much reduced by late Famennian extinctions. Only *Ufimia* and *Amplexocarinia* appear to have significant pre- and post-Famennian ranges, but further revisions may modify this.

3. *Tabulata*

Data for the Tabulata are much less reliable as little detailed modern work has been done on these corals at this level (Fig. 3.6). A diversity minimum in the Famennian is indicated, with most evolutionary activity concerning generic turnover in the Favositida. The genera that appear to have survived the late Devonian crises on present records were *Michelinia, Praemichelinia* (Favositida), *Syringopora, Tetraporinus, Thecostegites* (Syringoporida), *Aulopora,* and *Cladochonus* (Auloporida). The fact that the extinction events reflected by the Rugosa are not clearly mirrored by the Tabulata may only result from the limitations of the current data base.

Discussion

1. *Evolutionary patterns in Palaeozoic corals*

The polyps of solitary Rugosa rarely evolved the ability to bud. Overall, only some 36 per cent of rugose coral genera are colonial and of these, 84 per cent retain an epitheca around individual corallites. Even in amural

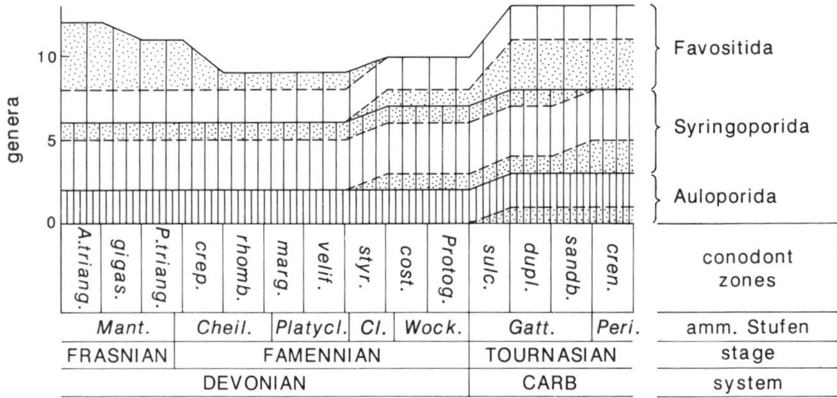

Fig. 3.6. Evolutionary history of Tabulata in the late Devonian and early Carboniferous. Cumulative standing diversity in numbers of genera of tabulate coral orders. Within each field, upper stippled area represents genera becoming extinct; lower stippled area represents newly appearing genera. conodont biozones arbitrarily of equal length. Data from various sources listed in text.

colonies, a holotheca covers the outer and under surface of the skeleton. The presence of the epitheca/holotheca appears to have been a great limiting factor in the evolutionary and ecological potential of the group. The relative inability to encrust limited the success of both colonial rugosans and tabulates in high energy environments, where stromatoporoids were more successful. Physiological limitations, together with the limited plasticity of growth form and probable lack of any algal symbiosis resulted in the poor performance of the colonial strategy in rugose corals (see, for example, Coates and Oliver 1973).

It is difficult to interpret the relative success of colonial form in tabulate corals. The perforate coenenchymal form so successful among scleractinians had no similar success in the Palaeozoic. Massive imperforate coenenchymal colonies appear ultimately to have been adaptively inferior to perforate cerioid forms and the perforate/communicate pattern dominates later tabulate faunas. The pores and tubules interconnecting corallites in this group could well have allowed gastric continuity in addition to colony-wide neural networks. Interpolypal gastric linkages were probably lacking in Palaeozoic imperforate coenenchymal colonies. However, they seem likely for perforate coenenchymal colonies which makes their relative lack of success difficult to understand. Most tabulate colonies also possess an external epitheca or holotheca, but had generally a much greater plasticity of growth form than rugosan colonies. This, and their greater number and variety of integrated colonial forms probably resulted in their more prominent contribution to high energy ecotopes.

The rise in diversity of the Rugosa and Tabulata broadly parallel the rise to dominance of the stromatoporoids as a principal component in mid-Palaeozoic reefs. Whereas both groups of corals may play major roles in biostromes and bioherms, they are subservient to stromatoporoids in fully diversified reef environments where more stable massive growth forms are the dominant contributors. The major extinction event in the late Givetian coincided with the beginning of the collapse of the mid-Palaeozoic reef ecosystem. Of the colonial Rugosa at this time, the more highly integrated amural forms were more successful and, together with cerioid perforate tabulates of flexible growth form, in places dominated deeper water bioherms on drowning reef platforms. With the almost complete elimination of the carbonate shelf environment at the end of the Frasnian, the survival of solitary and fasciculate genera only among the Rugosa suggests the possibility that increased movement of sediment on the shelves, with which these forms were more able to cope, may have been a significant factor in that extinction event. The syringoporoids and michelinids among the tabulate corals, the former fasciculate, and the latter with the largest corallites of any tabulate coral, were also adapted to cope with such conditions. Precise mechanisms are highly speculative, however, and other factors entrained by the important eustatic transgression correlated with this event (House 1975*a,b,* 1985) could have been as or more important. In addition, a meteorite impact has been postulated as implicated in this mass extinction (see below).

The apparently less severe effect of the end Frasnian extinctions on small, solitary, non-dissepimentate deeper water corals in contrast to their reaction to the late Famennian event suggests probable differences in the nature of the two crises. One possibility is a more significant role for temperature and/or circulatory fluctuations, which might disrupt the stability of deeper water environments, restricted to the latter.

The curious paucity of massive tabulate corals post-Devonian may relate to the disappearance of the stromatoporoids and the lack of reef environments comparable to those of the mid-Palaeozoic. Cause and effect are difficult to separate (Coates and Oliver 1973), but with the extensive carbonate shelves of the Carboniferous, it is difficult to avoid the conclusion that the poor development of reefs is a result of the lack of a suitable constructor. Late Palaeozoic shallow water reefs tend to be predominantly of algal and calcareous sponge construction, often associated with high stress environments in which corals are uncommon (James 1983). The disappearance of the stromatoporoids themselves is a curious phenomenon, with quite well diversified faunas having reappeared in the latest Famennian after a post-Frasnian gap. They were presumably further victims of the extinction event at the Devonian–Carboniferous boundary, although Mistiaen (1984) has ingeniously suggested that they were facultative calcifiers responding to Sandberg's (1983) aragonite-inhibiting cycles, reappearing in skeletal form in the Mesozoic. Whatever the reason, the rugose and tabulate corals, inhibited as encrusters by their epithecae or holothecae, possibly also limited in lacking algal symbionts, were incapable of assuming the stromatoporoid role. Rugose corals were the more successful massive colonial forms in the upper Palaeozoic, possibly because of the ecological flexibility conferred by their larger polyp size. It is interesting to note that corallite size in the contemporary massive tabulates reaches the highest attained in the group. Nevertheless, it is striking that the trend to massive and integrated colonies among the Rugosa is no more pronounced post-Devonian than before, despite the much reduced diversity of massive tabulate corals with which they might be expected to be in competition. This may be another reflection of the paucity of suitable niches caused by the disappearance of the stromatoporoids.

The steady decline in diversity in the Rugosa towards the end of the Palaeozoic presumably reflects the gradual diminution of habitat area resulting from the assembly of Pangaea and its associated eustatic and climatic changes (Valentine and Moores 1972). The effect is less marked in the Tabulata, already much reduced in diversity with the contraction of carbonate shelves in the later Carboniferous. Overall, the Palaeozoic corals mirror fairly closely the pattern of familial diversity of Sepkoski's (1984) Palaeozoic evolutionary fauna.

2. *Extinction events*

There is strong correlation between the late Devonian extinction events documented here and those established in the goniatite record by House (1985). In addition to a broad correlation between the late Givetian coral extinctions and House's Taghanic (Middle *varcus*-Zone) event, his Kellwasser (end *gigas*-Zone), Enkeberg (*marginifera*-Zone), and Hangenberg (late *costatus*- to early *sulcata*-Zones) events correlate well with peaks in the graph of percentage extinction in the coral record (Fig. 3.5A). Only his Annulata (early *styriacus*-Zone) event is undetected.

A wide variety of causal mechanisms have been postulated to explain extinction events in general and well documented mass extinction events like that at the Frasnian–Fammenian boundary in particular (McLaren 1982). Of these, various aspects of sea-level oscillations and climatic changes have been the most widely quoted 'conventional' factors (for example, Newell 1952, 1967; House 1975*b*; Copper 1977; Heckel and Witzke 1979). More recently, much has been written of the likely effects of extra-terrestrial causes such as meteorite (bolide) impacts, cometary showers, and cosmic radiation flux on the biosphere. Although a bolide has been proposed to explain the late Devonian extinctions (McLaren 1970, 1984), the only Ir anomaly so far documented in this interval occurs in the Canning Basin (Playford *et al.* 1984). Not only are the siderophile element ratios not compatible with those in meteorites but the anomaly is associated with microstromatolites of the iron cyanobacterium *Frutexites* which concentrates platinum group elements in its filaments. In addition, the anomaly occurs at the base of the upper *P. triangularis*-Zone, which assuming the accuracy of the biostratigraphic correlation, is clearly later than the Frasnian-Famennian extinction event at the end of the *gigas*-Zone.

Raup and Sepkoski (1984) have proposed a post-Palaeozoic 26 Ma mass extinction periodicity which has fostered speculation on forcing astronomical cycles (see review by Hallam 1984). They were unable to verify any Palaeozoic cyclicity and House (1985) points out that the mid-Palaeozoic extinction events he describes occur with a frequency much too short to support a cycle of the order of 26 Ma. They also differ in degree and effect, and appear to lack any regularity of interval.

House considers that these events, which often correlate with widespread shelf black shales, may be immediately related to eustatic sea level rises and euxinic conditions caused by the encroachment of the oxygen minimum zone onto the shelf. The interaction of tectonically driven global eustacy and the subtle environmental variations implied by sedimentary microrhythms facially mirroring the major sedimentary rhythms, and possibly forced by Milankovitch-type cycles, provides a mechanism which seems better to accommodate the differing characteristics of successive mid-Palaeozoic extinction events. Other extra-terrestrial factors such as major impacts, which undoubtedly occur and cannot be dismissed, may play a role in specific cases. Although precise causal effects are difficult to determine, eustatic and climatic perturbations of the environment are favoured here as the principal causes of extinction events in the coral record.

Conclusions

1. Rugose corals first appear in the Middle Ordovician and become extinct in the late Permian. Rates of diversification at the generic level decline from the Ordovician to the Upper Devonian, and from the Lower Carboniferous to the Permian following the near complete end Devonian extinctions. Subsidiary peaks of extinction and diversification occur close to all other system boundaries. Total generic diversity rises to a maximum in the Middle Devonian and declines from a second peak in the Lower Carboniferous to the end of the Permian.

2. Few rugose polyps had the ability to bud. The percentage of colonial genera rises slightly to the Middle Devonian, is reset by the end Devonian extinctions and rises again to a maximum in the Lower Permian. The percentage of amural genera (defined as colonial Rugosa lacking an epitheca separating constituent corallites) also rises with a similar pattern,

but peaks successively in the Upper Devonian and Upper Permian suggesting some enhanced survivorship over other colonial forms approaching major extinction events. Only 6 per cent of rugosan genera overall achieve the amural state, which is thought to reflect the limitation imposed by the epitheca on the evolutionary potential of the group.

3. Tabulate corals first appear in the Lower Ordovician and become extinct in the late Permian. Only pre-Silurian are they superior to the Rugosa in numbers of genera. Their patterns of extinction and diversification closely mirror those of the Rugosa, except that they suffer more significant end Ordovician extinctions, and their post-Devonian rediversification is markedly limited. This reduced success in the later Palaeozoic is thought to be linked to the disappearance of the stromatoporoids at the end of the Devonian. Stromatoporoids had created the niches in which many tabulate corals were particularly successful during the mid-Palaeozoic.

4. The full range of tabulate coral colonial organization evolved in the Ordovician. This diversity of structure was gradually reduced with time. Faunas were successively dominated by massive coenenchymal imperforate (Ordovician–early Silurian), massive cerioid perforate (Devonian) and fasciculate perforate (Permo–Carboniferous) colonies. There is a steady trend in growth form from dominantly massive towards more open erect (branching/fasciculate/cateniform) throughout the history of the group.

5. The Rugosa show a complex evolutionary history in the later Devonian at the generic level with important extinction events in the *varcus*- and *gigas*-Zones and less marked extinction peaks in the *marginifera*-Zone and the latest Famennian. Shallow water and deep water faunas seem to be affected differently. Some new genera evolve to give distinct if poorly diversified Famennian faunas. Early Tournaisian diversity is very low. Tabulate corals are less well documented over this interval but on present evidence appear to survive through only seven long ranging genera.

6. These extinction events match most of those reported over the same interval in the goniatite record and some are also reflected in other groups. This pattern of several, close spaced (*c.* 4 Ma apart on average) extinction events of varying severity and faunistic impact is considered to favour eustatic and climatic environmental perturbations as a causal mechanism.

Addendum

While this paper was in press, Sorauf and Pedder (1986) published an analysis of Upper Devonian rugose coral genera and species, with specific reference to the end Frasnian extinction event. Although their generic data base differs in some respects from that assembled here, the general pattern of events at that time that they describe agrees well with that outlined in this paper.

Acknowledgements

I thank Michael House, University of Hull, and Howard Armstrong, University of Newcastle upon Tyne, for comments on the manuscript. Christine Jeans (University of Newcastle upon Tyne) drafted text-figures 3.3–3.6.

References

Bambach, R. K. (1985). Classes and adaptive variety: the ecology of diversification in marine faunas through the Phanerozoic. In *Phanerozoic Diversity Patterns* (ed. J. W. Valentine), pp. 191–253. Princeton University Press, Princeton, New Jersey.

Coates, A. G. and Oliver, W. A., Jr (1973). Coloniality in zoantharian corals. In *Animal Colonies* (ed. R. S. Boardman, A. H. Cheetham, and W. A. Oliver, Jr), pp. 3–27. Dowden, Hutchinson and Ross, Stroudsberg, Pennsylvania.

Copper, P. (1977). Paleolatitudes in the Devonian of Brazil and the Frasnian-Famennian mass extinction. *Palaeogeogr. Palaeoclimat. Palaeoecol.* **21,** 165–207.

Fedorowski, J. (1981). Carboniferous corals: distribution and sequence. *Acta palaeont. polon.* **26,** 87–160.

Flügel, H. W. (1970). Die Entwicklung der rugosen Korallen im hohen Perm. *Verh. Geol. B.-A. Wien* **1970,** 1, 146–61.

Friakova, O., Galle, A., Hladil, J., and Kalvoda, J. (1985). A lower Famennian fauna from the top of the reefoid limestones at Mokra (Moravia, Czechoslovakia). *Newsl. Stratigr.* **15,** 43–56.

Hallam, A. (1984). The causes of mass extinctions. *Nature, Lond.* **308,** 686–7.

Harland, W. B., Cox, A. V., Llewellyn, P. G., Pickton, C. A. G., Smith, A. G., and Walters, R. (1982). *A Geologic Time Scale.* Cambridge University Press, Cambridge.

Heckel, P. H. and Witzke, B. J. (1979). Devonian world palaeogeography determined from distribution of carbonates and related lithic palaeoclimatic indicators. *Spec. Pap. Palaeont.* **23,** 99–123.

Hill, D. (1956). Rugosa. In *Treatise on Invertebrate Paleontology* (ed. R. C. Moore) Vol. F, pp. 233–324. Geological Society of America and University of Kansas Press, Boulder, Colorado, and Lawrence, Kansas.

Hill, D. (1981). Rugosa and Tabulata. In *Treatise on Invertebrate Paleontology* (ed. C. Teichert) Vol. F, Suppl. 1, pp. xl + 1–762. Geological Society of America and University of Kansas Press, Boulder, Colorado, and Lawrence, Kansas.

Hill, D. and Jell, J. S. (1970). Devonian corals from the Canning Basin, Western Australia. *Bull. Geol. Surv. W. Austr.* **121,** 1–158.

House, M. R. (1975*a*). Facies and time in Devonian tropical areas. *Proc. Yorks. Geol. Soc.* **40,** 233–88.

House, M. R. (1975*b*). Faunas and time in the marine Devonian. *Proc. Yorks. Geol. Soc.* **40,** 459–90.

House, M. R. (1985). Correlation of mid-Palaeozoic ammonoid evolutionary events with global sedimentary perturbations. *Nature, Lond.* **313,** 17–22.

James, N. P. (1983). Reef environment. In *Carbonate Depositional Environments* (ed. P.

A. Scholle, D. G. Bebout, and C. H. Moore), pp. 345–440. A.A.P.G. Memoir 33, Tulsa, Oklahoma.

Ilina, T. G. (1965). Chetyrekhluchevye korally pozdnei permi rannego triasa Zakavkazya. *Trudy pal. Inst.* **107,** 1–105.

McLaren, D. J. (1970). Time, life and boundaries. *J. Paleont.* **44,** 801–15.

McLaren, D. J. (1982). Frasnian–Famennian extinctions. *Spec. Pap. Geol. Soc. Am.* **190,** 477–84.

McLaren, D. J. (1984). Bolides and biostratigraphy. *Bull. Geol. Soc. Am.* **94,** 313–24.

McLean, R. A. (1984). Upper Devonian (Frasnian) rugose corals of the Hay River region, Northwest Territories, Canada. *Palaeontogr. Am.* **54,** 470–4.

McLean, R. A. and Pedder, A. E. H. (1984). Frasnian rugose corals of western Canada. Part 1: Chonophyllidae and Kyphophyllidae. *Palaeontographica A* **184,** 1–38.

Mistiaen, B. (1984). Disparition des Stromatopores paleozoïques ou survie du groupe: hypothèse et discussion. *Bull. Soc. géol. Fr.* **26,** 1245–50.

Newell, N. D. (1952). Periodicity in invertebrate evolution. *J. Paleont.* **26,** 371–85.

Newell, N. D. (1967). Revolutions in the history of life. *Spec. Pap. Geol. Soc. Am.* **89,** 63–91.

Newell, N. D. (1982). Mass extinctions—illusions or realities? *Spec. Pap. Geol. Soc. Am.* **190,** 257–63.

Oliver, W. A., Jr (1980). The relationship of the scleractinian corals to the rugose corals. *Paleobiology* **6,** 146–60.

Oliver, W. A. Jr, and Pedder, A. E. H. (1979). Rugose corals in Devonian stratigraphical correlation. *Spec. Pap. Palaeont.* **23,** 233–48.

Pedder, A. E. H. (1982). The rugose coral record across the Frasnian–Famennian boundary. *Spec. Pap. Geol. Soc. Am.* **190,** 485–9.

Playford, P. E., McLaren, D. J., Orth, C. J., Gilmore, J. S., and Goodfellow, W. D. (1984). Iridium anomaly in the Upper Devonian of the Canning Basin, Western Australia. *Science* **226,** 437–9.

Poty, E. (1985). A rugose coral biozonation for the Dinantian of Belgium as a basis for a coral biozonation of the Dinantian of Eurasia. *C.R. X Congr. Int. Strat. Géol. Carbonifère* **4,** 29–31.

Poty, E. (1986). Late Devonian to early Tournaisian rugose corals. *Annl. Soc. géol. Belg.* **109,** 65–74.

Raup, D. M. and Sepkoski, J. J., Jr (1984). Periodicity of extinctions in the geologic past. *Proc. nat. Acad. Sci. USA* **81,** 801–5.

Rosen, B. R. (1986). Modular growth and form of corals: a matter of metamers? *Phil. Trans. Roy. Soc. Lond.* **B313,** 115–42.

Rozkowska, M. (1969). Famennian tetracoralloid and heterocoralloid fauna from the Holy Cross Mountains (Poland). *Acta palaeont. polon.* **14,** 1–187.

Rozkowska, M. (1981). On Upper Devonian habitats of rugose corals. *Acta palaeont. polon.* **25,** 597–611.

Sandberg, P. A. (1983). An oscillating trend in Phanerozoic non-skeletal carbonate mineralogy. *Nature, Lond.* **305,** 19–22.

Sando, W. J. and Bamber, E. W. (1984). Coral zonation of the Mississippian System of western North America. *C.R. IX Congr. Int. Strat. Géol. Carbonifère, Washington, Urbana, 1979* **2,** 289–300.

Scrutton, C. T. (1968). Colonial Phillipsastraeidae from the Devonian of south-east Devon, England. *Bull. Br. Mus. nat. Hist.* (*Geol.*) **15,** 181–281.

Scrutton, C. T. (1979). Early fossil cnidarians. In *The Origin of Major Invertebrate Groups* (ed. M. R. House), Systematics Association Special Volume No. 12, pp. 161–207. Academic Press, London.

Scrutton, C. T. (1984). Origin and early evolution of tabulate corals. *Palaeontogr. Am.* **54,** 110–8.

Scrutton, C. T. (in press). Silurian coral and stromatoporoid biostratigraphy. In *A Global Standard for the Silurian System* (ed. C. H. Holland and M. G. Bassett). National Museum of Wales, Cardiff.

Sepkoski, J. J., Jr (1984). A kinetic model of Phanerozoic taxonomic diversity. III. Post-Palaeozoic families and mass extinctions. *Paleobiology* **10,** 246–67.

Shilo, N. A., *et al.* (1984). Sedimentological and palaeontological atlas of the late Famennian and Tournaisian deposits in the Omolon region (NE-USSR). *Annl. Soc. géol. Belg.* **107,** 137–247.

Sorauf, J. E. and Pedder, A. E. H. (1986). Late Devonian rugose corals and the Frasnian–Famennian crisis. *Can. J. Earth Sci.* **23,** 1265–87.

Valentine, J. W. and Moores, E. M. (1972). Global tectonics and the fossil record. *J. Geol.* **80,** 167–84.

Waterhouse, J. B. (1976). World correlations for Permian marine faunas. *Pap. Dept. Geol. Univ. Queensland* **7,** 2, 1–232.

Weyer, D. (1981). Korallen der Devon/Karbon Grenz aus hemipelagischer Cephalopoden-Fazies im mitteleuropäischen variszischen Gebirge–*Bathybalva* n.g., *Thuriantha* n.g. (Rugosa). *Freib. Forsch. C* **363,** 111–25.

Weyer, D. (1984). Korallen im Paläozoikum von Thüringen. *Hall. Jb. Geowiss.* **9,** 5–33.

Appendix

The following are the data used in the construction of Figs. 3.5 and 3.6. For the ranges, no entry indicates origin or termination outside the range of the diagrams. Conodont zone names in parentheses indicate maximum extensions of range used in the diagrams, but with an element of uncertainty, whilst names not in parentheses indicate ranges accepted as well authenticated. Sources refer to records *within* the range of the diagrams only. Abbreviations are: Fed, Fedorowski 1981; F, Friakova *et al.* 1985; H&J, Hill and Jell 1970; M, McLean 1984; M&P, McLean and Pedder 1984; P, Poty 1985; P′, Poty 1986; S&B, Sando and Bamber 1984; S, Shilo *et al.* 1984; W, Weyer 1981; W′, Weyer 1984.

Generic name	First record	Last record	Source
A. Rugosa			
Argutastraea	—	*A. triang*	P′
Scruttonia	—	*A. triang*	P′
Aristophyllum	—	*gigas*	P′
Charactophyllum	—	*gigas*	P′
Chonophyllum	—	*gigas*	P′ (?)
Fedorowskicyathus	—	*gigas*	P′
Frechastraea	—	*gigas*	M, P′
Hankaxis	—	*gigas*	P′
Hexagonaria	—	*gigas*	M, P′
Hunanophrentis	—	*gigas*	M
Iowaphyllum	—	*gigas*	P′
Marisastrum	—	*gigas*	P′
Mictophyllum	—	*gigas*	M, P′
Parasmithiphyllum	—	*gigas*	M, M&P
Peneckiella	—	*gigas*	P′
Phacellophyllum	—	*gigas*	P′
Phillipsastrea	—	*gigas*	P′
Piceaphyllum	—	*gigas*	P′
Tarphyphyllum	—	*gigas*	M&P
Wapitiphyllum	—	*gigas*	M&P, P′
Alaiophyllum	—	*gigas*. (*crep.*)	F
Disphyllum	—	*crep.*	F
Smithiphyllum	—	*marg.*	P′
Thamnophyllum	—	*marg.*	P′
Nalivkinella	—	*marg.* (*cost.*)	P′
Pterorrhiza	—	*marg.* (*cost.*)	P′
Temnophyllum	—	*gigas* (*Prot.*)	H&J, P′
Catactotoechus	—	*vel.* (*Prot.*)	H&J
Asthenophyllum	—	*cost.*	P′
Guerichiphyllum	—	*cost.*	P′
Metrioplexus	—	*cost.*	P′ (?)
Neaxon	—	*cost.*	P′
Pseudomicroplasma	—	*cost.*	P′
Syringaxon	—	*cost.*	H&J, P′
Famennelasma	—	*cost.* (*Prot.*)	W′
Metriophyllum	—	*cost.* (*Prot.*)	P′
Neaxonella	—	(*Prot.*)	P′, W′
Tabulophyllum	—	*sulc.*	P′, S
Laccophyllum	—	*sand.*	P′
Amplexocarinia	—	—	P′
Ufimia	—	—	P′
Bouvierphyllum	*gigas*	*gigas*	M&P
Ceciliaphyllum	*gigas*	*gigas*	M
Debnikiella	*gigas*	*gigas*	P′
Kakisaphyllum	*gigas*	*gigas*	M&P
Rachaniephyllum	*gigas*	*gigas*	P′
Smithicyathus	*gigas*	*gigas*	M, P′
Sudetia	*gigas*	*gigas*	P′
Trigonella	*gigas*	*gigas*	P′

Generic name	First record	Last record	Source
Petraiella	*P. triang.*	*vel.*	P′
Thecaxon	(*P. triang.*)	(*marg.*)	P′, W′
Kozlowskinia	(*crep.*)*marg.*	*marg.*	P′, W′
Campophyllum	*rhomb.*	—	P′, S
Kielcephyllum	*marg.*	*marg.*	P′
Gorizdronia	*marg.*	*cost.*	P′, S(?)
Cyathaxonia	*marg.*	—	P′, W′
Hillaxon	(*marg.*)*vel.*	*cost.*	P′
Famaxonia	(*styr.*)	(*Prot.*)	P′, W′
Pentaphyllum	(*styr.*)	—	P′
Czarnockia	*cost.*	*cost.*	P′
Friedbergia	*cost.*	*cost.*	P′
"*Clisiophyllum*" *omaliusi*	*cost.*	*sulc.*	P′, S(aff.)
"*Dibunophyllum*" *praecursor*	*cost.*	*sulc.*	P′, S(aff.)
Molophyllum	*cost.*	*dupl.*	P′
Euryphyllum	*cost.*	—	P′
Fasciculophyllum	*cost.*	—	P′
Melanophyllum	*cost.*	—	P′, S
Paleosmilia	*cost.*	—	P′
Pseudoclaviphyllum	*cost.*	—	P′
Siphonophyllia	*cost.*	—	P′, S
Pentaphyllum (*Commutia*)	(*cost.*)*sulc.*	*cren.*	P′
Saleelasma	(*cost.*)*sulc.*	—	P′
Antikinkaidia	(*cost.*)	(*cren.*)	P′
Dalnia	(*cost.*)	(*cren.*)	P′
Bradyphyllum	(*cost.*)	—	P′
Calophyllum	(*cost.*)	—	P′
Plerophyllum	(*cost.*)	—	P′
Soshkineophyllum	(*cost.*)	—	P′
Caninia	*Protog.*	—	P′, S
Dagmaraephyllum	(*Prot.*)	—	Fed
Pseudoendophyllum	(*Prot.*)	—	Fed
Uralinia	(*Prot.*)	—	Fed
Bathybalva	*sulc.*	*dupl.*	W
Drewerelasma	*sulc.*	*sand.*	P′, W′
Parasiphonophyllia	*dupl.*	*sand.*	P′, S
Thuriantha	*sand.*	*sand.*	W
Lophophyllum	*cren.*	—	P
B. Tabulata:			
Thamnopora	—	*gigas*	M&P
Natalophyllum	—	*crep.*	F
Scoliopora	—	*crep.*	F
Alveolites	—	*gigas* (*styr.*)	H&J, M&P
Roemeripora	—	*sulc.* (*sand.*)	S
Aulopora	—	—	W′
Cladochonus	—	—	W′
Michelinia	—	—	S, S&B, W′
Praemichelinia	—	—	—
Syringopora	—	—	F, S, S&B

Generic name	First record	Last record	Source
Tetraporinus	—	—	—
Thecostegites	—	—	S
Fuchungopora	*cost.*	—	S
Palaeacid	*cost.*	—	P′, S&B, W′
Rossopora	*dupl.*	—	W′
Sutherlandia	*dupl.*	—	W′
Yavorskia	*dupl.*	—	S
Ortholites	*sand.*	—	S

4. Extinctions and survivals in the Brachiopoda and the dangers of data bases

DEREK V. AGER

Department of Geology, University College of Swansea, Swansea, UK

Abstract

Undoubtedly there were major extinction eposides in the history of the brachiopods, for example at the end of the Frasnian and around the end of the Permian. There are also interesting examples of survivors — the spiriferids, the productids, the rhynchonellids. However, there is also a great danger that many of the alleged mass extinctions (for example, at the end of the Cretaceous) are artefacts produced simply by the way the information is recorded.

Introduction: the dangers of data bases

I do not deny for one moment that there were crises in the history of the brachiopods, with episodes of mass extinction and interesting examples of survival. In fact, the brachiopods are particularly useful for studying such phenomena in that they range from almost the simplest imaginable bivalve shells in the Early Cambrian to their sadly reduced state at the present day, when they cannot be far off their final extinction.

However, I am particularly concerned in this paper to draw attention to the man-made artefacts which make such episodes of mass extinctions more apparent than real and which also tend to put them in the wrong places in the stratigraphical record. This must be clarified before we start making deductions.

Extinction and Survival in the Fossil Record (ed. G. P. Larwood), Systematics Association Special Volume No. 34, pp. 89–97. Clarendon Press, Oxford, 1988.

We used to make observations and then suggest theories to explain them. Now we acquire data bases and formulate models. As a geologist talking to other scientists I now find it more convincing and persuasive if I say that I am sending my students into the field to upgrade their data bases. What they are doing is what geologists have always done, they are making geological maps. Apart from the phoney jargon which just covers up the obvious, it must be said that data are dangerous, especially in palaeontology. The greatest synthesis of palaeontological data ever produced was the multi-volumed *Treatise on Invertebrate Paleontology* inaugurated and edited by the late lamented Raymond C. Moore. He intended it to cover every known genus of fossil invertebrate and we are now reaching the stage of revising the volumes one by one. The *Treatise* is unbelievably useful as a source-book of information about the world's fossil faunas, but almost as soon as it was published the devotees of data descended on it and started juggling with figures. They started counting families and genera and making deductions about extinctions, quite regardless of how little this list of names really means in terms of what life was really like in the past. Though very useful in taxonomy, the *Treatise* largely reflects the spread of shallow shelf seas across particular continents and the distribution of palaeontologists around the world.

For example, a quick analysis of one brachiopod order in the Treatise (the Orthida) shows 53 type species coming from the USA, 44 from western Europe (mostly Britian), 24 from Estonia, 24 from Czechoslovakia, and only 18 from the rest of the world. This is obviously a wild distortion based on the chance of geological history and on the concentration of palaeontologists. There is also a bias produced by particular authors (in this case Cooper, Bancroft, Opik, and Barrande).

In the Mesozoic brachiopods, it is obvious to me that the number of nominal genera is disproportionate between the three systems of the Mesozoic, with the Jurassic far in the lead for purely monographic reasons, the Cretaceous second and the Triassic third (Ager *et al.* 1972, fig. 1). There is no reason to suppose that this reflects the reality of taxonomic diversity during the times in question. What is more the names are probably unevenly distributed between shallow water and deeper water forms, with the former exaggerated due to their greater accessibility (i.e. in non mountainous areas and thinner successions with less tectonic confusion).

It also seems to me that the brachiopods are slowly replaced, as one climbs up the Mesozoic succession, by the upstart molluscs. This happened first on muddy sea-floors, then on shallow water carbonate sea-floors, and finally as the bivalves and gastropods spread into deeper water. Of course there are exceptions, but this is the general pattern that is obvious to those who work on these things. Then the data jugglers moved in and proved by counting genera in the Treatise that it was not so. If they had been counting species it might have meant something, but what are genera? A living species is a rather clearly defined entity. There are problems, such as the continuously graded ring of seagull species around the North Atlantic, but generally speaking a species can be defined as a group of organisms capable of inter-breeding and of producing fertile off-spring. In palaeontology it is more difficult, as we cannot breed our fossils or cross the time planes, but we can at least look for a statistically significant unity of characters. Every specialist tells stories of the excessive creation of new supposed species in his own favourite group. Mine is of 19 species of a certain zeilleriid brachiopod in one thin bed of rock in one quarry in Somerset.

However, genera have even less validity. There cannot be any reality to a genus. Most genera relate to monographic bursts rather than to evolutionary ones. They can be nothing more than pragmatic conveniences in the mind of the palaeontologist. I tell my students when they are struggling with the literature that the ultimate criterion can only be that of usefulness. If it is useful to have a name then have one. I suppose that the cladists would say that it was the branching before that of the non-interbreeding species, but that can mean nothing to a palaeontologist.

However, the devotees of data dive in and do things with named genera, or what is worse, with families, which are even further divorced from reality. Thus, Cherfas (1985) plotted 25 000 'genera', 20 000 of which are said to be extinct, on a graph showing the extinctions through time with marked peaks at certain 'moments' in time. This follows Sepkoski's well-known claim of a 26-million-year cycle of family extinctions blamed on a dark companion of our Sun, aptly named Nemesis. As a palaeontologist I am extremely sceptical of all these deductions, since I know only too well the lack of reality in named fossil species, genera, and families. It seems to me extremely naive to claim something meaningful on the basis of names given by scores of different palaeontologists.

Thus, when I wrote my section of the *Treatise* (Ager 1965) our distinguished editor added some words about the great changes that occurred in my group between the Palaeozoic and the Mesozoic, but I rejected these comments because I just did not believe them. Certainly, the names change, certainly the classification changes, even the names of parts of the anatomy change. However, this was all a study of the psychology of palaeontologists rather than a study of palaeontology. The harsh facts are that in the brachiopods hardly any Palaeozoic specialists work also on the Mesozoic and vice versa. So inevitably the names used for things in the Palaeozoic do not get carried through into the next era. The same control operates at all levels.

Though I am something of a 'catastrophist' in my interpretation of the geological record and have claimed a 60 million year cyclicity (Ager 1976), I am not convinced about the 26 million year extinction events that are now fashionable. I certainly do not see such a cycle in the history of the brachiopods though I have deduced something of that order (i.e. 30 million years) in the pattern of regression, transgressions and climatic optima through Mesozoic times (Ager 1981).

I cannot see how anyone can be dogmatic about fossil species and genera when they are based on such subjective criteria. Thus, in the Mesozoic rhynchonellids we find sharp peaks in the Pliensbachian and Bajocian, due almost entirely to the abundance of brachiopods at those levels in the west of England and the work of one arch splitter — S. S. Buckman (see Ager *et al.* 1972, figs 6 and 7).

What is more it is easy to produce peaks if you compartmentalize your data. It all depends on the class interval. Very often in accumulating information about brachiopods from around the world, I have no stratigraphical range more accurate than 'Cretaceous' or even 'Mesozoic'. I am particularly lucky if I get 'Upper Cretaceous'. I cannot help, therefore, but produce an evident mass extinction of genera at the end of that era. That is not to say that I do not believe that something happened then, but certainly it is vastly exaggerated. Thus, no major taxonomic category of brachiopods became extinct at that time and if you go to New Zealand (as I just have) you would think that this was where the brachiopods really began!

We see the same thing even more blatantly in extinctions at the end of the Devonian Period. Cherfas was much better about this than most data analysers. He emphasized that the major extinctions happened at the end of the penultimate stage of the Devonian — the Frasnian — rather than at the end of the period itself. Nevertheless, his graph still shows the biggest drop in numbers between the Devonian and the Carboniferous, presumably because of the class interval fixation mentioned above.

It may be an unfashionable doctrine, but I believe in experience in palaeontology rather than digested data. After more than 30 years sweating (or soaking) over the Mesozoic rocks of the world looking (among other things) for their brachiopods, I will say dogmatically that I think I know what was going on. If someone tells me that he has collected brachiopods in the Upper Triassic rocks of the eastern Alps, or the Jurassic of Kutch in north-west India, or the Cretaceous of the Mississippi Embayment, then I have a pretty good idea what he has found. I cannot often quantify it and if I did I would not trust the results. I certainly do not necessarily trust other people's fossil names. Still less do I trust them when they are put in the literature and other people then use them to prove something. Thus, for some inexplicable reason certain binomens seem to have got into the collective palaeontological subconscious, *Atrypa reticularis,* for example, which is allegedly found everywhere.

So to hell with data bases and let us trust our eyes! What other computer could recognize a friend's face on the other side of the world out of its five billion inhabitants? And do not let us forget, as the Dutch geologist R. Hooykaas wrote recently, 'we have to keep in mind that our ancestors' eyes were as good as ours'. I tend to believe that old species therefore are best. In the case of my brachiopods that means the names given by the Sowerbys and Davidson.

It is, however, worth considering in more detail the specific moments in geological time when something significant seems to have happened in brachiopod history.

Major events in brachiopod history

1. The end of the Frasnian event

I have long drawn attention to the tragedy recorded at the top of the Frasnian Stage of the Upper Devonian with the world-wide extinction of the atrypids, well demonstrated by Copper (1966). With the atrypids went the pentamerids and many other groups. Among the survivors were the productids, which were to flourish in the Late Palaeozoic, the spiriferids (s.s.) and the rhynchonellids which I suggest (Ager 1968) took over at least one of the ecological niches previously occupied by the atrypids and flourished as a result. But this was a decline not an execution.

McLaren (1970) suggested, long before the days when iridium became fashionable, that the mass extinctions at the end of the Frasnian may have been caused by an extra-terrestrial body falling into one of the world's oceans. Sure enough in due course an iridium anomaly was found at this level, though I tend to suggest cynically that this was because they looked for it there. Personally, I prefer to blame Pluto stoking his subterranean fires (and thereby producing plate movements) rather than Jupiter sending down his heavenly thunderbolts, but that hardly matters in the present context.

What does matter is that the end of the Frasnian was a bad time for brachiopods, but many of them survived and flourished in the absence of competition.

2. The end Permian event

It has long been presumed that, when the trumpet sounded at the end of the Palaeozoic, all of the familiar Palaeozoic brachiopods became extinct and only a few brachiopod faunas, of low diversity and little interest, survived into the Mesozoic. Up to a point this would be the American view, whereas in Europe we see abundant and diverse Mesozoic brachiopod faunas, and tend to take a similarly disdainful view of

Caenozoic faunas. New Zealanders' or Japanese views would be different again.

In the *Treatise* one sees no fewer than thirteen brachiopod superfamilies becoming extinct at the end of the Permian, but again this is something of an artefact, resulting from excessive taxonomic splitting. It should also be borne in mind that certain groups, such as the richthofeniids and the stenocsimatids only had a brief spell of glory in the perhaps peculiar conditions of Permian seas. What is more, Palaeozoic brachiopods did not exit suddenly. Some, such as the pentamerids, had gone long before and even within the Permian there was a gradual decline rather than a mass exit. There were also far more survivors than we generally think.

One of the most interesting facts that has emerged about brachiopod history in recent years is the Palaeozoic-like appearance of Early Triassic brachiopod faunas.

Generally speaking, the Early Triassic was not a good time for brachiopods (or for many other groups). It may have been, as Fischer has suggested, a time of global salinity crisis far more serious and widespread than the better-publicized Messinian one at the end of the Miocene, which was restricted to Mediterranean waters.

In several widely separated parts of the world, however, we now have Lower Triassic brachiopod faunas including such typically Palaeozoic groups as the productids. These were first recognized by Trumpy (1961) in East Greenland, later in the southern part of the Soviet Union (Ruzhentsev and Sarycheva 1965) and most recently they have been described in southern China (Liao 1980). Naturally, this has been disputed, but that is all part of the fun. Certainly, particularly typical 'Palaeozoic' groups, such as the athyrids continued to flourish in the Triassic. It is just that, these faunas are not well known to the generality of palaeontologists.

3. *The end Triassic event*

It is not generally appreciated how abundant and diverse Triassic brachiopod faunas were after their difficult time at the beginning of that period. This appears to be because they are largely confined to deeper water environments and to sediments that later became involved in major tectonism, such as the Southern Alps of Europe and the Southern Alps of New Zealand. Thus, locally the athyrid *Tetractinella* is fantastically abundant in northern Italy. Usually, however, not only is the terrain difficult of access, but the sediments are highly deformed, as is the Torlesse super-group in New Zealand, where Milne and Campbell (1969) found the southern European Norian form *Halorella* (which I earlier overlooked myself in a paper on the genus).

Pearson (1977) described and revised the diverse Rhaetian brachiopod faunas of Europe and Dagis (1965) described those of Siberia, though without using the stage-name Rhaetian. It is evident from both of these works, and from the classic work of Bittner (1890), that the brachiopods of the Late Triassic were very diverse and still included many forms generally associated with the Palaeozoic. However, they were limited to certain parts of the world and like all locally endemic faunas, were vulnerable. Most of them became extinct at the end of the period.

Only a few rhynchonellid and terebratulid stocks survived into the Jurassic, plus one genus — *Spiriferina* — of all the vast diversity of spiriferids (in the widest sense) that had existed earlier. I have suggested elsewhere (Ager, in press) that the rhynchonellids survived and the spiriferids became extinct because the latter were constrained by rigid calcareous spiralia, whereas the rhynchonellids were able to extend their lophophores outside the shell for feeding and reproduction. Whereas most of the spire-bearers became extinct at the end of the Triassic (with no evidence of either heavenly trumpets or heavenly missiles), *Spiriferina* lingered on, with fewer and fewer species, until the early Toarcian.

The success story among the articulate brachiopods, in terms of longevity, were the rhynchonellids, which ranged from the Ordovician to the present day. The terebratulids, which only started in the Devonian had a longer, slower climb to fame. They came slowly to dominate during the Mesozoic, and in their twin groups of short and long-looped forms, ruled the brachiopod world throughout the Caenozoic. They appear to have survived each catastrophe almost untouched.

4. *The end Cretaceous event*

The most publicized mass extinction in the whole fossil record is that at the end of the Cretaceous. All that I can say about it here is that no-one told the brachiopods. Certainly, as I have pointed out earlier, by counting names around the world one would get the impression that the brachiopods suffered in the same way as the dinosaurs. However, no taxonomic group above the family became extinct at this point in time and if one is to believe the 'Treatise' only two of those. I would also doubt if many true genera died out in this particular apocalypse. The Danian Stage, now included in the Caenozoic and generally accepted as 'post-Nemesis', contains abundant brachiopods closely related to their Maastrichtian forebears.

Though we generally think of the Tertiary in Europe as (to adapt a rude remark about England and music) 'das Lande ohne Brachiopoden', this was certainly not my impression on a recent visit to New Zealand. Indeed, one gets the impression that the Tertiary was where brachiopods really started — sometimes in rock-forming abundance. Even in Europe, for example in the Vienna Basin extending over into Hungary, brachiopods are often abundant in Miocene shallow-water carbonates.

However, it must be emphasized that the brachiopods were adapted to fewer ecological niches than their obvious competitors, the benthonic molluscs. It was this competition, coupled with changes in sea-level, that I blame for the traumas in the brachiopod life-story.

5. *Other events*

There were, of course, other events, or supposed events, in the history of the brachiopods, but many of these are more apparent than real. Thus, in a discussion of the Jurassic–Cretaceous boundary, I showed (Ager 1975) that the most important event in brachiopod history thereabouts happened at a higher level than the conventional boundary. If forced to draw a line on the basis of my second favourite animals (after women) I would be forced to put it, not at the base of the Berriasian, but within the Valanginian, where the pygopid terebratulids declined markedly and the rhynchonellid cyclothyrids took off in an orgy of explosive evolution. However, this would be a foolish exercise and no sort of argument in the business of defining stratigraphical limits, which to me should be a matter of pure pragmatism, as is argued in that paper. The peak in Cherfas's graph at this point (1985, p. 49) seems to me highly artificial.

There were probably other important events in strata that I know less about. From the literature these would seem to be new appearances rather than old departures. In each case one must presume that 'Death makes way for life'. At the beginning of the Ordovician, for example, there is said to have been a sudden burst of new stocks such as the Enteletacea and the Strophomenacea. Similarly the Spiriferacea are said to have begun abruptly at the beginning of the Silurian and the Productacea at the beginning of the Devonian. However, all these things may be, in part at least, artefacts, like the mass extinction of thirteen superfamilies at the end of the Permian discussed above.

Conclusions

In their long history, the brachiopods have suffered a series of severe bouts of extinction. Fewer and fewer major groups have survived, and these have been replaced ecologically more and more by the benthonic molluscs. Though they continue to flourish in certain parts of the world, they may be close to final extinction. *Lingula* soup is eaten locally in Japan, but I am told that brachiopods generally are not very palatable and they appear to have few predators. Perhaps the evolution of a new group of organisms even more addicted to mollusc-eating than I am myself, will wipe out their chief opposition, or perhaps a nuclear winter will give the brachiopods another chance to conquer the world. On that cheerful note I will end.

References

Ager, D. V. (1965). Mesozoic and Cenozoic Rhynchonellacea. In *Treatise on Invertebrate Palaeontology* (ed. R. C. Moore), Geol. Soc. Am. and Univ. Kansas Press, Pt. H, H597–625.

Ager, D. V. (1968). The Famennian takeover. *Circ. Palaeont. Assoc.* No. **54a,** 1.

Ager, D. V. (1975). Brachiopods at the Jurassic–Cretaceous boundary. *Mem. Bureau Recherches geol. min.*, No. **86,** 150–62.

Ager, D. V. (1976). The nature of the fossil record. *Proc. Geol. Ass.* **87,** 131–60.

Ager, D. V. (1981). Major marine cycles in the Mesozoic. *J. geol. Soc. Lond.* **138,** 159–66.

Ager, D. V. Why the rhynchonellids survived and the spiriferids did not—a suggestion. *Palaeontology* **30,** 853–7.

Ager, D. V., Childs, A., and Pearson, D. A. B. (1972). The evolution of the Mesozoic Rhynchonellida. *Geobios, Lyon* No. **5,** 157–233.

Bittner, A. (1890). Brachiopoden der Alpinen Trias, I. *Abhandl. k.k. geol. Reichsanst.* **14,** 1–35.

Cherfas, J. (1985). Extinction and the pattern of evolution. *New Scientist* No. 1476, 48–51.

Copper, P. (1966). The *Atrypa zonata* brachiopod group in the Eifel, Germany. *Senckenbergiana Lethaea* **47,** 1–55.

Dagis, A. S. (1965). *Triassic brachiopods from Siberia.* Izdat. 'Nauka', Akad. Nauk. SSSR (In Russian.)

Liao, Z.-t. (1980). Brachiopod assemblages from the Upper Permian and Permian-Triassic boundary beds south China. *Can. J. Earth Sci.* **17,** 289–95.

McLaren, D. (1970). Time, life and boundaries. *J. Paleont.* **44,** 801–15.

Milne, J. D. G. and Campbell, J. D. (1969). Upper Triassic fossils from Oroua Valley, Ruahine Range, New Zealand. *Trans. Roy. Soc. NZ* **6,** 247–50.

Pearson, D. A. B. (1977). Rhaetian brachiopods of Europe. *Neue Denkschrift. naturhist. Mus. Wien* **1,** 1–70.

Ruzhentsev, V. E. and Sarycheva, T. G. (1965). Distribution and changes of marine organisms at the boundary between the Palaeozoic and Mesozoic. *Paleontol. Inst. Akad. Sci. SSSR* **108,** 1–431. (In Russian.)

Trumpy, R. (1961). Triassic of East Greenland. In *Geology of the Arctic* (ed. G. O. Raasch), pp. 248–58. University of Toronto Press, Toronto.

5. Mass extinctions and the pattern of bryozoan evolution

P. D. TAYLOR

Department of Palaeontology, British Museum (Natural History), London, UK

G. P. LARWOOD

Department of Geological Sciences, University of Durham, Durham, UK

Abstract

The pattern of bryozoan evolution is examined in relation to the five mass extinctions recognized to have effected the marine realm. At familial level the diversity curve for the Bryozoa is remarkably similar to that for marine animals in total. The most conspicuous features are a Palaeozoic diversity plateau, a late Permian diversity crash, and a Mesozoic–Recent climb in diversity to levels above those of the Palaeozoic.

Neither familial nor generic data reveal high levels of extinction at the times of the end-Ordovician or late Devonian mass extinctions, though there is some evidence for local or regional extinctions of bryozoan faunas. The Permian extinction was spread over the two terminal stages of the period and resulted in the removal of many infraordinal taxa. However, contrary to the 'textbook' view, only one order seems to have become extinct. The depauperate Triassic bryozoan fauna was effected by the end-Triassic mass extinction which possibly removed two orders of Palaeozoic 'holdovers'. Most post-Palaeozoic bryozoans are closely-related to two orders which existed as minor and 'weedy' components of Palaeozoic bryozoan faunas. Bryozoan extinctions at the end of the Cretaceous are

Extinction and Survival in the Fossil Record (ed. G. P. Larwood), Systematics Association Special Volume No. 34, pp. 99–119. Clarendon Press, Oxford, 1988.

more evident at low taxonomic levels and seem to have effected cheilostomes more than cyclostomes. Major faunal turnovers (from trepostome- to fenestrate-dominated faunas in the Palaeozoic, and cyclostome- to cheilostome-dominated faunas in the post-Palaeozoic) bear no obvious relationship to mass extinctions.

Introduction

Mass extinctions may have played a major role in determining evolutionary patterns (e.g. Gould 1985). In an important paper, Raup and Sepkoski (1982) were able to distinguish five times of mass extinction when familial extinction rate for marine animals stood out above background levels. These major mass extinctions occurred during the late Ordovician, late Devonian, late Permian, late Triassic, and late Cretaceous. Mass extinctions may be other than simple intensifications of background extinction rates (Jablonski 1986*a,b*). The purpose of the current paper is to examine the effects (if any) of the five mass extinctions on the pattern of bryozoan evolution. Did the standing diversity of bryozoans change after a mass extinction event, and have mass extinctions had any appreciable effect on the taxonomic composition of bryozoan faunas?

The Bryozoa are a mainly marine phylum of colonial metazoans (see Ryland 1970, for an introduction). Colonies consist of multiples of suspension-feeding zooids which are variously arranged to give a wide variety of different colony forms (e.g. F. K. McKinney 1986). Most species of bryozoans are sessile, inhabit the continental shelf, and have colonies with calcareous hard parts. Skeletal morphology is employed extensively in bryozoan taxonomy (Boardman *et al.* 1983), and living species are generally distinguished using the same morphological features as fossils. The Bryozoa have a rich fossil record extending back to the Lower Ordovician (Larwood and Taylor 1979; Taylor and Curry 1985). Unfortunately, however, the study of fossil bryozoans has not matched their abundance, and their evolutionary history is not well-known (cf. Schopf 1977; Morozova and Viskova 1977). With such a poor data base we are largely restricted to a review of evolutionary pattern in relation to mass extinctions; it would be premature to address problems of the processes causing extinctions.

As with any study of extinction pattern, the data available for the Bryozoa can be grouped according to its position within taxonomic, geographical, and stratigraphical hierarchies. The ideal would be to have data on global species extinctions with a fine level of stratigraphical control (e.g. zonal). Data from other hierarchical levels may, for various reasons, lead to a different picture of the pattern of extinction. Therefore, we are careful to distinguish the hierarchical levels of our data bases.

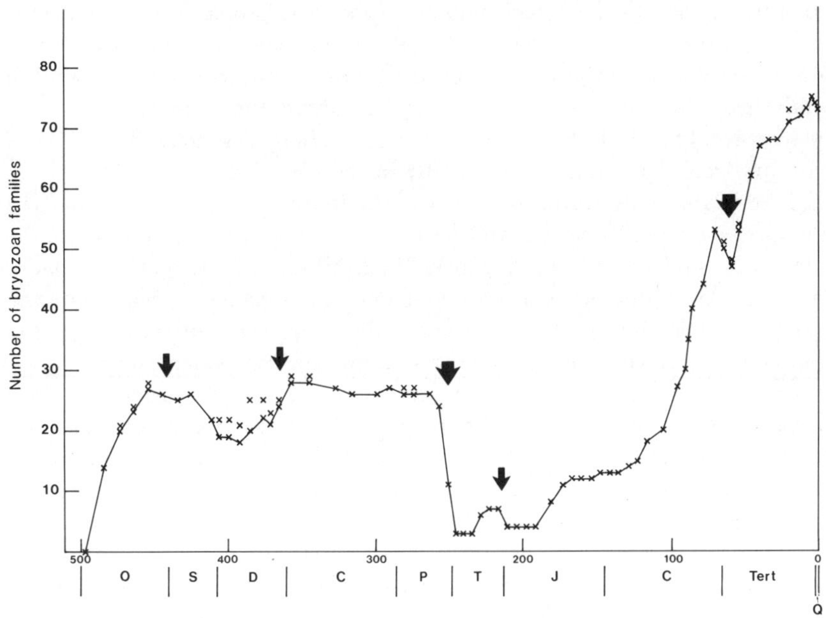

Fig. 5.1. Bryozoan family diversity curve. Stage-by-stage plot using a modified version of Sepkoski's (1982) data base and the Harland *et al.* (1982) time scale. Points above the curve are due to uncertainties in ranges; times of mass extinction in the marine record are arrowed.

Broad diversity pattern

The Ordovician to Recent family level diversity curve (Fig. 5.1) for the Bryozoa, constructed using an up-dated and modified version of Sepkoski's (1982) compilation, is a useful starting point. Personal experience of bryozoan assemblages through the Phanerozoic suggests that this curve may be a reasonable first-order approximation of relative species diversity. Following a dramatic rise in diversity during the early and mid Ordovician, Palaeozoic diversity levelled-off at about 20–30 families, with a Silurian–Devonian diversity trough superimposed. Most of the Palaeozoic diversity of the Bryozoa is accounted for by four stenolaemate orders (Cystoporata, Trepostomata, Fenestrata, and Cryptostomata); a fifth stenolaemate order (Cyclostomata) and a gymnolaemate order (Ctenostomata) were less important. Assemblage diversities of Palaeozoic age seldom reach 100 species. A precipitous drop in diversity occurred during the late Permian. Diversities remained low throughout the Triassic and early Jurassic before rising in the mid-Jurassic with the radiation of cyclostomes. During the late Jurassic the Order Cheilostomata appeared, probably from a ctenostome ancestor (see Boardman *et al.* 1983), and began a major diversification in the mid-Cretaceous. Bryozoan diversity rose dramatically to attain family levels which at the present day are about three times those of the Palaeozoic. This is reflected in significantly higher species diversities in fossil bryozoan assemblages, especially in the Neogene.

There is a striking similarity between the bryozoan family diversity curve (Fig. 5.1; see also Morozova and Viskova 1977, fig. 2) and that for marine animals in total (Raup and Sepkoski 1982, fig. 2). The two curves share a Palaeozoic diversity plateau, an end Palaeozoic decline, and a continuing post-Palaeozoic rise towards the Recent. The bryozoan curve differs mainly in having a lower base level, lacking a Cambrian component, and having a delayed post-Palaeozoic diversity recovery. If the Bryozoa epitomize marine diversity in general, then perhaps the same causal mechanism(s) should be sought to explain these patterns, and any focus on bryozoans specifically should be directed at the areas in which they differ from the curve of total marine diversity. Other estimates of marine diversity using several different methods (Sepkoski *et al.* 1981) also broadly resemble the bryozoan family curve. Marine diversity patterns have been explained by changes in provinciality (Schopf 1979) and species packing in communities (Bambach 1985). The latter seems a more attractive hypothesis for bryozoan family diversity in view of the cosmopolitan distribution of most families and the changes in assemblage diversity which appear to more-or-less parallel the family diversity trend.

The two-way taxonomic split of bryozoan diversity into a Palaeozoic component made up largely of stenolaemates and a post-Palaeozoic component largely of gymnolaemates corresponds to Sepkoski's (1981) 'Paleozoic Fauna' and 'Modern Fauna', respectively. These evolutionary faunas may relate to the dominant types of marine benthic communities (Sepkoski and Miller 1985). In this respect, Palaeozoic stenolaemate faunas are associated mainly with soft-bottom, muddy sediments, whereas post-Palaeozoic gymnolaemates are more often found in coarser shell sands and gravels.

Ordovician extinction

An important extinction has been recognized in the uppermost Ordovician. Brenchley (1984) noted that this extinction effected most elements of the biota, and attributed a climatic origin related to the growth and decay of the Gondwana ice cap in Hirnantian (late Ashgill) times. According to Brenchley and Newall (1984), the extinction was preceeded by a late Caradoc–early Ashgill reduction in species diversity, and the timing of the extinction itself was variable.

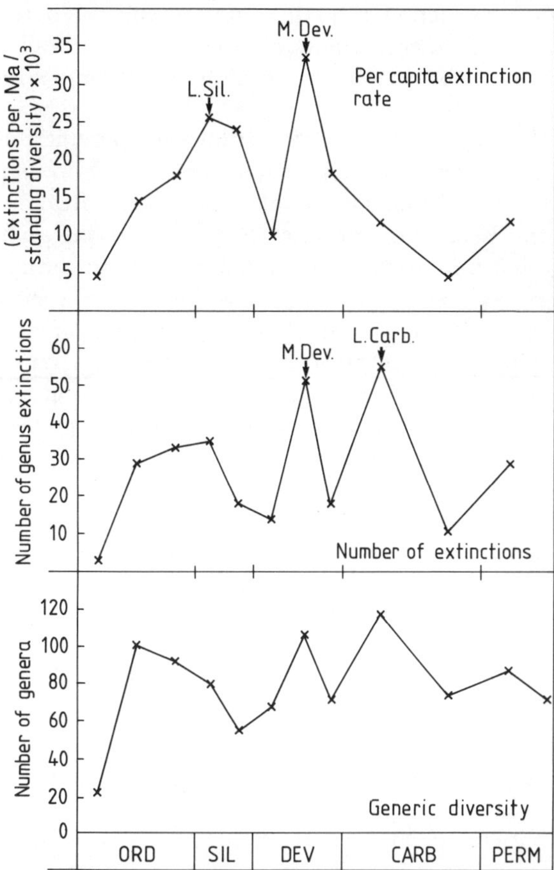

Fig. 5.2. Bryozoan genus diversity, number of extinctions and per genus extinction rate (per Ma) for series of the Palaeozoic. Main sources: Astrova (1970), Boardman *et al.* (1983), Cuffey and McKinney (1979), Dzik (1981), Lavrentjeva (1985), Morozova (1974), and Ross (1978, 1981).

The bryozoan family diversity curve (Fig. 5.1) shows no significant decline in diversity between the Ashgill and Llandovery. M. L. McKinney (1985), using Sepkoski's (1982) data base, calculated that only 11 per cent of bryozoan families became extinct at this time. A new data set (Fig. 5.2) of bryozoan generic diversity through series of the Palaeozoic also fails to detect any major drop in diversity. However, changes in diversity result not only from the negative effects of extinction, but also the additive effects of origination. If the extinction component is separated out, the Upper Ordovician can be seen to have experienced a greater number of bryozoan genus extinctions than the Middle Ordovician, but fewer than the Lower Silurian. The Lower Silurian also stands out as a peak above the Upper

Ordovician when extinction rate (per Ma) is calculated relative to the standing diversity of genera. Therefore, this stratigraphically coarse data provides no evidence for a global mass extinction of bryozoans in the late Ordovician.

Anstey (1986), in a paper on late Ordovician provincialism in North American bryozoan genera, recognized a terminal Ordovician extinction which affected the endemic taxa of the Laurentian Realm, notably the Cincinnati and Reedsville-Lorraine Provinces. Long-surviving, morphologically complex bryozoans were differentially lost during this extinction. They were replaced in the early Silurian by immigrants from other realms. Spjeldnaes (1982) noted the disappearance of most bryozoan faunas in the Upper Ashgill and commented that this represents a true extinction in the trepostomes, but that some families and even genera of cystoporates and cryptostomes reappear in the Upper Llandovery or Lower Wenlock (i.e. they are Lazarus taxa).

Unfortunately, there are no modern studies of changes in the bryozoan fauna across Ordovician–Silurian boundary sequences. Bryozoans from the relatively complete section on Anticosti Island were last studied by Bassler (*in* Twenhofel 1928), and only the ptilodictyine cryptostomes have subsequently been revised (Ross 1960, 1961). There appears to be no single horizon of high species extinction, and it is notable that stratigraphical pseudoextinctions (45) exceed true extinctions (36) in the two latest Ordovician and two earliest Silurian formations. However, standing diversity of species in these two Ordovician formations (26 and 25 spp.) is noticeably greater than that of the two basal Silurian formations (7 and 9 spp.).

A similar pattern is evident in Baltoscandia where Brood (1981) identified 59 species of Hirnantian bryozoans and commented on the low diversity of the overlying Silurian strata. There is a very clear need for further research to establish the geographical extent and quantitative and qualitative aspects of bryozoan faunal change at the end of the Ordovician.

Devonian extinction

Raup and Sepkoski (1982) found higher than usual rates of family extinction distributed over two or three stages of the Middle and Upper Devonian. Most attention has been paid to the so-called Kellwasser event about the level of the Frasnian–Famennian boundary (House 1985). According to McGhee (1982), this extinction had its greatest effects on epifaunal filter-feeding animals, while both Copper (1977) and McLaren (1982) stress the extinction of reefs and their associated biotas. Climatic cooling (Copper 1977) and bollide impact (McLaren 1970) have been proposed as possible causal processes.

The mid-late Devonian falls within a general trough of low bryozoan family diversity and it is not possible to distinguish any times of outstanding reduction in diversity (Fig. 5.1). M. L. McKinney (1985) calculated that only three of 32 bryozoan families became extinct during the two terminal stages of the Devonian. Series-level data for bryozoan genera (Fig. 5.2) reveals a diversity high in the Middle Devonian followed by a low in the Upper Devonian. The number of generic extinctions also peaks in the Middle Devonian, is low in the Upper Devonian, and peaks again in the Lower Carboniferous. The Lower Carboniferous peak is largely due to the long duration of the system and disappears when per genus extinction rate is plotted (Fig. 5.2), but the Middle Devonian peak remains. A high proportion of genera that became extinct during the Middle Devonian were endemic to eastern North America and are known only from the Middle Devonian; it seems possible that a monographic bias may exist as a result of the relatively intensive study of these often well-preserved bryozoan assemblages.

In western Europe and eastern North America bryozoan faunas disappear just before the end of the Frasnian (Bigey 1986), possibly relating to widespread black shale accumulation. Elsewhere in shelf regions of the Palaeotethys, Famennian bryozoans may be of common occurrence (e.g. Troitskaya 1968). Any global effects of a Frasnian–Famennian bryozoan extinction seem to have been slight at generic and higher taxonomic levels.

Palaeozoic extinctions and faunal turnover

Cuffey and McKinney (1979), and others have commented on the change from trepostome-ceramoporoid dominated bryozoan faunas of the early Palaeozoic to fenestrate-rhomboporoid dominated faunas of the late Palaeozoic. The main feature of this faunal turnover is an apparent replacement of trepostomes by fenestrates in broadly similar sedimentary facies. An obvious question to pose is whether this change was related to either the Ordovician or Devonian mass extinctions (cf. Gould and Calloway 1980 on the effect of the end Permian extinction on bivalves vs. brachiopods; see also Benton 1983).

The proportion of fenestrate genera relative to trepostome genera for the series of the Palaeozoic is shown in Fig. 5.3. There is a fairly even and continuous increase in fenestrates throughout the Palaeozoic; most successive series have a higher proportion of fenestrate genera than the preceding series. The largest increase occurs between the Upper Silurian and the Lower Devonian (e.g. the diverse fenestrate fauna described by McKinney and Kříž 1986). Neither the Ordovician nor Devonian mass extinctions have any appreciable effect at this level of taxonomic and

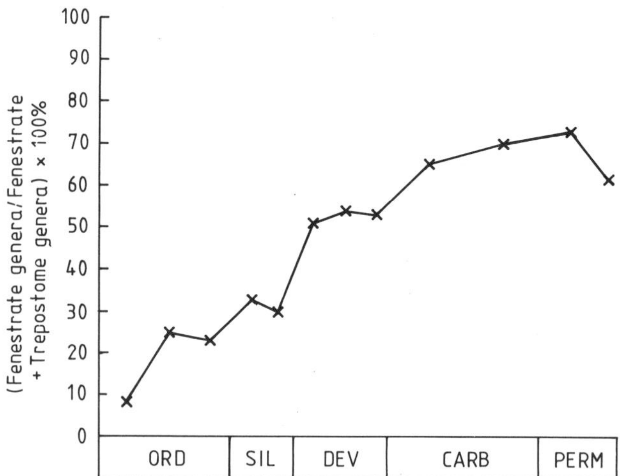

Fig. 5.3. Changing proportion of fenestrate:trepostome genera during series of the Palaeozoic. Same data base as Fig. 5.2.

stratigraphical resolution; there is no evidence for either mass extinction re-setting the trepostome:fenestrate balance.

Permian extinction

A massive extinction was experienced by the marine realm in the late Permian (Raup 1979). This extinction has been ascribed a variety of causes (see Dickins 1983, for a review) amongst which the most popular is the reduction in area of the shallow water marine continental shelves (Schopf 1974; Simberloff 1974; but see Stanley 1984; Jablonski 1985). Detailed examination of the end Permian extinction is made difficult by problems of correlation in the Upper Permian and the common occurrence of unconformities separating marine sequences of Permian and Triassic rocks. Nevertheless, it is clear that a gradual decline in overall marine diversity occurred through a lengthy period of late Permian time.

The bryozoan family diversity curve (Fig. 5.1) shows a huge drop in diversity during the Upper Permian, spread out over the two terminal stages (Guadalupian and Dzhulfian). A 76 per cent extinction of bryozoan families was calculated by M. L. McKinney (1985) for these two stages, one of the highest of any invertebrate group. Ross (1978) pointed out that, although late Permian bryozoan faunas are not well-known (see Morozova 1970), most Permian 'lineages' became extinct either before or early in the Dzhulfian in nearly all non-Tethyan regions.

Generic extinctions (Fig. 5.4) in the Upper Permian amount to about 52 genera in the Guadalupian (68 per cent of standing diversity) and 14 genera in the Dzhulfian (64 per cent of standing diversity). No bryozoan genera are known to have originated in the Dzhulfian (Schäfer and Fois-Erickson 1986) and low origination rates undoubtedly contributed to the dramatic drop in diversity of bryozoans and other marine animals (Hüssner 1983).

The only detailed study of bryozoans across a relatively complete Permo–Trias boundary sequence was undertaken by Morozova (1965) in Trans-Caucasia. Here the Guadalupian is divided into a lower Gnishik Horizon and an upper Khachik Horizon, overlain by Dzhulfian, and supposed Induan (Lower Triassic) deposits. Twenty-eight species were recorded in the Gnishik Horizon, only one of which ranged into the Khachik Horizon, and six different species in the Dzhulfian Stage. Most of the Dzhulfian species occur in the lower half of the stage. A single species, the fenestrate *Polypora darashamensis* Nikiforova, survives into the supposed Induan Stage. However, these Lower Triassic deposits are now thought to be of latest Permian age and have been included in a new stage (Dorashamian) by Rostovtsev and Azaryan (1973).

Extinction and survival of higher taxa of bryozoans following the Permian and Triassic extinctions are discussed together below (p. 108).

Triassic extinction

Relatively little attention has been paid to the mass extinction which occurred at the end of the Triassic, although this event was recognized in 1967 by Newell. More recently, Hallam (1981) has reviewed the end-Triassic extinction with particular regard to the bivalves which, like ammonites, brachiopods, and conodonts, were affected significantly. Hallam implicates regression and widespread anoxia during the succeeding Hettangian transgression as possible causal factors.

Sepkoski's (1982) family-level compilation fails to reveal a late Triassic bryozoan extinction. However, knowledge of Triassic bryozoans has increased enormously during the last few years (Bizzarini and Braga 1982; Hu 1984; Sakagami 1985; Morozova 1986; Schäfer and Fois-Erickson 1986) and the modified family curve (Fig. 5.1) shows an appreciable Upper Triassic diversity peak followed by a decline. Schäfer and Fois-Erickson (1986) recognized the extinction of about 10 genera during the terminal Rhaetian Stage of the Triassic which followed a modest late Triassic diversification, especially of trepostomes (Fig. 5.4). Although this extinction is small in terms of numbers of genera, it is large relative to standing diversity.

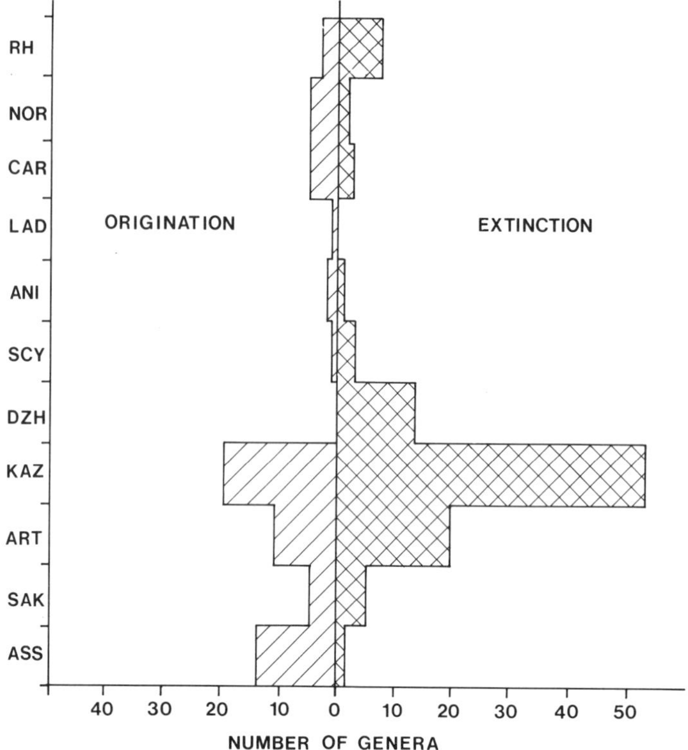

Fig. 5.4. Bryozoan generic originations and extinctions during stages of the Permian and Triassic. After Schäfer and Fois-Erickson (1986).

Extinction and survival of Palaeozoic bryozoan orders

A traditional view of bryozoan evolutionary history (Fig. 5.5A) includes the extinction at the end of the Permian of the four typically Palaeozoic orders of stenolaemates (Trepostomata, Cryptostomata, Cystoporata, Fenestrata), and the continuation into the Mesozoic of the stenolaemate order Cyclostomata and the gymnolaemate order Ctenostomata, the latter the eventual ancestors of the Cheilostomata. New finds of Triassic and Jurassic bryozoans, together with new interpretations of phylogenetic relationships, have led to this view being challenged. Two modern views can now be presented, the first (Fig. 5.5B) using a conservative hypothesis and the second (Fig. 5.5C) a radical hypothesis of phylogenetic relationships.

The Fenestrata are the only order without certain representation in the Triassic; they appear to have been a victim of the end-Permian mass

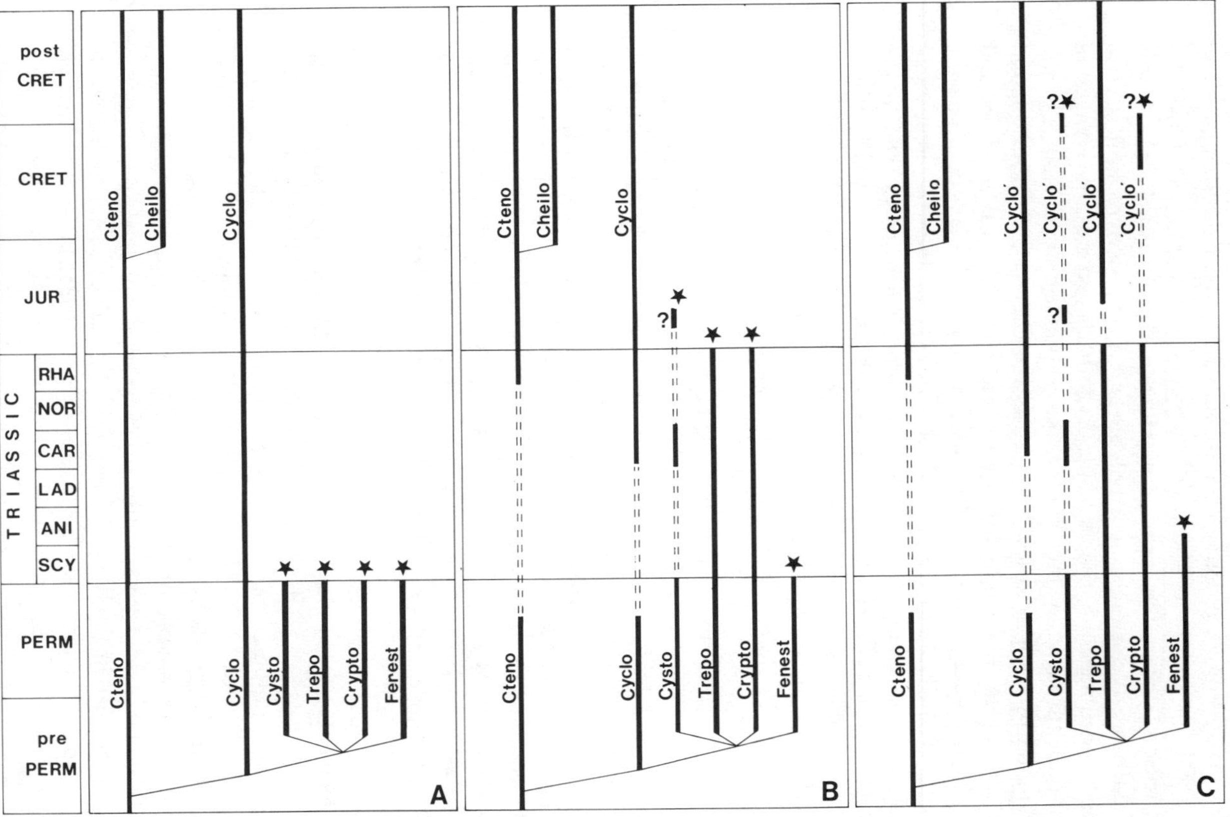

Fig. 5.5. Three alternative views ordinal ranges, extinctions (asterisks), and phylogenetic relationships across the critical Palaeozoic–Mesozoic transition. (A) Traditional 'text book' view with the extinction of four orders at the end of the Permian. (B) Conservative modern view with the extinction of one order at the end of the Permian and two at the end of the Triassic. (C) radical modern view with one end-Permian and two possible end-Cretaceous ordinal extinctions.

extinction. According to the conservative hypothesis, cystoporates range into the Carnian and possibly the Lower Jurassic (Boardman 1984), whereas cryptostomes and trepostomes both range into the Rhaetian and may have been victims of the end-Triassic mass extinction (Schäfer and Fois-Erickson 1986). The radical hypothesis (Boardman 1984) regards post-Palaeozoic cyclostomes as a polyphyletic group derived from several Palaeozoic stenolaemate orders. This hypothesis extends the ranges of trepostomes through to the Recent (as cerioporid 'cyclostomes'), the cryptostomes through to at least the end of the Cretaceous (as petaloporid and related 'cyclostomes'), and the cystoporates similarly (as *Semicrescis* and related 'cyclostomes'). Whereas the conservative hypothesis necessitates the evolution of several morphological features (e.g. lunaria, hemisepta, annular wall thickenings) in post-Palaeozoic cyclostomes which are convergent with Palaeozoic stenolaemate orders, the radical hypothesis demands parallel evolution of brood chambers, interzooidal pores, and pseudoporous exterior walls in three or four separate groups of stenolaemates during the post-Palaeozoic. Rigorous phylogenetic analysis using more morphological evidence than is currently available is needed to test the two hypotheses. However, both hypotheses alter the traditional view of bryozoan extinction by reducing ordinal extinctions at the end of the Permian from 4 to 1, and introducing possible ordinal extinctions during either the end-Triassic or end-Cretaceous. Although Triassic bryozoans are known from many parts of the world (see Schäfer and Fois-Erickson 1986), knowledge of Jurassic and to a lesser extent Cretaceous bryozoans is virtually confined to north-west Europe. It is likely that ordinal ranges will be extended significantly through the discovery of bryozoan faunas from other parts of the world, e.g. the two Lower Jurassic occurrences of probable cystoporates (Boardman 1984) are from Canada and Chile.

Whichever hypothesis of relationships is correct, the majority of post-Triassic bryozoans were derived from two orders which had been extremely minor elements of bryozoan faunas in the Palaeozoic and survived the extinction at its end. The Ctenostomata and the Cyclostomata are paraphyletic groupings of primitive gymnolaemates and stenolaemates, respectively. Ctenostomes are soft-bodied and are known in the Palaeozoic only as borings in shells and other hard substrata (Pohowsky 1978); non-boring forms are presumed to have been present but have left no fossil record (Larwood and Taylor 1979). An encrusting, runner-like ctenostome probably gave rise in the late Jurassic to the first of the cheilostomes which today dominate marine bryozoan faunas (Taylor 1981, table 2). Although possessing calcareous skeletons, cyclostomes are rare in the Palaeozoic and most consist of small encrusting or erect colonies with delicate branches (see Brood 1975; Dzik 1981). It is thought that all or most post-Palaeozoic cyclostomes were most closely-related to a group (crownoporids) that have been found only in the Ordovician (see Taylor 1985). Palaeozoic ctenostomes and crownoporid cyclostomes share simple, weed-like morphologies, and may have been distributed mainly in nearshore habitats (Anstey 1981, 1986). Their evolutionary pattern fits the classic picture of initially subordinate, morphologically simple groups surviving mass extinction events (Permian and Triassic) that brought about the demise of previously dominant, complex groups, and subsequently radiating into vacant habitats.

Cretaceous extinction

The terminal Maastrichtian Stage of the Cretaceous has long been recognized as a time of mass extinction for both marine and terrestrial biotas. Interest in this extinction was rekindled in 1980 when Alvarez *et al.* proposed their bollide impact theory. Although some groups of invertebrates were in decline immediately before the extinction, many became suddenly truncated at the level of the Ir-rich boundary clay (Alvarez *et al.* 1984).

The bryozoan family diversity curve (Fig. 5.1) shows a slight, but conspicuous and abrupt drop in diversity between the Maastrichtian and Danian, and another drop between the Danian and Thanetian. These declines interrupt the otherwise continuous and steep climb in diversity which began in the mid-Cretaceous. M. L. Mckinney (1985) using Sepkoski's (1982) data base calculated an extinction of eight of 66 (12 per cent) bryozoan families in the Maastrichtian.

The extinction is more marked at generic level. For example, Viskova (1980) noted that about 60 genera of cyclostomes became extinct, almost 50 per cent of the standing crop. Voigt (1981, 1986) has criticized the use of taxonomic compilations in studying Maastrichtian–Danian changes in bryozoans because the systematics are so poorly-known and many of the older monographs are unreliable. However, Voigt (1986, table 1) was able to recognize the Maastrichtian extinction of 33 cheilostome genera and 26 cyclostome genera (to which must be added a further five cheilostomes and six cyclostomes; E. Voigt, *pers. comm.* 1986).

Bryozoans have special importance because they are the dominant macrofossil group in most Maastrichtian and Danian sediments of the North Sea region, over 500 species occurring in Denmark alone (Håkansson and Thomsen 1979). Two studies at the species level with different stratigraphical and geographical resolutions are pertinent in relation to extinctions in this region. Brood (1972) monographed cyclostome bryozoans from the Upper Cretaceous and Danian of Scandanavia. His table 4 gives the zonal ranges of 110 species. Only two of 56 species (3.5 per cent) become extinct at the end of the Maastrichtian; this compares with 23 of 51 species (45 per cent) that become extinct at the end of the Campanian. Such a low level of species extinctions is not easily reconcilable with data on generic extinctions and may reflect differences in taxonomic opinion.

Working in Denmark, Håkansson and Thomsen (1979) noted that about 25 per cent of Maastrichtian cyclostomes fail to occur in the Danian and over 80 per cent of cheilostomes. Maastrichtian and Danian cyclostome faunas have about the same species diversity, whereas cheilostomes are less diverse in the Danian. Superimposed on this taxonomic difference

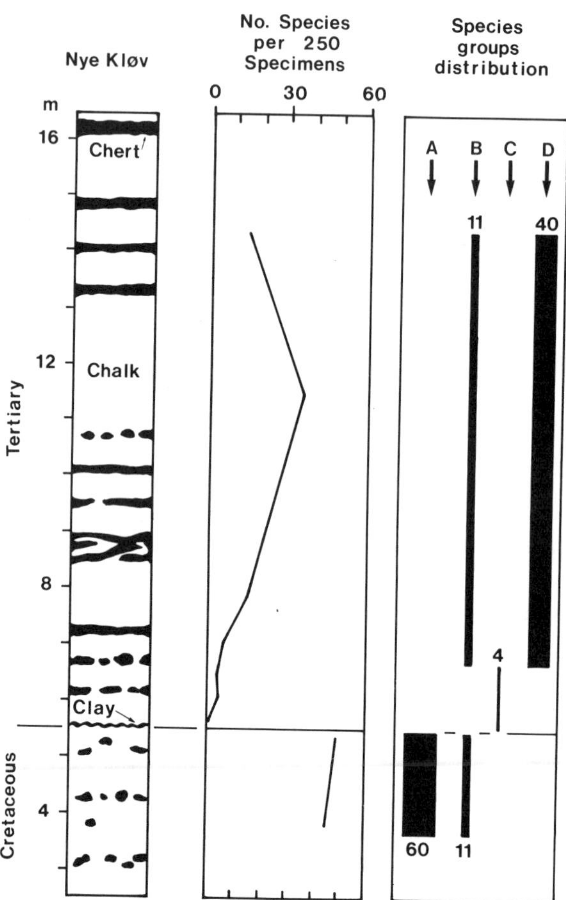

Fig. 5.6. Distribution of cheilostome bryozoan species across the Maastrichtian–Danian boundary sequence at Nye Kløv, Denmark (after Håkansson and Thomsen 1979). See text for explanation.

Håkansson and Thomsen also drew attention to facies-related differences between Maastrichtian and Danian bryozoan faunas. Whereas pelagic chalks contain high diversity (commonly over 100 spp.), but low density bryozoan faunas in the Maastrichtian, Danian chalks virtually lack bryozoans and have very low diversities (less than 10 spp.). By contrast, bryozoan limestones are of very similar diversity (typically over 75 spp.) in the Maastrichtian and Danian, but cheilostomes tend to dominate by weight in the Maastrichtian and cyclostomes in the Danian. This pattern of greater extinction in pelagic chalks may relate to the paucity of hard substrata (e.g. echinoid tests) for encrustation in the Danian. A detailed analysis of cheilostome distribution across the boundary section at Nye Kløv (Fig. 5.6) is instructive. About 70 spp. occur in the Maastrichtian here, only four spp. in the basal Danian chalk, rising to over 40 spp. in the overlying Danian bryozoan limestone. Based on their stratigraphical occurrence, Håkansson and Thomsen distinguished four groups of cheilostomes at Nye Kløv: group A comprising 60 spp. recorded only from the Maastrichtian; group B of 11 spp. found in both the Maastrichtian and the upper parts of the Danian; group C of 4 spp. found only in the basal Danian; and group D recorded only in the higher parts of the Danian (though found in the Maastrichtian elsewhere in the basin). The low diversity group C fauna has been recognized in several localities, and comprises rooted and free-living species adapted to the soft chalk sea-bed.

In summary, evidence from Denmark shows that cheilostomes were affected more than cyclostomes by the end-Cretaceous extinction, and pelagic chalk species more than bryozoan limestone species. Notable among the extinctions of very common Upper Cretaceous bryozoans were all of the bifoliate species of *Onychocella* and all but one of the species of *Lunulites* (Voigt 1986). It is unclear to what extent this regional extinction pattern typifies the global pattern.

Cretaceous extinction and faunal turnover

An important feature of post-Jurassic bryozoan evolution is the change from cyclostome-dominated assemblages to cheilostome-dominated assemblages (e.g. Voigt 1959). Data has been compiled (Fig. 5.7) on the relative composition of mid-Cretaceous to Recent bryozoan faunas to see whether the Maastrichtian extinction had any role in this faunal turnover. This reveals a gradual trend of increasing proportions of cheilostomes through the stages of the Upper Cretaceous, levelling-off approximately in the Maastrichtian. The end-Cretaceous extinction causes a slight perturbation in favour of cyclostomes, but the pre-extinction level is quickly re-established. Therefore, the end-Cretaceous extinction seems to have had no lasting effect with regard to the ordinal composition of bryozoan faunas. On

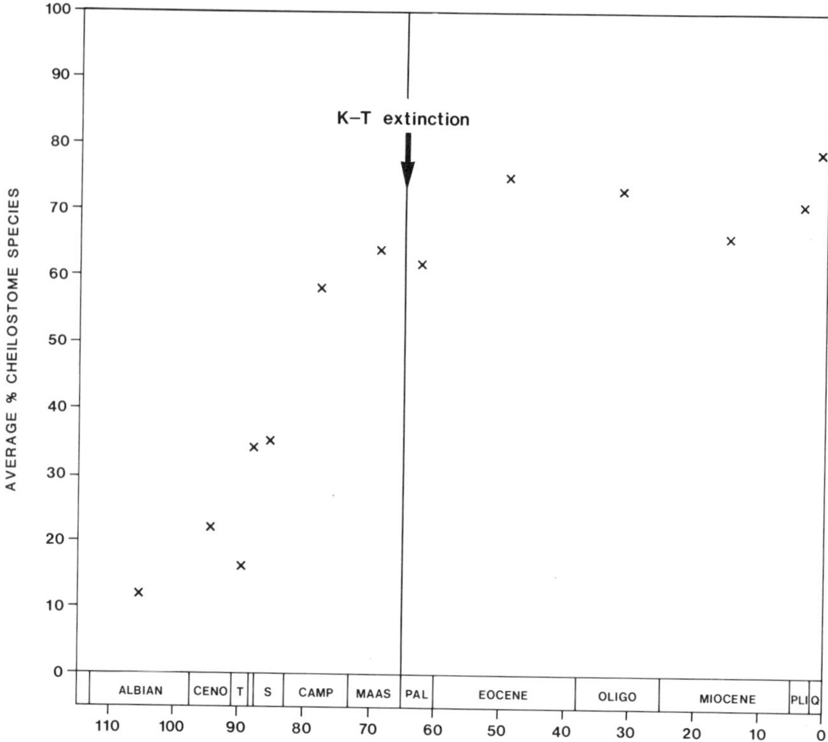

Fig. 5.7. Changing proportion of cheilostome bryozoan species in mixed cheilostome:cyclostome assemblages from the mid-Cretaceous to the Recent.

a global level it seems unlikely that the extinction discriminated significantly between cyclostomes and cheilostomes.

Conclusions

The quantitative and qualitative effects of the five mass extinctions on bryozoans are varied. Although local and regional evidence point to the existence of end-Ordovician and late Devonian extinctions, they are not detectable as significant events in global family and order level data; neither did they have any role in the major taxonomic turnover from trepostome- to fenestrate-dominated faunas. The end-Permian extinction had a massive effect on the Bryozoa. Only one order (Fenestrata) may have become extinct at the end of the Permian, but a large number of familial, generic, and species extinctions occurred during Guadalupian and Dzhulfian stages, and Triassic bryozoan faunas are markedly depauperate relative to those of the Permian. New data (Schäfer and Fois-Erickson 1986) suggests the existence of an end-Triassic bryozoan extinction. Depending upon the taxonomy employed, this may have involved the extinction of up two orders or none at all. The end-Cretaceous extinction of bryozoans occurred mostly at low taxonomic levels and effected cheilostomes more than cyclostomes in the only region for which good data is available (Håkannson and Thomsen 1979).

In order to clarify patterns of bryozoan extinction and to address the causal processes much more information is needed on: (a) faunal changes across boundary sections; (b) faunas from hitherto little known geographical provinces; and (c) phylogenetic relationships and recognition of truly monophyletic groups.

Acknowledgements

We are grateful for discussions with various members of the International Bryozoology Association during the association's 7th conference, and to R. A. Fortey for his constructive comments on the manuscript.

References

Alvarez, L. W., Alvarez, W., Asaro, F., and Michel, H. V. (1980). Extraterrestrial cause for the Cretaceous–Tertiary extinction. *Science* **208,** 1095–1108.

Alvarez, W., Kauffman, E. G., Surlyk, F., Alvarez, L. W., Asaro, F., and Michel, H. V. (1984). Impact theory of mass extinctions and the invertebrate fossil record. *Science* **223,** 1135–41.

Anstey, R. L. (1981). Geographical distribution and evolutionary potential of extant and extinct bryozoan taxa in the late Ordovician of North America. *Geol. Soc. Am. Abstr. Progams* **13,** 398.

Anstey, R. L. (1986). Bryozoan provinces and patterns of generic evolution and extinction in the late Ordovician of North America. *Lethaia* **19,** 33–51.

Astrova, G. G. (1970). The history of development, system and phylogeny of the Bryozoa order Trepostomata. *Trudy paleont. Inst.* **169,** 1–240. (In Russian.)

Bambach, R. K. (1985). Classes and adaptive variety: the ecology of diversification in marine faunas through the Phanerozoic. In *Phanerozoic Diversity Patterns* (ed. J. W. Valentine), pp. 191–253. Princeton University Press, Princeton.

Benton, M. J. (1983). Large-scale replacements in the history of life. *Nature, Lond.* **302,** 16–7.

Bigey, F. P. (1986). Biogeography of Devonian Bryozoa. In *Bryozoa: Ordovician to Recent* (ed. C. Nielsen and G. P. Larwood), pp. 9–23. Olsen & Olsen, Fredensborg.

Bizzarini, F. and Braga, G. (1982). The Triassic Bryozoa of the western Tethyan basin. *Boll. Soc. palaeont. ital.* **21,** 223–34.

Boardman, R. S. (1984). Origin of the post-Triassic Stenolaemata (Bryozoa): a taxonomic oversight. *J. Paleont.* **58,** 19–39.

Boardman, R. S., *et al.* (1983). Bryozoa. In *Treatise on Invertebrate Paleontology, Part G, Revised,* pp. 1–625. Geological Society of Amerca and University of Kansas, Boulder, Colorado.

Brenchley, P. J. (1984). Late Ordovician extinctions and their relationship to the Gondwana glaciation. In *Fossils and Climate* (ed. P. J. Brenchley), pp. 291–315. Chichester.

Brenchley, P. J. and Newell, G. (1984). Late Ordovician environmental changes and their effect on faunas. In *Aspects of the Ordovician System.* Palaeontological Contributions from the University of Oslo, No. **295,** (ed. D. L. Bruton), pp. 65–79. Universitetsforlaget, Oslo.

Brood, K. (1972). Cyclostomatous Bryozoa from the Upper Cretaceous and Danian in Scandinavia. *Stockholm Contrib. Geol.* **26,** 1–464.

Brood, K. (1975). Cyclostomatous Bryozoa from the Silurian of Gotland. *Stockholm Contrib. Geol.* **28,** 45–119.

Brood, K. (1981). Hirnantian (Upper Ordovician) Bryozoa from Baltoscandia. In *Recent and Fossil Bryozoa* (ed. G. P. Larwood and C. Nielsen), pp. 19–27. Olsen & Olsen, Fredensborg.

Copper, P. (1977). Paleolatitudes in the Devonian of Brazil and the Frasnian–Famennian mass extinction. *Palaeogeogr. Palaeoclimatol. Palaeoecol.* **21,** 165–207.

Cuffey, R. J. and McKinney, F. K. (1979). Devonian Bryozoa. *Spec. Pap. Palaeont.* **23,** 307–11.

Dickins, J. M. (1983). Permian to Triassic changes in life. *Mem. Ass. Aust. Palaeontol.* **1,** 297–303.

Dzik, J. (1981). Evolutionary relationships of the early Palaeozoic 'cyclostomatous' Bryozoa. *Palaeontology* **24,** 827–61.

Gould, S. J. (1985). The paradox of the first tier: an agenda for paleobiology. *Paleobiology* **11,** 2–12.

Gould, S. J. and Calloway, C. B. (1980). Clams and brachiopods—ships that pass in the night. *Paleobiology* **6,** 383–96.

Håkansson, E. and Thomsen, E. (1979). Distribution and types of bryozoan communities at the boundary in Denmark. In *Cretaceous–Tertiary Boundary Events. I. The Maastrichtian and Danian of Denmark* (ed. T. Birkelund and R. G. Bromley), pp. 78–96. University of Copenhagen, Copenhagen.

Hallam, A. (1981). The end-Triassic bivalve extinction event. *Paleogeogr. Palaeoclimatol. Palaeoecol.* **35,** 1–44.

Harland, W. B., Cox, A. V., Llewellyn, P. G., Pickton, C. A. G., Smith, A. G., and Walters, R. (1982). *A Geologic Time Scale.* Cambridge University Press, Cambridge.

House, M. R. (1985). Correlation of mid-Palaeozoic ammonoid evolutionary events with global sedimentary perturbations. *Nature, Lond.* **313,** 17–22.

Hu, Z. X. (1984). Triassic Bryozoa from Xizang (Tibet) with reference to their biogeographical provincialism in the world. *Acta palaeont. sin.* **23,** 568–77.

Hüssner, H. (1983). Die Faunenwende Perm/Trias. *Geol. Rdsch.* **72,** 1–22.

Jablonski, D. (1985). Marine regressions and mass extinctions: a test using the modern biota. In *Phanerozoic diversity patterns* (ed. J. W. Valentine), pp. 335–54, Princeton University Press, Princeton.

Jablonski, D. (1986*a*). Causes and consequences of mass extinctions: a comparative approach. In *Dynamics of Extinction* (ed. D. K. Elliott), pp. 183–229. Wiley, New York.

Jablonski, D. (1986*b*). Background and mass extinctions: the alternation of macroevolutionary regimes. *Science* **231,** 129–33.

Larwood, G. P. and Taylor, P. D. (1979). Early structural and ecological diversification in the Bryozoa. In *The origin of major invertebrate groups* (ed. M. R. House), Systematics Association Special Volume No. 12, pp. 209–34. Academic Press, London.

Lavrentjeva, V. D. (1985). Bryozoa Suborder Phylloporina. *Trudy palaeont. Inst.* **214,** 1–101. (In Russian.)

McGhee, G. R. (1982). The Frasnian–Famennian extinction event: a preliminary analysis of Appalachian marine ecosystems. *Spec. Pap. Geol. Soc. Am.* **190,** 491–500.

McKinney, F. K. (1986). Historical record of erect bryozoan growth forms. *Proc. Roy. Soc.* series B, **228,** 133–49.

McKinney, F. K. and Kříž, J. (1986). Lower Devonian Fenestrata (Bryozoa) of the Prague Basin, Barrandian Area, Bohemia, Czechoslovakia. *Fieldiana Geol.* New Series, **15,** 1–90.

McKinney, M. L. (1985). Mass extinction patterns of marine invertebrate groups and some implications for a causal phenomenon. *Paleobiology* **11,** 227–33.

McLaren, D. J. (1971). Time, life and boundaries. *J. Paleont.* **44,** 801–15.

McLaren, D. J. (1982). Frasnian–Famennian extinctions. *Spec. Pap. Geol. Soc. Am.* **190,** 477–84.

Morozova, I. P. (1965). Bryozoa. In *The development and change of marine organisms at the Palaeozoic-Mesozoic boundary. Part* 1 (ed. V. E. Ruzhenstev and T. G. Sarycheva), Transactions of the Palaeontological Institute, Vol. 108, pp. 54–8. (In Russian.)

Morozova, I. P. (1970). Late Permian Bryozoa. *Trudy palaeont. Inst.* **122,** 1–347. (In Russian.)

Morozova, I. P. (1974). Revision of the bryozoan genus *Fenestella. Paleont. J.* **1974,** 167–80.

Morozova, I. P. (1986). Bryozoa. In *Parastratigraphic Groups of Flora and Fauna of the Triassic* (ed. A. N. Oleinikov and A. I. Zhamoida). *Trudy vses. nauchno-issled. geol. Inst.* **334,** 67–78. (In Russian).

Morozova, I. P. and Viskova, L. A. (1977). Historical development of marine Bryozoa (Ectoprocta). *Paleont. J.* **1977,** 393–408.

Newall, N. D. (1967). Revolutions in the history of life. *Spec. pap. geol. Soc. Am.* **89,** 63–91.

Pohowsky, R. A. (1978). The boring ctenostomate Bryozoa: taxonomy and paleobiology based on cavities in calcareous substrata. *Bull. Am. Paleont.* **73,** 1–192.

Raup, D. M. (1979). Size of the Permo-Triassic bottleneck and its evolutionary implications. *Science* **206,** 217–8.

Raup, D. M. and Sepkoski, J. J. (1982). Mass extinctions in the marine fossil record. *Science* **215,** 1501–3.

Ross, J. R. P. (1960). Larger cryptostome Bryozoa of the Ordovician and Silurian, Anticosti Island, Canada — Part I. *J. Paleont.* **34,** 1057–76.

Ross, J. R. P. (1961). Larger cryptostome Bryozoa of the Ordovician and Silurian, Anticosti Island, Canada — Part II. *J. Paleont.* **35,** 331–44.

Ross, J. R. P. (1978). Biogeography of Permian ectoproct Bryozoa. *Palaeontology* **21,** 341–56.

Ross, J. R. P. (1981). Biogeography of Carboniferous ectoproct Bryozoa. *Palaeontology* **24,** 313–41.

Rostovtsev, K. O. and Azaryan, N. R. (1973). The Permian-Triassic boundary in Transcaucasia. *Mem. Can. Soc. Petrol. Geol.* **2,** 89–99.

Ryland, J. S. (1970). *Bryozoans.* Hutchinson, London.

Sakagami, S. (1985). Paleogeographic distribution of Permian and Triassic Ectoprocta (Bryozoa). In *The Tethys* (ed. K. Nakazawa and J. M. Dickins), pp. 171–83. Tokyo University Press, Tokyo.

Schäfer, P. and Fois-Erickson, E. (1986). Triassic Bryozoa and the evolutionary crisis of Paleozoic Stenolaemata. In *Global Bio-Events* (ed. O. Walliser), Lecture Notes in Earth Sciences, Vol. 8, pp. 251–5. Springer-Verlag, Berlin.

Schopf, T. J. M. (1974). Permo-Trias extinction: relation to sea-floor spreading. *J. Geol.* **82,** 129–43.

Schopf, T. J. M. (1977). Patterns and themes of evolution among the Bryozoa. In *Patterns of Evolution* (ed. A. Hallam), pp. 159–207. Elsevier, Amsterdam.

Schopf, T. J. M. (1979). The role of biogeographic provinces in regulating marine faunal diversity through geologic time. In *Historical Biogeography, Plate Tectonics and the Changing Environment* (ed. J. Gray and A. J. Boucot), pp. 449–57, Oregon State University Press, Corvallis.

Sepkoski, J. J. (1981). A factor analytic description of the Phanerozoic marine fossil record. *Paleobiology* **7,** 36–53.

Sepkoski, J. J. (1982). A compendium of fossil marine families. *Milwaukee Publ. Mus. Contrib. Biol. Geol.* **51,** 1–125.

Sepkoski, J. J., Bambach, R. K., Raup, D. M., and Valentine, J. W. (1981). Phanerozoic marine diversity and the fossil record. *Nature, Lond.* **293,** 435–7.

Sepkoski, J. J. and Miller, A. I. (1985). Evolutionary faunas and the distribution of Paleozoic marine communities in space and time. In *Phanerozoic Diversity Patterns* (ed. J. W. Valentine), pp. 153–90. Princeton University Press, Princeton.

Simberloff, D. S. (1974). Permo-Trias extinctions: effects of area on biotic equilibrium. *J. Geol.* **82,** 267–74.

Spjeldnaes, N. (1982). Ordovician bryozoan faunas. *4th International Symposium on the Ordovician System, Oslo,* Abstr. p. 49.

Stanley, S. M. (1984). Marine mass extinctions: a dominant role for temperature. In *Extinctions* (ed. M. H. Nitecki), pp. 69–117. University of Chicago Press, Chicago.

Taylor, P. D. (1981). Functional morphology and evolutionary significance of differing modes of tentacle eversion in marine bryozoans. In *Recent and Fossil Bryozoa* (ed. G. P. Larwood and C. Nielsen), pp. 235–47, Olsen & Olsen, Fredensborg.

Taylor, P. D. (1985). Carboniferous and Permian species of the cyclostome bryozoan *Corynotrypa* Bassler, 1911 and their clonal propagation. *Bull. Br. Mus. nat. Hist.* (*Geol.*) **35,** 359–72.

Taylor, P. D. and Curry, G. B. (1985). The earliest known fenestrate bryozoan, with a short review of Lower Ordovician Bryozoa. *Palaeontology* **28,** 147–58.

Troitskaya, T. D. (1968). *Devonian Bryozoa of Kazakhstan.* Nedra Press, Moscow. (In Russian.)

Twenhofel, W. H. (1928). Geology of Anticosti Island. *Can. geol. Survey Mem.* **154,** 1–481.

Viskova, L. A. (1980). Bryozoa. In *Development and change of invertebrates on the boundary of the Mesozoic and Cenozoic. Bryozoa, arthropods and echinoderms*, Paleontologicheskii Institut, Akademiya Nauk SSSR, Moscow. (In Russian.)

Voigt, E. (1959). La significance stratigraphique des Bryozoaires dans le Crétacé supérieur. *Congr. Soc. savantes* **84,** 701–7.

Voigt, E. (1981). Critical remarks on the discussion concerning the Cretaceous–Tertiary boundary. *Newsl. Stratigr.* **10,** 92–114.

Voigt, E. (1986). The Bryozoa of the Cretaceous–Tertiary boundary. In *Bryozoa: Ordovician to Recent*, pp. 329–42. Olsen & Olsen, Fredensborg.

6. Extinction and survival in the Bivalvia

A. HALLAM

Department of Geological Sciences, University of Birmingham, Birmingham, UK

A. I. MILLER

Department of Geology, University of Cincinnati, Cincinnati, Ohio, USA

Abstract

For a variety of reasons, the Class Bivalvia is well suited for an evaluation of its evolutionary history. In this paper, Phanerozoic global diversification and extinction patterns among bivalves are examined in some detail.

On the basis of analyses of diversification in the class as a whole, as well as in constituent orders and life habit groupings, it is apparent that patterns of diversification established following the group's initial radiation in the Ordovician continued through the Palaeozoic and post-Palaeozoic. Mass extinctions, including the Late Permian event, interrupted, but did not permamently alter these patterns.

Following their Ordovocian radiation, bivalves were most prominent in nearshore terrigenous regions, but as the Palaeozoic progressed they became increasingly important offshore and in carbonate habitats. Pterioids and nuculoids were notable contributors to faunal diversity in deepwater during the latter portion of the Palaeozoic.

While Palaeozoic mass extinctions did not affect bivalves as greatly as they did many other marine organisms, Mesozoic bivalves were notably affected by mass extinctions at the ends of the Norian, Pliensbachian, Tithonian, and Maastrichtian; the Pliensbachian and Tithonian events are thought to have been regional, rather than global in scale. Additional, bivalves suffered noteworthy extinctions during the late Eocene and

Extinction and Survival in the Fossil Record (ed. G. P. Larwood), Systematics Association Special Volume No. 34, pp. 121–38. Clarendon Press, Oxford, 1988.

Plio-Pleistocene. It is suggested that these extinctions resulted from climatic fluctuations, bottom-water anoxia, and/or reductions in amount of habitable area.

Introduction

For a number of reasons the Bivalvia are an unusually favourable group for the study of extinction and survival. Because of the high abundance and preservational potential of their shells, the fossil record can be expected to give a reliable indication of their history, and indeed, there is a record of a rich and diverse fauna extending back through almost the whole of the Phanerozoic. Unlike many fossil groups, much can be deduced about mode of life from shell form (Stanley 1970), because the vast majority of the wide array of adaptive types still survive. Therefore, extinction/survival analysis can be conducted with respect to life habits as well as taxonomic subdivisions, though there is of course to some extent a relationship between the two.

The origin and early evolution of the Bivalvia has been thoroughly discussed by Morris (1979), Pojeta (1978, 1985), and Runnegar (1985). The group apparently evolved from rostroconch molluscs in the early Cambrian, with two genera, *Fordilla* and *Pojetaia*, being widely distributed at that time; no bivalves have yet been recognized in younger Cambrian strata. Bivalve radiation started in the early Ordovician and became substantial in the middle of the period. Thus, Pojeta (1985) points out that the number of recorded species rises from about 60 in the early, to 360 in the mid, and 420 in the late Ordovician. The radiation was accompanied by a significant increase in the maximum size attained by individual organisms. The largest known Cambrian species is 5 mm while the maximum size for Ordovician species rises from 17 mm in the early to 110 mm in the late Ordovician (Pojeta 1985).

Phyletic size increase continued through the Phanerozoic in many groups, to culminate in giant rudists and inoceramids in the late Cretaceous and the extant giant clam *Tridacna*. Two adaptive breakthroughs that led to major radiations are recognized by Stanley (1968, 1972, 1977). The first was the paedomorphic retention into the adult of the larval byssus, which took place polyphyletically in the Pterioida and gave rise to a significant epifaunal radiation in the Palaeozoic. The second was mantle fusion, which produced siphons and provided a tight seal for the mantle cavity during the hydraulic process of burrowing, which thus became more efficient. The evolutionary consequence was a striking post-Palaeozoic radiation of heterodonts following an earlier pulse of diversification of deep-burrowing anomalodesmatids in the Carboniferous. The rise to dominance as a result of this adaptive breakthrough is evidenced by the fact that nearly 90 per cent of the living species of suspension-feeding bivalves are siphonate (Boss 1971).

Diversity changes through the Phanerozoic

Patterns of diversity change through the Phanerozoic are presented in Figs 6.1–6.4. Further details on how these curves were constructed are given in Miller (1986). Figure 6.1 presents two curves of bivalve family diversity from the Ordovician to the present. Figure 6.1b shows more oscillations than Fig. 6.1a because it was compiled at the stage, rather than the series, level of resolution. Both curves are, however, closely comparable in showing a progressive rise in diversity through the Phanerozoic that was

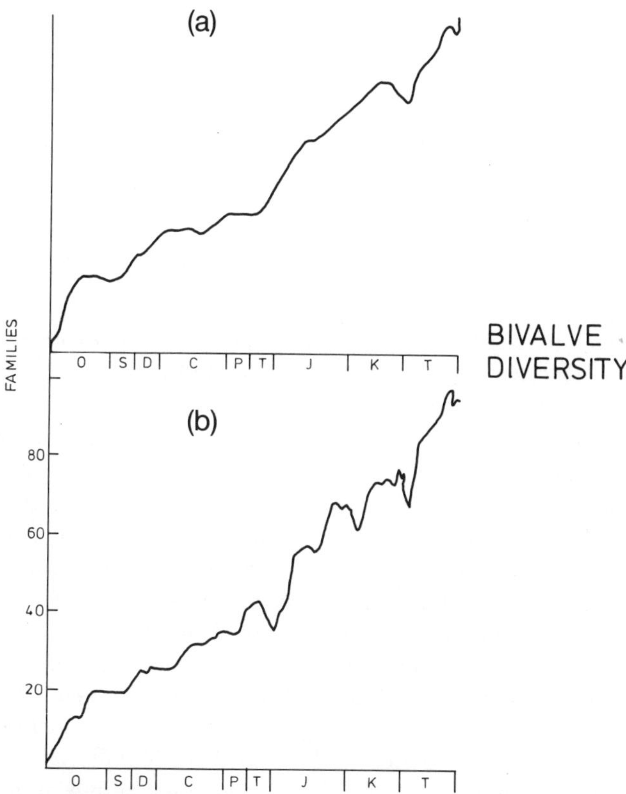

Fig. 6.1. Bivalve family diversity through the Phanerozoic, Cambrian excepted. (a) Adapted from Newell and Boyd (1978, fig. 1). (b) Based on data in Sepkoski (1982). O, Ordovician; S, Silurian; D, Devonian; C, Carboniferous; P, Permian; T, Triassic; J, Jurassic; K, Cretaceous; T, Tertiary.

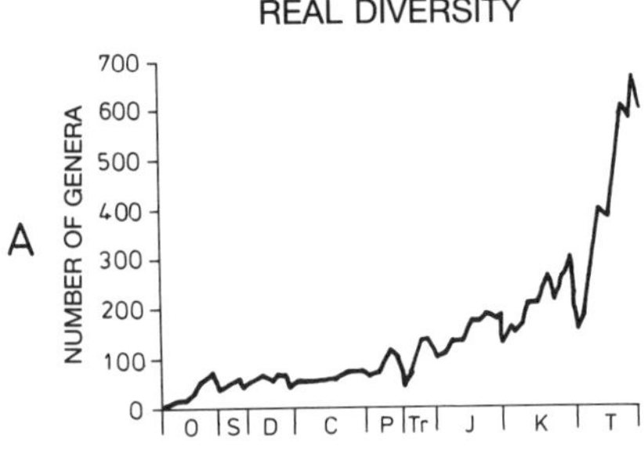

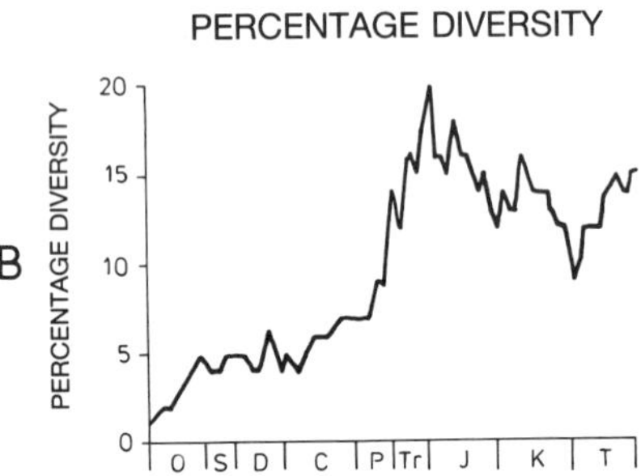

Fig. 6.2. Bivalve generic diversity through the Phanerozoic, Cambrian excepted. Bottom curve shows bivalve diversity as a percentage of whole faunal marine diversity. Based on unpublished data of J. J. Sepkoski.

little disturbed by mass extinction events, with even the major end-Palaeozoic event having little long-term effect. According to Sepkoski's (1982) compilation, on which Fig. 6.1b is based, the great majority of bivalve families are extant. Of the 55 families that have gone extinct, 22 did so at the five major extinction events generally recognized at or near the end of the Ordovician, Devonian, Permian, Triassic, and Cretaceous, viz. the Ashgillian, Frasnian, Djulfian, Norian, and Maastrichtian. Five more went extinct in the Tithonian and four in the Guadalupian, which should probably be included in the late Permian extinction event. Thus, 56 per cent of family extinctions are concentrated at six mass extinction horizons.

We present generic diversity data in Fig. 6.2. These curves were constructed using data from an unpublished compendium of fossil marine genera compiled by J. J. Sepkoski Jr. The upper curve presents information in the same way as in Fig. 6.1, and shows a radiation through the Ordovician, subsequently stabilizing before a further notable diversity rise in the Permian that was interrupted by the late Permian mass extinction, following which there was no dramatic long-term increase in diversification in the Mesozoic. The sharp Tertiary increase almost certainly reflects a real phenomenon, but it might be exaggerated somewhat by the 'pull of the Recent' (Raup 1979), whereby extant genera are likely to be overemphasized relative to those that are known only as fossils.

The lower curve of Fig. 6.2 shows bivalve diversity as a percentage of the total marine faunal diversity. It is an attempt to construct a diversity curve that is uniquely characteristic of bivalves, because many of the fluctuations in the upper curve might be expected also in the diversity curve of other higher taxa, for instance the decrease at major extinction horizons, increase because of the 'pull of the Recent' and changes related to temporal changes in rock volume (Raup 1976) or 'palaeontologist interest' (Sheehan 1977).

The most important feature to observe in the lower curve is that bivalve percentage diversity had reached half the average Mesozoic level by Carboniferous times. There is a striking spike at the end of the Permian, which indicates that bivalves fared much better than most other marine organisms during the late Permian mass extinction. Indeed, bivalves are seen to have fared better than the pool of all other marine organisms during every interval of overall declining generic diversity during the Palaeozoic and Triassic. Gould and Calloway (1980) have argued that the replacement of brachiopods by bivalves as the dominant shelly benthic macroinvertebrates was a consequence of differential extinction in the late Permian rather than displacive competition. It is evident from the percentage curve that attention should not be confined to the late Permian event in attempting to account for bivalve diversification through time, but it should also be noted that this pattern of differential success starts to break down in the Jurassic.

Figure 6.3 offers a taxonomic breakdown into orders. It presents plots of familial diversity for all those bivalve orders that ever achieved a standing familial diversity of five or more. The only order in which Mesozoic patterns are dramatically discontinuous is the Hippurtitoida. Prominent post-Palaeozoic orders (most importantly veneroids and pterioids) had become well established during the Palaeozoic. Hippuritoids excepted, there were no post-Palaeozoic orders that had not been of some importance in the Palaeozoic.

A breakdown of bivalves with respect to life habits is presented in Figure 6.4, which shows generic diversity curves (diversity plotted as the percentage of total bivalve generic diversity) for the five major life-habit groups. These curves were based on life-habit assignments made to nearly 2000 of the ~2200 genera included in the generic diversity curve of Fig. 6.2, as detailed by Miller (1986). The curves are somewhat misleading in that post-Palaeozoic decreases in epibyssate, endobyssate, and deposit-feeding bivalves are directly attributable to the increasing importance of free-burrowing suspension feeders. The real diversity of all these groups

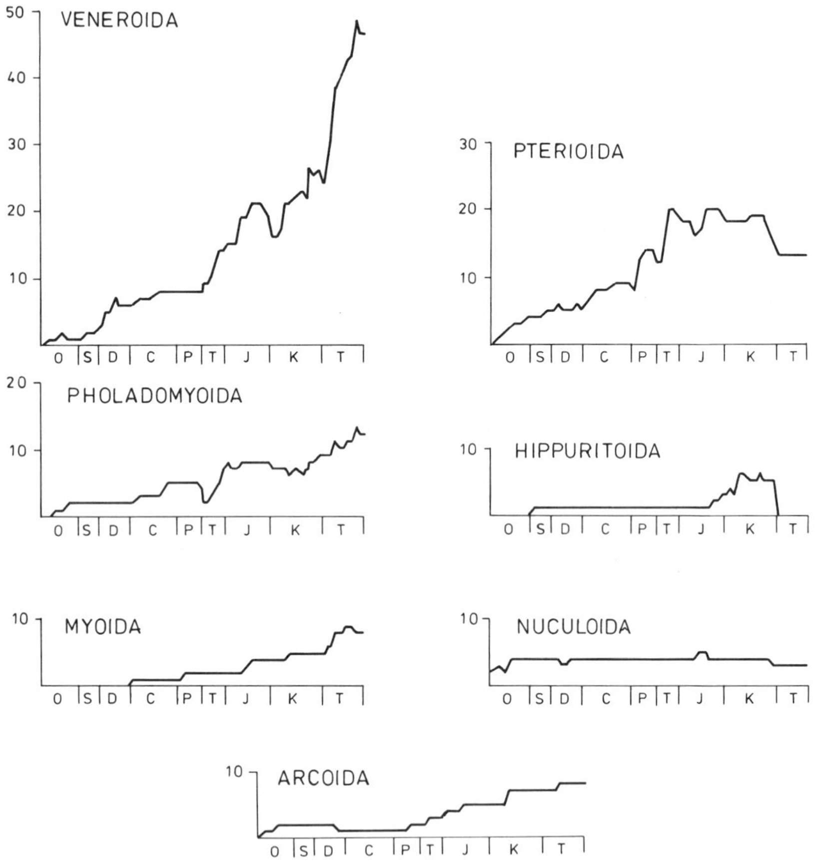

Fig. 6.3. Familial diversity curves of bivalve orders that ever achieved a standing familial diversity of five or more.

increased after the Palaeozoic, but not nearly as dramatically as free-burrowing suspension feeders, so that their percentage share of bivalve genera has decreased. In any case, it is clear that the patterns established in the Palaeozoic were once more carried over into the post-Palaeozoic, with free-burrowing suspension feeders beginning their climb to prominence in the Silurian after an Ordovician percentage drop, which is a by-product of radiations of other life-habit groups. Again, there were no major, long-term shifts as a result of the late Permian mass extinction.

The only life-habit group that was of limited importance in the Palaeozoic, but rose to prominence in the post-Palaeozoic, is the cementers. Their percentage share of bivalve genera began to increase in the Jurassic and expanded rapidly in the Cretaceous. With the extinction of rudists and other cementers at the end of the Cretaceous the group's

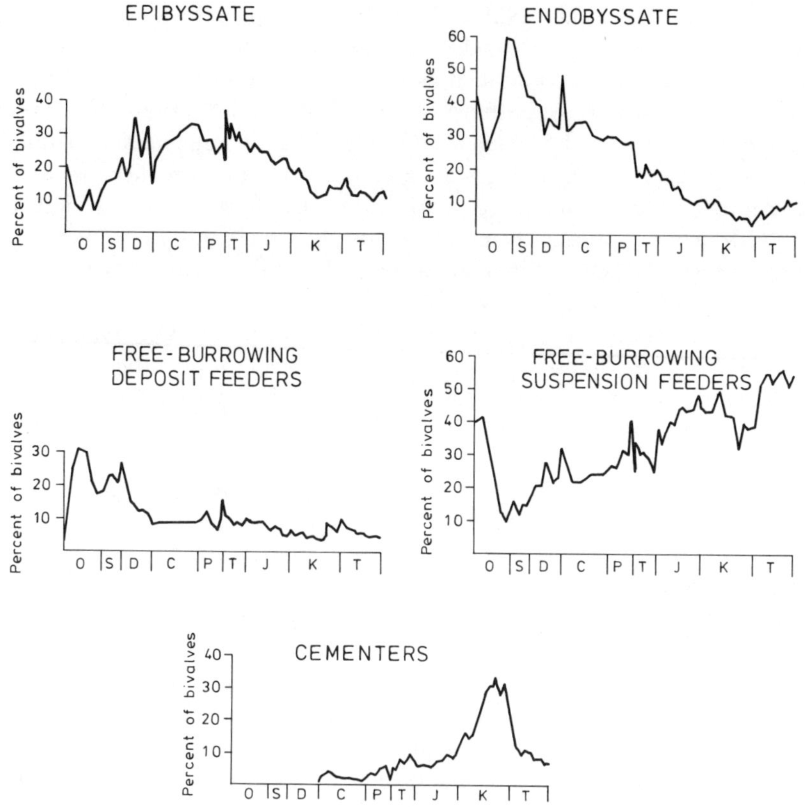

Fig. 6.4. Generic diversity curves, with diversity plotted as the percentage of total bivalve generic diversity, in terms of life habits.

percentage share decreased subsequently to early Mesozoic levels. Given the diversification pattern of cementers, it is clear that the late Permian mass extinction was not the primary agent in their diversification.

The results of this analysis of bivalve life habits are consistent with those of Stanley (1972), for the Palaeozoic, and Thayer (1983) for the entire Phanerozoic.

Patterns of change within each era

1. *Palaeozoic*

By late Ordovician times there had become established in the shallow neritic zone a moderate diversity of bivalves dominated by infaunal burrowing nuculoids and epifaunal to semi-infaunal byssate mytiloids and pterioids (Pojeta 1971, 1985). Two families, the Cycloconchidae and Colpomyidae, went extinct in the Ashgillian, but the end-Ordovician mass extinction could not be described as a traumatic event for the bivalves as a whole. Kriz (1984) records that the event had a more serious effect on the pteriomorphs than the palaeotaxodonts (mainly nuculoids), with 25 Ashgillian genera being reduced to 9, compared with 15 reduced to 7. This differential effect is tentatively attributed by Kriz to the widespread establishment of bottom-water anoxia in Llandovery seas, which might have affected byssate epifaunal suspension feeders more severely than free-burrowing deposit feeders.

Bivalve diversity was substantially higher in the late Devonian than in late Ordovician times but only two families, the Antipleuridae and Ambonychiidae, went extinct during the Frasnian mass extinction event, which had a negligible effect on bivalves as a whole, in marked contrast to some other groups. Not surprisingly, the late Permian event was more significant. Sepkoski (1982) records five bivalve families going extinct during the Guadalupian–Djulfian interval—the Modiomorphidae, Pterineidae, Edmondiidae, Megadesmidae, and Grammyssidae. To these must be added the Alatoconchidae, a family of free-living recliners retricted to the Old World Tethys and attaining the largest size of any Palaeozoic bivalve—up to 100 cm long (Yancey and Boyd 1983).

Nevertheless, the bivalves belong with the groups least affected by the end-Palaeozoic crisis, as pointed out by Nakazawa and Runnegar (1973), who report a significant decline in generic and familial diversity throughout the late Permian, but no evidence for abrupt extinction of numerous bivalve taxa at the end of the period. The Pterioida and Pholadomyoida experienced the most decline, with the other orders being little affected (cf. Fig. 6.3). The pholadomyoids are important components of Australian, Siberian, and South American faunas, but are rare elsewhere. Only the Edmondiidae and Megadesmidae are known from the latest Permian. Nakazawa and Runnegar also note that a number of well-known Mesozoic families, such as the Pteriidae, Bakevellidae, Inoceramidae, Isognomonidae, Terquemiidae, Anomiidae, Oxytomidae, Entoliidae, and Ostreaidae, made their first appearance in the Permian. Since many of these are poorly represented, if at all, in early Triassic faunas, the apparent disappearance of many taxa across the Permian–Triassic boundary must be attributable to what Jablonski (1986*b*) has termed the Lazarus Effect—at least some of the extinctions are more apparent than real.

General patterns of change through the Palaeozoic have been addressed by several workers. Bretsky (1973) argued that following major extinction events bivalve replacement occurred in two distinct waves separated by a period of biotic stability, and that cosmopolitan genera were less affected by these events than endemics, with the notable exception of the end-Palaeozoic event. He did not, however, clearly define what he meant by endemic genera. Bretsky (1969) was also the first to draw attention to an apparent recurrent pattern in marine benthic communities, with bivalves dominant, in association with linguloid brachiopods, in the inshore shallow neritic zone, and articulate brachiopods and other groups dominant in the offshore deeper neritic zone. The latter groups were considered to be more stenotopic and, therefore, extinction-vulnerable.

Although Bretsky's work has had a seminal influence, further study suggests that it requires some revision. Multivariate analysis of faunal abundance data collected from a variety of palaeogeographic settings in the Upper Ordovician reveals that Ordovician bivalves were most abundant in both onshore and offshore regions of high terrigenous influx, and were relatively scarce in carbonate-dominated settings. By the late Palaeozoic, bivalves had become important faunal components in a variety of carbonate environments as well. This expansion was at least partly related to the diversification of infaunal suspension feeders (Miller 1985).

Factor analysis of Palaeozoic marine communities indicates a succession of three major evolutionary faunas in the Palaeozoic (Sepkoski and Miller 1985). All three faunas are clearly manifested in local community compositions and may be characterized as (1) trilobite-rich, (2) brachiopod-rich, and (3) mollusc-rich. In post-Cambrian times (3) was concentrated in nearshore, shallow water environments and (1) concentrated in relatively deeper water environments, with (2) occupying an intermediate position. Changes in dominance involved onshore-offshore expansion of new community types, successively (2) and (3). By late Palaeozoic times, the brachiopod-dominated fauna was more or less restricted to the middle and outer shelf as a consequence of expansion of the 'modern' mollusc-dominated fauna. The great majority of bivalves occurred with fauna (3), but others, including several pterioid and nuculoid genera, provide notable exceptions, being important faunal components in more offshore, deeper water settings (Boardman *et al.* 1984; Kammer *et al.* 1986).

2. *Mesozoic*

Whereas late Permian bivalve faunas exhibited provinciality to a high degree, the earliest Triassic faunas were cosmopolitan, and were dominated overwhelmingly by nuculoids and pterioids. Diversity was initially low and did not exceed the highest Permian value until the middle of the period (Fig. 6.2A, and Nakazawa and Runnegar 1973).

The next major extinction event affecting the Bivalvia took place at the end of the Triassic (Hallam 1981*b*). At family level, the event was not particularly marked, with only six families (Cassianellidae, Megalodontidae, Monotidae, Myophoricardiidae, Mysidiellidae) going extinct out of a total of 52, but at generic level the term mass extinction is certainly apposite. It is also clear that extinction was concentrated at or very close to the end of the period. Thus, the vast majority of Norian genera (87 out of 95) persisted into the late Norian or Sevatian (including the 'Rhaetian') and include most of the distinctive, common, and characteristic Triassic genera that did not survive into the Jurassic. A minimum of 33 out of a total of 40 genera that became extinct, survived until the end or almost the end of the stage. Information is also available at species level for the relatively well studied continent of Europe. Of a total of at least 60 European Upper Norian species only five definitely survived into the Jurassic, a 92% extinction rate at the end of the Triassic implying an almost complete turnover of organisms at the system boundary.

For five of the nine bivalve orders the generic extinction rate ranged between 40 and 62 per cent. The Hippuritoida sample of two late Triassic genera is too small for the 100 per cent extinction rate of be held to be especially significant, but it is noteworthy that the Unionoida (80 per cent) and Trigonioida (85 per cent) were strongly affected by the extinction event whereas the Pholadomyoida (11 per cent) were only slightly affected at generic level. With regard to ecology, a higher proportion of apparently euryhaline (and no doubt more generally eurytopic) genera survived into the Jurassic, but because of the high rate of turnover, all ecological categories were affected by the extinction event.

Two relatively minor events, which nevertheless represent mass extinction of bivalves at species level, are recognizable in the Jurassic of Europe though not as yet elsewhere, pointing to a regional rather than a global control (Hallam 1986). The first event, recognizable in nearly all fossil groups, occurred in end Pliensbachian — beginning Toarcian times, and was responsible for an 84 per cent extinction rate in bivalve species, but does not appear significant at the generic level. The second event, at the end of the Jurassic, coincides with the global disappearance of five families, the Aviculopectinidae, Hippopodiidae, Quenstedtiidae, Ceratomyidae, and Sowerbyidae, but it should be noted that all these families are genus-poor. Trigonioids and hippuritoids were the most notably affected by the late Tithonian extinction event, but mainly at species level.

Cretaceous radiations and extinctions among the Bivalvia have been fully dealt with by Kauffman (1984). An extinction event among some other fossil groups has been recognized at the Cenomanian–Turonian boundary, but the bivalves were evidently not affected, at least above the species level. The end of the period is a different matter, and the Maastrichtian marks the last appearance of the following ten families: Buchiidae, Inoceramidae, Dicerocardiidae, Icanotiidae, Tancrediidae, Caprinidae, Hippuritidae, Monopleuridae, Radiolitidae, Requienidae. Half of these families are rudists, the most spectacular of Cretaceous bivalves, which occupied tropical reef habitats. Two groups must be distinguished. The Requienidae and Monopleuridae maintained a low generic diversity throughout the Cretaceous, whereas the Caprinidae and Hippuritidae underwent a moderate, and the Radiolitidae a pronounced radiation in the latter part of the period (Fig. 6.5). The radiolitid and hippuritid radiations are related by Skelton (1985) to a major adaptive breakthrough, as ligament invagination permitted the uncoiled growth of extended umbones to produce tubular valves for elevated and recumbent growth forms.

According to Kauffman the final extinction of the rudists did not take place abruptly at the end of the Maastrichtian, but was preceded by a substantial diversity reduction at the end of the mid-Maastrichtian, along with other warm-water bivalves such as large trigoniids and oysters. Similarly, that other important Cretaceous group, the inoceramids, was also in decline before the end of the stage. These was apparently, however, an extinction event at the end of the Maastrichtian in northern Europe at least, as recorded in the well known Danish section at Stevns Klint (Alvarez *et al.* 1984). Species of the rudist *Gyropleura* and the inoceramid *Tenuiptera* persists until immediately below the K–T boundary Fish Clay. Of 12 bivalve genera in the top Maastrichtian hardground that are definitely calcite-bearing, four became extinct and eight have persisted to the present. There remains a problem concerning genera with aragonitic shells, whose proven stratigraphic ranges are probably unreliable because of possible preservational failure. Relatively complete sections across the K–T boundary in shallow-water facies with bivalves are scarce compared with deep-water sections containing only micro- and nannoplakton, and it would be premature to generalize too much in the present state of knowledge from Stevns Klint alone.

Kauffman considers that extinction among temperate zone bivalves was only moderate compared with the tropical zone, with some 50 per cent of lineages crossing the K–T boundary with only species-level breaks.

Jablonski (1986*a*) has studied late Cretaceous bivalve and gastropod faunas of the Atlantic and Gulf coastal plains of the USA. He claims that during times of what he calls background extinction, planktotrophic larval development, broad geographical range, and high species richness enhanced the survivorship of species and genera. In contrast, during the end-Cretaceous mass extinction these factors were ineffectual, but broad geographic deployment of entire lineages, regardless of the ranges of constituent species, served to enhance survivorship. Therefore, geographic range at species level apparently had little influence on the survival of genera.

With regard to general patterns of change through the Mesozoic, the most striking was the marked increase in infaunalization in the late Cretaceous, persisting into the Cainozoic, that was probably a defensive response to the radiation of benthic predators in the form of neogastropods, crabs and teleost fish (Vermeij 1977). At the present day, predation is

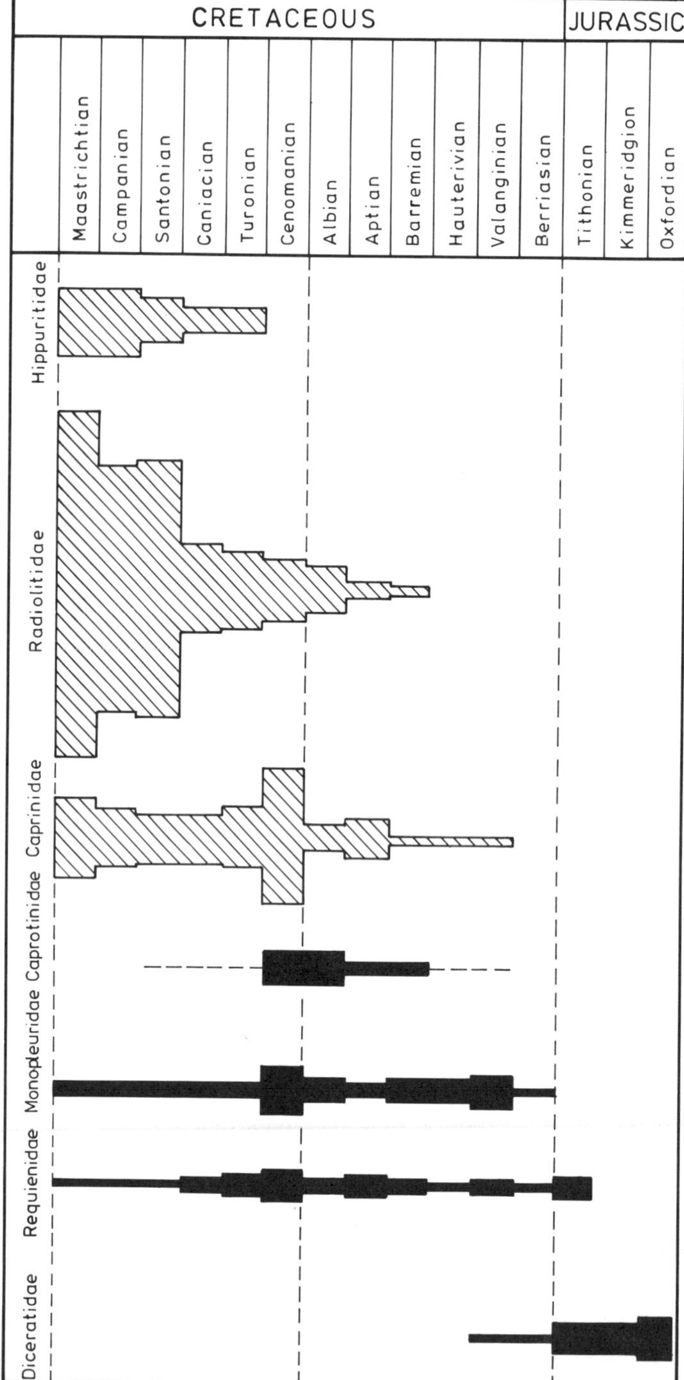

Fig. 6.5. Diversity variation through the late Jurassic and Cretaceous of rudist families. Adapted from Kauffman (1984, fig. 8.2).

regarded as by far the most important limiting factor on subtidal, suspension-feeding bivalves (Stanley 1973) and in at least some tropical seas population densities diminish drastically for this reason at depths of only a few metres compared with the shallower, more inshore areas (Jackson 1972). The situation must have been substantially different in the early Mesozoic because of the relatively low predation pressure at that time. This is borne out by the fact that Jurassic bivalve faunas have a much higher proportion of epifauna than Recent faunas (compare the data in Stanley 1970 and Hallam 1976). If individual rather than taxa are counted the contrast appears even more striking. In an unpublished study by A. Hallam 20 polytypic assemblages, and a few monotypic assemblages at a wide range of stratigraphic horizons within the Jurassic of western Europe were collected and subjected to diversity analysis. The epifaunal Pterioida accounted for an astonishingly high proportion of the total number of individuals counted in the polytypic assemblages, no less than 58 per cent. This figure would be even higher if the monotypic assemblages, such as shell-packed oyster beds, were also taken into account.

While there appears to be at least some correlation between endemicity and extinction vulnerability, well evidenced, for example, by the Trigonioida, a notable exception is provided by certain Pterioida, namely the pectinaceans *Daonella, Monotis, Halobia, Aulacomyella, Pectinula, Bositra,* and *Buchia,* and the pteriacians *Inoceramus, Retroceramus,* and *Pseudomytiloides.* Like their Palaeozoic precursors, these genera tended to occupy more offshore, deeper water habitats than other bivalves. They also tend to have relatively cosmopolitan species with short stratigraphic ranges. For this reason, species of Triassic *Daonella, Halobia,* and *Monotis,* Jurassic *Buchia* and *Retroceramus,* and Cretaceous *Inoceramus* have been widely used as stratigraphic indices. That the genera may also be usually wide ranging and short-lived compared with other bivalves is shown for Jurassic examples in Fig. 6.6. A short stratigraphic range implies a relatively high vulnerability to extinction, so the combination of cosmopolitanism and high extinction vulnerability appears paradoxical in the light of the relationship pointed out by Boucot (1975) between taxon longevity and geographic distribution. Why this should be so remains a mystery in the light of the more stable environments experienced in deeper water offshore.

3. Cainozoic

The most dramatic Cenozoic extinction event according to Stanley (1984) took place along the margin of western North America in the late Eocene, with taxa having warm-water affinities being more severely affected. In the southern circum-Pacific region relatively few species survived into the Oligocene. The only other extinction event of note took place in late Pliocene to mid Pleistocene times, when 65–75 per cent of benthic molluscs

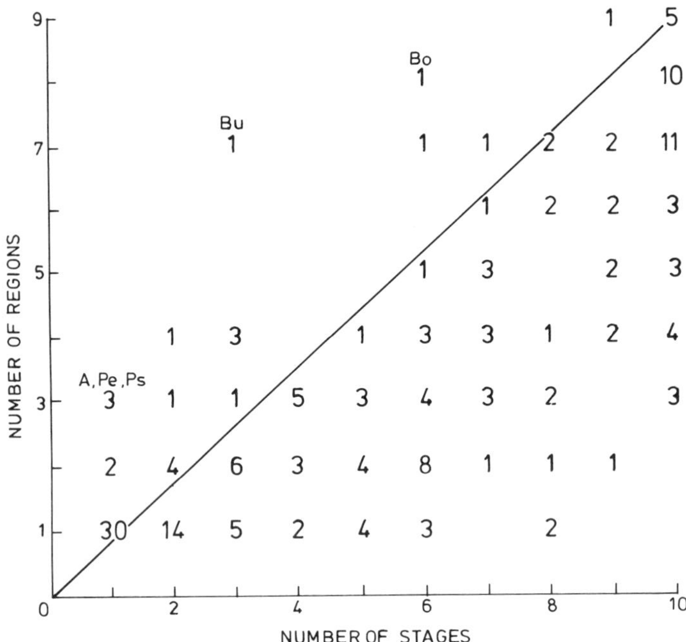

Fig. 6.6. Plot of geographic against temporal range for Jurassic bivalve genera (the numbers on the plot refer to numbers of genera). A, *Aulacomyella*; Bo, *Bositra*; Bu, *Buchia*; Pe, *Pectinula*; Ps, *Pseudomytiloides*; R, *Retroceramus*. Note that these pectinacean and pteriacian pterioids all lie well above the oblique line, which indicates that they had a relatively cosmopolitan distribution. After Hallam (1977, fig. 3).

went extinct in the western North Atlantic and Caribbean. Casualties in the Mediterranean and Red Sea were significantly fewer and there was little or no excessive Plio-Pleistocene extinction off Japan and California. The event was therefore regional rather than global in extent and was caused according to Stanley (1984) by severe reduction of water temperatures in an oceanic cul-de-sac, coincident with the onset of continental glaciation in the northern hemisphere. All purely tropical species disappeared and small-sized siphonate burrowers experienced higher rates of extinction than large-sized species, probably because the small species tended to be more endemic and therefore more vulnerable to regional temperature changes (Stanley 1986*a*).

Stanley (1986*b*) also studied relative survivorship in North Pacific Neogene bivalves, predominantly the faunas off Japan and California. Among non-protobranch faunas, siphonate burrowers have a much higher survivorship than non-siphonates, whose species suffer heavier predation and have smaller population size. Among siphonate burrowers, survivorship and population size are inversely proportional to body size. Pectinaceans have a much lower survivorship than other epifauna, with small populations suffering heavy predation. Other epifaunal bivalves are less vulnerable because of larger populations, in turn related to small body size and intertidal or cryptic habitats. In contrast to the findings of Jablonski and Lutz (1983) there is no evidence that species with non-planktotrophic larvae are either narrowly distributed or of brief geological duration. In general, high extinction rates promote high speciation rates and fragmented distributions, a phenomenon that Stanley terms the fisson effect.

Conclusions

Compared with other groups of marine invertebrates, the bivalves were less vulnerable to mass extinction events, at least until the early Mesozoic. This fact is shown up especially clearly for the most important mass extinction event of all, in the late Permian. This event, while provoking an increased extinction rate among the Bivalvia, and hence some diversity reduction, did not seriously disturb patterns of change in the constituent groups that had become established in the late Palaeozoic, and which continued into the Mesozoic after experiencing what was evidently only a modest setback.

With regard to relative extinction vulnerability within the Bivalvia, groups whose ecological requirements are inferred to have included warm-water conditions were evidently always more at risk. This is shown most clearly for the hippuritoids, which were severely affected by the extinction events at the end of the Triassic, Jurassic, and Cretaceous. Endemics were also relatively vulnerable, as well shown by the fate of trigonioids during the same events. The pterioids contain some Mesozoic groups, however, that paradoxically combined cosmopolitan distribution with short species durations, for reasons that are not understood. Studies on Neogene bivalves suggest that small population size and large body size (usually well correlated with small population size) have served to increase extinction vulnerability, as has an epifaunal habitat unprotected from benthic predators such as crabs, neogastropods, and teleosts. One should, however, resist the temptation to seek too many generalizations in the present limited state of knowledge, and each extinction event should be studied on its own merits. Thus, it is by no means clear why pholadomyoids were apparently relatively vulnerable to extinction during the late Permian event, but relatively invulnerable during the late Triassic event.

An adequate discussion of the possible cause of causes of the mass extinctions would demand a full discussion of other fossil groups and geological evidence, and is beyond the scope of this article. The subject is well reviewed by Jablonski (1986*b*), who argues the merits and demerits of the more plausible of the suggested explanations. Extrapolating to the whole Phanerozoic from his Neogene work, Stanley (1984) has stressed the dominant role of water temperature as a causal factor. While there is a good case to be made for declining temperatures being a major factor in the Plio-Pleistocene and late Eocene extinctions, the case is much weaker for earlier times in Phanerozoic history. This is especially true for the late Permian extinctions that took place at a time of global climatic amelioration. The relatively high vulnerability to extinction of tropical groups could, at least to some extent, be the consequence of their being more generally stenotopic. For events prior to the Cainozoic, severe reduction of neritic habitat as a consequence of either regression or bottom-water anoxia, both phenomena bound up with sea-level change, seems the likeliest cause (Hallam 1981*a*,*b*, 1986).

Acknowledgment

We thank Jack Sepkoski for allowing us to use his unpublished generic compendium.

References

Alvarez, W., Kauffman, E. G., Surlyk, F., Alvarez, L. W., Asaro, F., and Michel, H. V. (1984). Impact theory of mass extinctions and the invertebrate fossil record. *Science* **223,** 1135–41.

Boardman, D. R., Mapes, R. H. Yancey, T. E., and Malinkey, J. M. (1984). A new model for depth-related allogenic community succession within North American Pennsylvanian cyclothems and implications on the black shale problem. In *Limestones of the Mid-Continent,* Tulsa Geological Survey Special Publication No. 2 (ed. N. J. Hynes), pp. 141–82.

Boss, K. J. (1971). Critical estimate of the number of Recent Mollusca. *Occ. Pap. Mollusks* **3,** 81–135.

Boucot, A. J. (1975). *Evolution and Extinction Rate Controls.* Elsevier, Amsterdam, 427 pp.

Bretsky, P. W. (1969). Evolution of Paleozoic benthic marine invertebrate communities. *Palaeogeogr. Palaeoclimatol. Palaeoecol.* **6,** 45–59.

Bretsky, P. W. (1973). Evolutionary patterns in the Paleozoic Bivalvia: documentation and some theoretical considerations. *Bull. geol. Soc. Am.* **84,** 2079–95.

Gould, S. J. and Calloway, C. B. (1980). Clams and brachiopods — ships that pass in the night. *Paleobiology* **3,** 23–40.

Hallam, A. (1976). Stratigraphic distribution and ecology of European Jurassic bivalves. *Lethaia* **9,** 245–59.

Hallam, A. (1977). Jurassic bivalve biogeography. *Paleobiology* **3,** 58–73.

Hallam, A. (1981*a*). *Facies Interpretation and the Stratigraphic Record.* W. H. Freeman, Oxford.

Hallam, A. (1981*b*). The end-Triassic bivalve extinction event. *Palaeogeogr. Palaeoclimatol. Palaeoecol.* **35,** 1–44.

Hallam, A. (1986). The Pliensbachian and Tithonian extinction events. *Nature* **319,** 765–8.

Jablonski, D. (1986*a*). Background and mass extinctions: the alternation of macroevolutionary regimes. *Science* **231,** 129–33.

Jablonski, D. (1986*b*). Causes and consequences of mass extinctions: a comparative approach. In *Dynamics of Extinction* (ed. D. K. Elliott), pp. 183–229. Wiley, New York.

Jablonski, D. and Lutz, R. A. (1983). Larval ecology of maringe benthic invertebrates: palaeobiological implications. *Biol. Rev.* **58,** 21–89.

Jackson, J. B. C. (11972). The ecology of the molluscs of *Thalassia* communities, Jamaica, West Indies. II. Molluscan population variability along an environmental stress gradient. *Mar. Biol.* **14,** 304–37.

Kammer, T. W., Brett, C. E., Boardman, D. R., and Mapes, R. H. (1986). Ecologic stability of the dysaerobic biofacies during the late Palaeozoic. *Lethaia* **19,** 109–22.

Kauffman, E. G. (1984). The fabric of Cretaceous marine extinctions. In *Catastrophes and Earth History* (ed. W. A. Berggren and J. A. Van Couvering), pp. 151–246. Princeton University Press, Princeton.

Kriz, J. (1984). Autecology and ecogeny of Silurian Bivalvia. *Spec. Pap. Palaeont.* **32,** 183–95.

Miller, A. I. (1985). The spatio-temporal development of the Class Bivalvia during the Paleozoic era. *Geol. Soc. Am. Abstr. Programs* 17, 663.

Miller, A. I. (1986). Spatio-temporal development of the Class Bivalvia during the Paleozoic Era. PhD dissertation, University of Chicago (unpublished).

Morris, N. J. (1979). On the origin of the Bivalvia. In *The Origin of Major Invertebrate Groups* (ed. M. R. House), pp. 381–413. Academic Press, London.

Nakazawa, K. and Runnegar, B. (1973). The Permian-Triassic boundary: a crisis for bivalves? In *The Permian and Triassic Systems and their Mutual Boundary* (ed. A. Logan and L. V. Hills), pp. 608–21. Canadian Society of Petroleum Geology, Calgary.

Newell, N. J. and Boyd, D. W. (1978). A palaeontologist's view of bivalve phylogeny. *Phil. Trans. Roy. Soc. Lond.* **B284,** 203–15.

Pojeta, J. (1971). Review of Ordovician pelecypods. *US geol. Surv. Profess. Pap.* **695,** 1–46.

Pojeta, J. (1978). The origin and early taxonomic diversification of pelecypods. *Phil. Trans. Roy. Soc. Lond.* **B284,** 225–46.

Pojeta, J. (1985). Early evolutionary history of diasome mollusks. In *Mollusks: Notes for a Short Course* (ed. T. W. Broadhead), University of Tennessee Department of Geological Sciences Studies in Geology 13, pp. 102–21.

Raup, D. M. (1976). Species diversity in the Phanerozoic: an interpretation. *Paleobiology* **2,** 289–97.

Raup, D. M. (1979). Biases in the fossil record of species and genera. *Bull. Carnegie Mus. Nat. Hist.* **13,** 85–91.

Runnegar, B. (1985). Origin and early history of mollusks. In *Mollusks: Notes for a Short Course* (ed. T. W. Broadhead), University of Tennessee Department of Geological Sciences Studies in Geology 13, pp. 17–32.

Sepkoski, J. J. (1982). A compendium of fossil marine families. *Milwaukee Public Mus. Contr. biol. Geol.* **51,** 1–125.

Sepkoski, J. J. and Miller, A. I. (1985). Evolutionary faunas and the distribution of Paleozoic benthic communities in space and time. In *Phanerozoic Diversity*

Patterns (ed. J. W. Valentine), pp. 153–90. Princeton University Press, Princeton.

Sheehan, P. W. (1977). Species diversity in the Phanerozoic: a reflection of labor by systematists? *Paleobiology* **3,** 325–9.

Skelton, P. W. (1985). Preadaptation and evolutionary innovation in rudist bivalves. *Spec. Pap. Palaeont.* **33,** 159–74.

Stanley, S. M. (1968). Post-Paleozoic adaptive radiation of infaunal bivalve molluscs — a consequence of mantle fusion and siphon formation. *J. Paleont.* **42,** 214–29.

Stanley, S. M. (1970). Relation of shell form to life habits in the Bivalvia. *Geol. Soc. Am. Mem.* **125,** 1–296.

Stanley, S. M. (1972). Functional morphology and evolution of bysally attached bivalve molluscs. *J. Paleont.* **46,** 165–212.

Stanley, S. M. (1973). Effects of competition on rates of evolution, with special references to bivalve mollusks and mammals. *System. Zool.* **22,** 486–506.

Stanley, S. M. (1977). Trends, rates and patterns of evolution in the Bivalvia. In *Patterns of Evolution as Illustrated by the Fossil Record* (ed. A. Hallam), pp. 209–50. Elsevier, Amsterdam.

Stanley, S. M. (1984). Marine mass extinctions: a dominant role for temperature. In *Extinctions* (ed. M. Nitecki), pp. 69–118. Chicago University Press, Chicago.

Stanley, S. M. (1986*a*). Anatomy of a regional mass extinction: Plio-Pleistocene decimation of the western Atlantic bivalve fauna. *Palaios* **1,** 17–36.

Stanley, S. M. (1986*b*). Population size, extinction and speciation: the fission effect in Neogene Bivalvia. *Paleobiology* **12,** 89–110.

Thayer, C. W. (1983). Sediment-mediated biological disturbance and the evolution of marine benthos. In *Biotic Interactions in Recent and Fossil Benthic Communities* (ed. M. J. S. Tevesz and P. L. McCall), pp. 479–625. Plenum, New York.

Vermeij, G. J. (1977). The Mesozoic marine revolution: Evidence from snails, predators and grazers. *Paleobiology* **3,** 245–58.

Yancey, T. E. and Boyd, D. W. (1983). Revision of the Alatoconchidae: a remarkable family of Permain bivalves. *Palaeontology* **26,** 497–520.

7. Extinction and survival in the Cephalopoda

MICHAEL R. HOUSE

Department of Geology, University of Hull, Hull, UK

Abstract

The major groups of the Cephalopoda result from three main innovative phases in their evolution. The late Cambrian saw the introduction of a chambered shell with a linking siphuncle and then followed the radiation of the Actinoceratoidea, Endoceratoidea, and Nautiloidea, in the early Ordovician. Only *Nautilus* survives to the present day. In the Lower Devonian the Ammonoidea appear showing a protoconch and ammonitella giving bouyancy advantage and novel larval strategies; the Ammonoidea survived to the end Cretaceous. From the same stock evolved the endocochleate Coleoidea many of the fossil groups of which are poorly known, but substantial radiation(s) must have produced the Belemnitida and the living teuthoids, cuttle fish, and octopods. To these factors must be added the impressive diversification in soft part organization for which the fossil record supplies no chronology.

Extinction in the varied groups is complex. Pseudoextinction in evolving lineages must be distinguished from true extinction at all taxonomic levels. At the species and subspecies level morphologies due to phenotypic modification are difficult to distinguish from true evolution. Comments are made using examples mostly from the Ammonoidea on how extinction and survival patterns change with taxon grade. At a familial level there is often a 'package' evolution in which there is a relay-like replacement of one group by another. The major extinctions of several independent stocks and unrelated groups, best seen at familial and higher taxon levels, bear witness to a common external cause. These are particularly noteworthy for the ammonoids at the end of the Devonian, the mid-Carboniferous, end-

Extinction and Survival in the Fossil Record (ed. G. P. Larwood), Systematics Association Special Volume No. 34, pp. 139–154. Clarendon Press, Oxford, 1988.

Permian, end-Triassic, and at the final end-Cretaceous event which the ammonoids did not survive.

Introduction

The Cephalopoda have been an abundant and successful, but exclusively marine group, from the late Cambrian ot the present day, that is for about 500 million years. The fossil and present-day evidence suggests a quite complex evolution (Fig. 7.1) in which major structural innovations may be expressed by the appearance of:

(a) a chambered shell with a siphuncle which defines almost all the Cephalopoda;

(b) a siphuncle with endocones giving a poise adjustment mechanism (Endoceratoidea);

(c) a siphuncle with beaded form between the septa which encloses complex ray-shaped cavities perhaps giving both poise and hydrostatic control mechanisms (Actinoceratoidea);

(d) groups with simple, ancestral type, siphuncles, but with a range of shell form adaptations (Nautiloidea);

(e) protected fertilization leading to an encapsulated egg for the developing embryo and the concomitant loss of a larval ciliated stage as in other molluscs;

(f) forms with egg-shaped protoconchs and an ammonitella suggesting exploitation of a pelagic larva stage, a marginal siphuncle, and more folded septa (Ammonoidea); and

(g) the endocochleate groups, probably arising from simple orthoconic Bactritida or Orthoceratida which gave rise to the only forms well represented at the present day (Coleoidea or Endocochlea). This radiation of internal-shelled forms may also be associated with the achievement of mantle pump respiration, skin camouflage and chromatophore pigmentation, and the replacement of the pin-hole by a complex eye.

Soft-part organization and evolution

A quite unknown factor in the evolution of the Cephalopoda is the role of the changes in soft tissues for these can rarely be deduced. The importance of these has been discussed elsewhere (House 1987), but may be summarized. The initiation of a buoyant shell brought with it higher energy requirements and this resulted in selections which favoured greater respiratory efficiency. Gills became more advanced and there was a replacement of an ancestral cilial flow respiration by first a chamber-pump and then a mantle-pump respiration mechanism. There will have been accompanying musculature changes. The ancestral blood systems appear

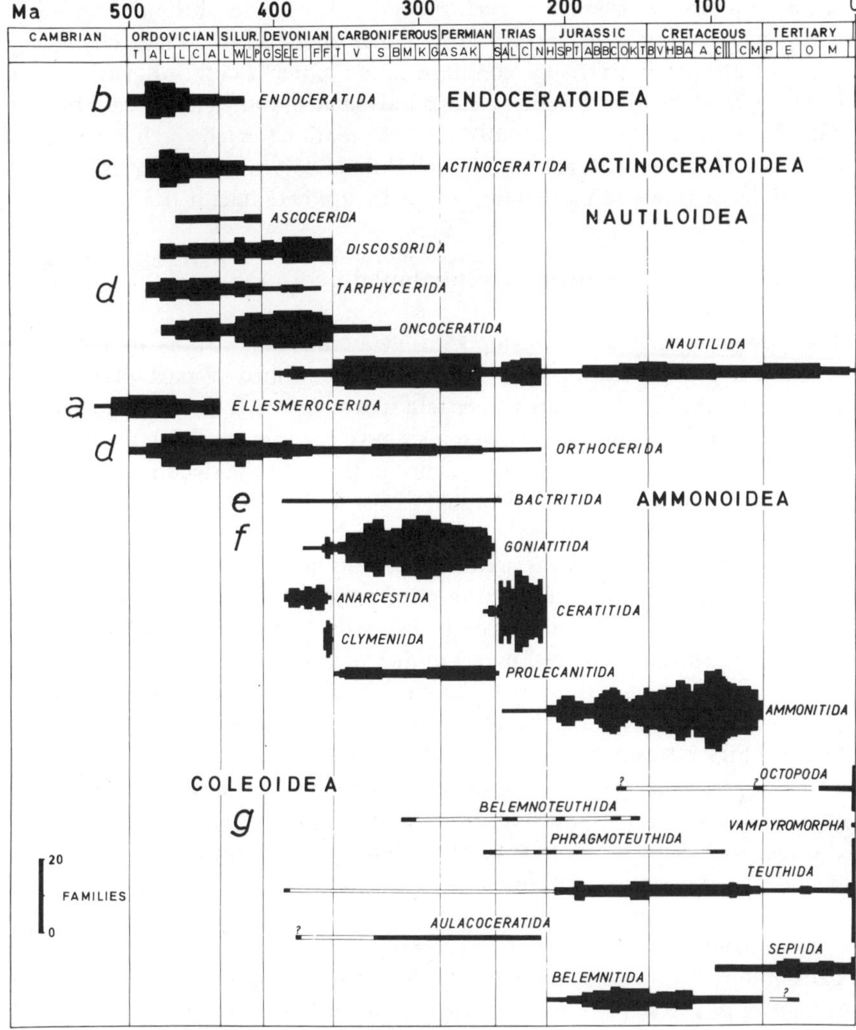

Fig. 7.1. Diagram illustrating the fossil record of the Cephalopoda. The number of families recognized in the past is indicated by the width of the columns. Important points of structural innovation are indicated by letters and discussed in the text (after House 1987).

to have been of open vacuities and weakly developed pumps. By the living cephalopods this had been replaced by a 'closed' system of vessels with capillaries, and arterial and venous vessels, and well developed systemic and branchial hearts. The exhalent respiratory flow from the mantle became developed as a propulsive mechanism. Very complex nervous systems were achieved often giving the ability for learning skills. The origin of the complex eye may have been related to the origin of the internal shell and associated chrondrophore ability to give sudden camouflage changes to the skin and perhaps visual communication skills. These innovations will have resulted in different life styles, and offence and defence capabilities. The fossil record gives virtually no information on the achievement of advances in these characters, either in their role as initiating successful radiations or in enabling certain groups to survive difficult times.

Setting of cephalopod evolution

The broad setting of cephalopodan evolution in relation to other invertebrate groups shows how the nautiloids formed a part of the very widespread radiation in many invertebrates in the early Ordovician (House 1967). The groups mostly involved also have calcareous skeletons, supports or shells. The rise in these groups is antipathetic to the decline in trilobites and chitinophosphatic shelled brachiopods. This radiation seems to result from an extension of broad carbonate shelves in a reasonably equable climatic setting. Perhaps the nautiloid radiation reflects the opening up of new niches at this time in a setting of generally low selection pressure. It will be notable that the subsequent post-early Ordovician decline towards the late Palaeozoic is shared by a number of groups and hence may have a common environmental control showing itself as a restriction in diversity (House 1967, p. 44). The rise of the ammonoids in the Devonian and their history (Fig. 7.1 and 7.3) show detailed fluctuations, especially at the end-Permian and end-Trias. Changes at these levels are shown also by other groups and witness to diversity-constraining times. The final extinction of the ammonoids corresponds to extinction of many groups at the K–T boundary, especially the rudist bivalves, sponges, and most belemnites among the larger invertebrates.

This contribution is thus essentially biased by the fact that the palaeontological evidence is partial. Notwithstanding this, it is possible to recognize in the sequence of fossil cephalopod faunas patterns of evolution which may be said to represent styles of the evolutionary process which may have more general applicability. This is, perhaps, more true of the evidence regarding the Ammonoidea than for other groups because they have been studied so intensively to take advantage of the age-dating which their rapid evolution provides.

Nautiloids

This term loosely and conveniently embraces the external shelled cephalopods without the specialized protoconch and ammonitella which characterizes the Ammonoidea.

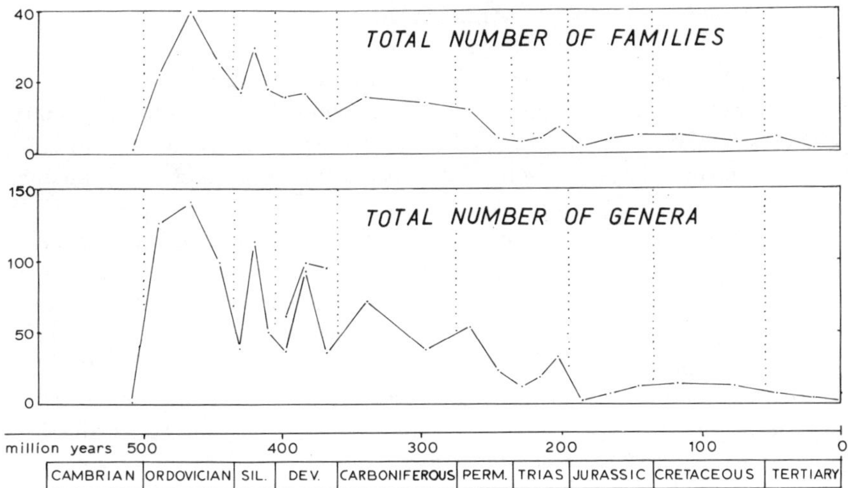

Fig. 7.2. Diagrams showing the numbers of nautiloid genera and families in the fossil record (data from Teichert *et al.* 1964).

When analysed in detail the progressive diversification of the nautiloids, as indicated by the number of genera described, is as follows (Crick 1981): Franconian (1), Trempealeauan (31), Tremadocian (44), Arenigian (172), Llanvirnian (130), Llandeilian (119), Caradocian (120), and Ashgillian (116). The Cambrian–Ordovician boundary is drawn at the base of the Tremadocian stage. By the end of the Arenigian (Fig. 7.2) there is a substantial decline in generic numbers, and Crick (1981) has shown that this is a decline both by extinction and a marked reduction in innovation. In rapidly evolving groups, generic evolutionary replacement shows in taxon tots as both high new taxa and high numbers of taxa becoming extinct. This is the feature of the Arenigian record and was demonstrated by the author (House 1967, p. 51) to be a general feature in the invertebrate groups undergoing radiation at this time.

The record (Figs. 7.1 and 7.2) shows that by the early Ordovician all the major high taxon groups of the nautiloids were established. No higher taxa of nautiloids have evolved since. This is thought to be an example of adaptive radiation in which the major problems associated with exploitation of the cephalopodan innovation were overcome and led to the establishment of distinct groups based to a large extent on functionally significant characters. Subsequent extinction may successively have removed forms unable to cope with the environmental stresses encountered. Later extinctions are not particularly patterned and groups become extinct from time to time (Fig. 7.1). Accompanying this is a real decline in abundance of faunas except in particular facies.

Ammonoidea

Because the ammonoids have been subject to considerable detailed biostratigraphic analysis, they shed interesting light on evolutionary processes in the groups at all levels. The current view of the evolution in the ammonoids, bringing together the work of many specialists, is indicated in Fig. 7.3.

It has been stressed (House 1985*b*) that the different taxonomic levels illustrate different evolutionary processes with regard to innovation, extinction, and survival. Thus, this discussion is conveniently divided into various taxon levels and types of evolutionary process to give illustrations of this as seen in practice.

Attention should be drawn to the fact that the delimitation of present-day species and genera is relatively simple because they are constrained in space and time, and because their genetic, social, and distribution data are at least knowable. These last factors are usually unknown in the fossil record and, in addition, the fossil record is analysed in *time* so that another dimension is added to the problem of taxon definition. Since the definition of taxa is subjective, and based mainly on criteria that are not universally acknowledged, terminology becomes very confusing. It is here that use of taxon tots by the uninformed, however good their computers, can be very misleading unless it uses the data of one specialist alone using constant criteria.

1. Extinction in phenotypic variety

The morphological variety of a conspecific fauna illustrates the diversity permitted by the sum of the genetic variation in the population and the phenotypic environmental controls on growth. The diversity this can produce in ammonites is illustrated by a classic study by Reeside and Cobban (1960) of *Neogastroplites* from a single level in which a very great variety of shell forms and ornamented types occur. Generally, when large faunas are available from particular horizons a wide diversity is recognized; other examples have been discussed by Kennedy and Cobban (1976).

Names are applied by ammonoid specialists to distinctive forms rather freely, but with the scientific intent not to hide in a blanket terminology forms which may have age-dating value. Thus, terminology is made subservient to practical chronostratigraphical usefulness. The result may appear on taxon tots to indicate the extinction of many subspecies or species, but this may be a false statement of the real situation.

2. Phenotypic change in time

Environments change with time, and it is to be expected that, even with no change in the genetic composition of a population, a succession of slightly

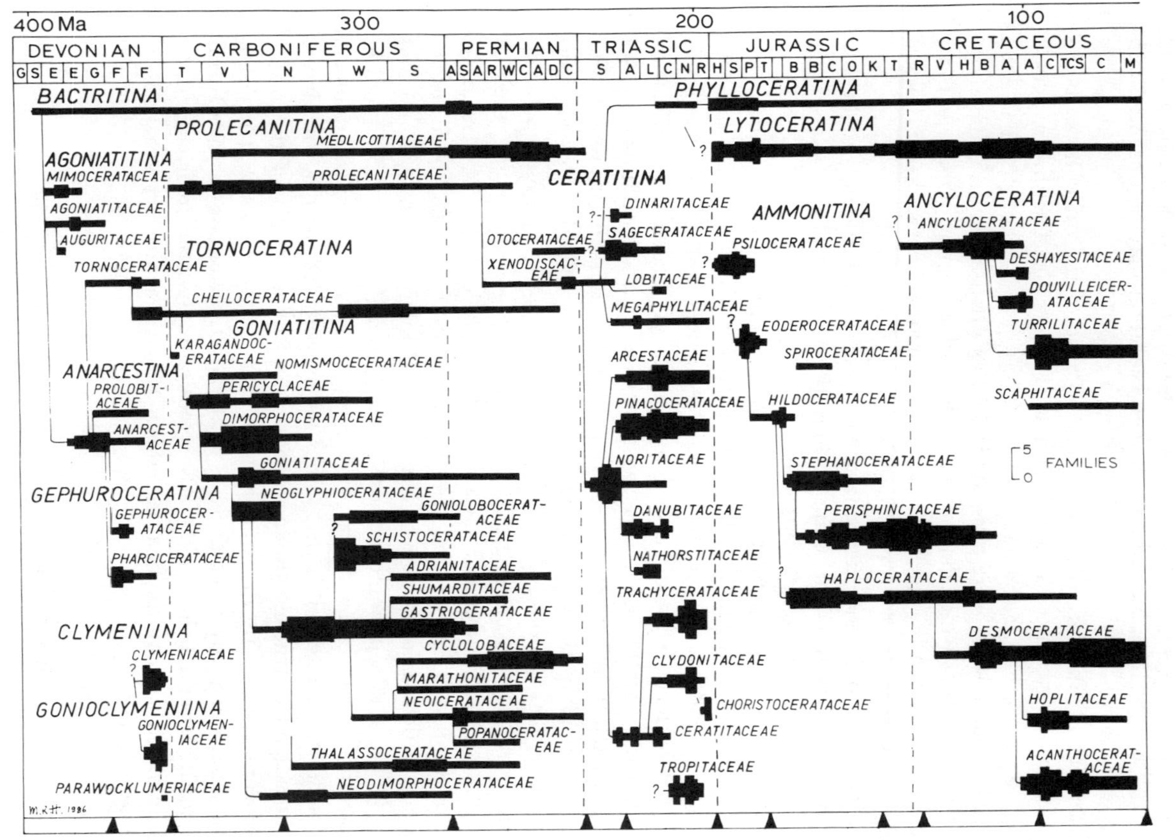

Fig. 7.3. Diagram illustrating the current view of the evolution of the Ammonoidea and the variety of families in the group through time. Data from the work of many specialists (in House and Senior 1981, after House 1987).

different morphologies might result from changes in environmental controls during growth alone. These may result in a succession of different names. Since the palaeontologist has, at present, no means of determining genetic composition, the fact that this is not true evolution may be obscured and extinction inferred.

Examples of this sort raise problems because of the difficulty of being sure that a single successive population is being dealt with. In a case involving the genus *Tornoceras* (House 1965) it was thought that shell changes in one lineage were phenotypic as illustrated by statistical analysis of successive faunas which demonstrated the shell ontogeny of many of them as being significantly different. Other characters in the same lineage seemed to indicate long-term trends which were more likely to indicate real evolution. Had only the shell characters been known, names might still have been applied to elements in the lineage on statistical grounds yet this might be taken to indicate extinctions.

This type of 'false' extinction problem is probably widespread at the lower levels of evolutionary diversification. It is emphasized by the fact that ammonoids tend not to occur throughout sequences, but to appear at levels separated by levels without them. Thus, forms at particular levels can be distinguished from those at other levels by small characters. In some cases this is linked with small-scale sedimentary rhythmicity in which ammonites are only preserved in certain environments. Whether this rhythmicity is caused by sea level changes or climatic changes is perhaps debated, but the effect on a succeeding population given no genetic change could be to produce a rather irregular succession of morphologically different forms which might well receive different names. This might be the cause of statistically definable change in *Kosmoceras* (Brinkmann 1929), Raup and Crick (1981, 1982), or some of those in *Cardioceras* (Callomon 1985) and in *Tornoceras* already discussed. Part of the reason why punctuated equilibria cannot be tested in detail is that minor phenotypic fluctuations will always obscure detailed true evolution. This was, of course, one of the major conclusions reached by Brinkmann in his classic detailed work at Peterborough (1929), a study he undertook of ammonite evolution in the Oxford Clay in the hope that he could study evolution in a stable environment. In fact, he found the environment as subject to sedimentary perturbations as at other levels.

3. *Small-scale evolutionary change*

Small-scale taxonomic turnover at subspecies or species level may involve sequential evolution. Names change as one species gives rise to another. The number of name changes and their taxon grade depends fortuitously on the whim of the palaeontologist given few agreed rules on taxon definition. Thus, whilst real evolution is involved, novelty is indicated by the disappearance of certain named species and their replacement by new ones as the lineage proceeds.

In practice, good examples of continuously changing morphology in evolution are rare. Rather populations as a whole change. Thus, from the Devonian, in *Sporadoceras* there is a progressive deepening of the first lateral lobe, and in *Maenioceras* a progressive closing of the umbilicus. In the Jurassic the progressive introduction of a corded keel in the *Amaltheus* lineages and again in those of *Cardioceras* are also of this type. Again the name changes in successive faunas of these groups are subjectively defined and the apparent loss of taxa represents a 'false' extinction.

4. *Generic change in evolutionary lineages*

Rather similar problems are met in the replacement of one generic name by another along a gradually evolving lineage. The boundaries between these may be clearly defined using specific morphological criteria (for example the differentiation of a keel in the *Quenstedtoceras* to *Cardioceras* lineage), but may result in a named ancestral genus occurring with a descendant genus in a single population. This is part of the problem the time factor gives to fossil terminology. Along the lineage it may be taken to indicate an extinction, but this would not be true. Single lineages may in this way have several successive generic names, for example in the Middle Devonian, *Cabrieroceras, Maenioceras, Pharciceras,* and *Petteroceras*. A simple taxon tot for the stage would suggest a high extinction at the close, but this would be false.

5. *Generic extinction during adaptive radiation*

As indicated in Fig. 7.2 for the nautiloids, generic turnover is highest during adaptive radiation when much novelty is produced, often in relation to the expolitation of new environments. For the Ammonoidea this is less clearly seen (Fig. 7.4). The Devonian has some seven of the fourteen ammonoid suborders (using the classification in House and Senior 1981). This presumably represents a radiation concerned with the efficient exploitation of whatever advantage the novel characters of an ammonitella and more complex sutures provided the ammonoids. However, this does not show greatly at generic level (Fig. 7.4). On the other hand, the short-term increase of genera associated with the early stocks of the Goniatitina and Ceratitina may represent generic diversification of the same type as suggested for the early Nautiloidea. Whilst this will include replacement taxa in evolving lineages as discussed in the previous section it will also include taxa of limited during and true extinction. These could only be disentangled by specialists familiar with the details and probably at the generic level this would still be a somewhat subjective exercise.

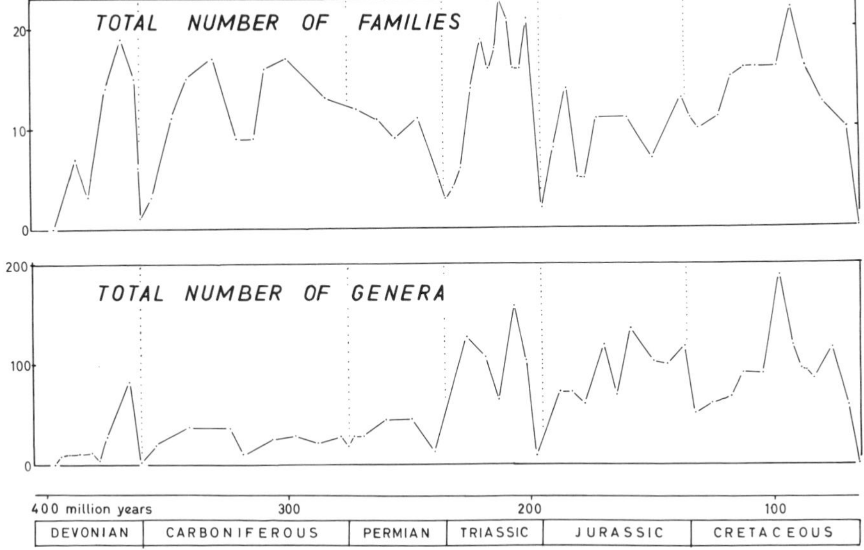

Fig. 7.4. Number of families and genera of ammonoids through time. Data for families from the work of many specialists (in House and Senior 1981), generic data from Arkell *et al.* (1957, after House 1985*b*).

6. *Common extinctions of generic lineages*

When several stocks thought to have been unrelated for a considerable while become extinct at the same time then real extinction is reasonably deduced, especially if testing internationally gives the same result. Of the many examples available, the events at the Devonian–Carboniferous boundary are illustrated here (Fig. 7.5). It has been argued (House 1985*a*) that for the mid-Palaeozoic this pattern is seen on at least eight occasions and the associated extinctions follow a pattern. First, there is a slight decline in generic diversity; secondly, there is a marked extinction; thirdly, there is a low point with very limited fauna at all; fourthly, there is a return of diversity; fifthly, associated with the renewed diversity, or first seen as it develops, are quite novel forms, often initiating new groups. It has been argued that this type of extinction results from environmental perturbations often shown by sedimentological characteristic, notably anoxia, and regressive or transgressive events.

In the Devonian cases (House 1985*a*) the critical factor seems to be related to the early stages. The new groups arising often show their novelty within the first few millimetres of growth, and this suggests that their survival or initiation may result from characters advantageous in early growth. That such changes may have fundamental changes for the adult is clear. It is tornoceratids which survive several of the Devonian extinction

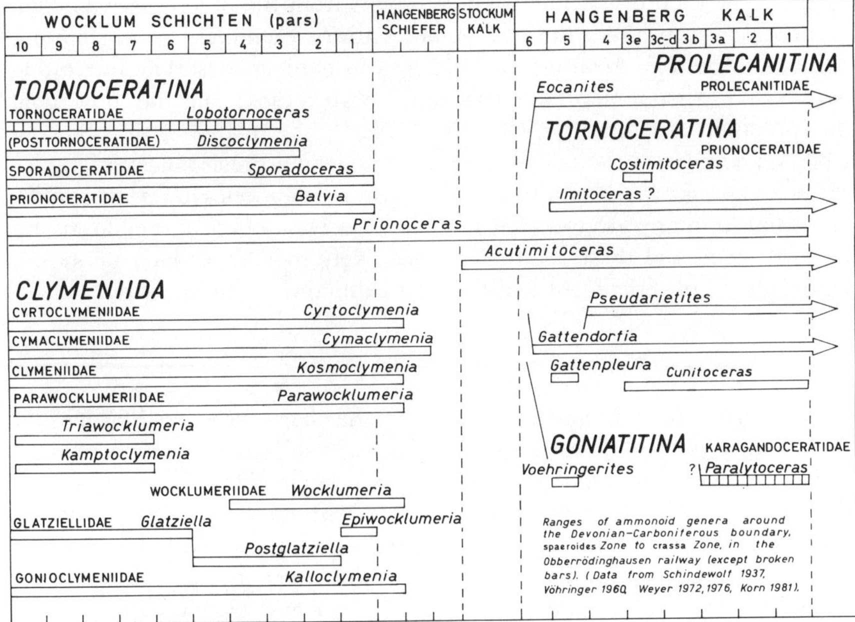

Fig. 7.5. Ranges of ammonoid genera around the Devonian/Carboniferous boundary and the Hangenberg Event (for sources see House 1985*b*).

events and this group may have lived in colder or deeper waters, and periodically produced successful diversifications, as after the Kellwasser and Hangenberg Events (House 1985*a*). The role of early stages in family innovation at younger levels has not been documented.

7. *Extinctions in adaptive radiation*

Extinction during adaptive radiation has been commented on at generic levels and this is reflected at family level (Figs. 7.2 and 7.4) where where is an approximately sympathetic relationship. At the higher taxon levels of family and superfamily, the radiations often follow extinction events (as illustrated in Fig. 7.3 by the Viséan (V) radiation of the early Carboniferous or that of the Ceratitina in the early Triassic. Often a period of build-up of diversity is evidenced, and this is indicated for the early Carboniferous on Fig. 7.5 which shows how two new groups enter (the Goniatitida and Prolecanitida) as minor elements, yet soon come to dominate ammonoid faunas (Fig. 7.3) in the later Carboniferous. The Carboniferous data especially illustrates how family diversity is often great early in the radiation of a stock and declines later, presumably indicating, as at generic level, that a higher level of diversity was sustained during the early exploitation of the novelty of survivors from the preceding extinction events. However, there are also some contrary examples. So much work still has to be done relating such changes to habit modes. The interesting work of, for example, Saunders and Swan (1984) on the functional interpretation of goniatite shell form still has to be related to the observed facies occurrence of the different types, and given a precise biostratigraphical tie to specific levels and place in the group evolution. However, it seems clear that for many groups, after innovation, a wide range of shell forms are often attained and these are then progressively reduced as later periods of taxonomic impoverishment and the next extinction event approaches.

8. Family and superfamily level packet evolution

One feature of the ammonoid record which has been recognized for most of this century is that there are rather compact and clearly defined faunas which become extinct and are replaced by other replacement faunas. The ammonite 'ages' of a former generation bear witness to their recognition of this (House 1986). This has been referred to as 'relay' evolution. This is well illustrated at family and superfamily level by a succession of taxonomic 'packets' in time by which a group may slowly or quickly be replaced by another. A succession of relay-like faunal groups thus appear sequentially. This is illustrated for the Triassic and early Jurassic by Fig. 7.6. A similar analysis for the late Jurassic and Cretaceous has been published by Ward and Signor III (1983). Notable is the way many of these are spindle-shaped, and show a slow rise in diversity to a maximum and then a decline. The important event may be the initiation of the decline in diversity rather than the final demise of the family. Usually, other groups are associated with the replacements suggesting that this type of innovation and extinction results from competition within the ammonoid group as a whole rather than from a cause external to the group. The change may have been externally triggered, but the actual relay-type exchange is thought to result from the succeeding group being more able to cope with the change. Partly, these patterns result from the fact that a surviving group will differ from the group it replaces and hence a taxonomically discrete 'package' will result.

For the Devonian (House 1985*a*), it is been argued that some of these changes are triggered by extinction events associated with anoxic events. For the Jurassic, which has a more complex palaeogeography associated with continental break-up, it seems that migration from Boreal or Tethyan seas (Enay and Mangold 1982) and sea level effects (Bayer and McGhee 1984) gave fluctuating fortunes. The use of the faunal record as a bioseismograph to investigate such palaeogeographical changes has hardly begun.

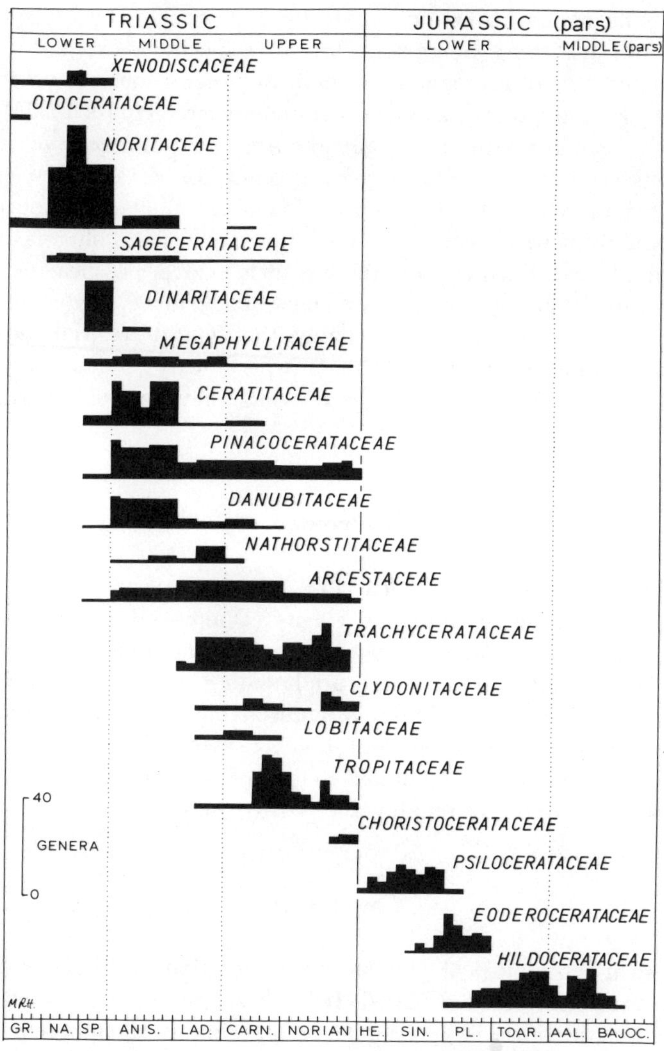

Fig. 7.6. Packet evolution in the Ammonoidea illustrated for the Trias and early Jurassic. Data from the contributors to House and Senior (1981, after House 1985*b*).

9. Common extinctions at high taxon level

Periods of extinction at family or superfamily level where several separate stocks are involved occurred especially near the end of the Middle Devonian, Upper Devonian, early Namurian (E_2), late Permian, late Triassic, and late Cretaceous. These are marked on Fig. 7.3 by a black triangles and some rather lesser extinction events are also marked. They represent the major extinction events for the group. The cause of these events is still in dispute. Some seem linked to specific anoxic events, others seem related to transgressions, others to regressions. A preliminary comment on such causes has been made elsewhere (House 1985*b*), much, very much, remains to be done, and the argument of recent years on the Cretaceous–Tertiary boundary gives ground for caution. When these events were analysed from the point of view of taxon change (House 1985*a*, fig. 9) each event was preceded by a decline in diversity, often lasting over a substantial time. Furthermore, there is little evidence of clear periodicity. This indicates that the changes are more likely to be related to palaeogeographic and perhaps, climatic change than extraterrestrial causes such as a Nemesis Star, a companion star, or a periodicity related to oscillations about the orbit giving the solar year which have excited journalists in recent years.

Coleoidea

The fossil record of the Coleoidea (Fig. 7.1) is too sparse to be discussed profitably in relation to extinction events (Donovan 1964, 1977; Jeletsky 1964). The contrast between known living taxa, and fossil representative in the Teuthoidea and Octopoda, indicate how in groups in which the shell is reduced, the fossil record becomes very unreliable. The Belemnitida may or may not survive the Cretaceous, but in any case it is clear that they suffered near, if not complete, extinction at the K–T boundary as did the ammonoids and were presumably affected by the same cause.

Conclusions

The physical and biological controls of the evolutionary failure or success of organisms are complex and varied. It is, therefore, to be expected that the patterns of extinction will be also. To invoke simple, all embracing models, such as punctuated equlibria, is naive. Environmental stresses grade from slight changes of temperature, winds, tidal flow, food availability, or predator activity which may effect the survival of an individual or a group to the major and permanent palaeogeographical, climate, or food chain patterns which may change the survival parameters for major groups. This need only happen once.

So far, as here, analysis of 'taxon tots', has figured largely in the discussion. We are only beginning to analyse such data in the detail the biostratigraphic scale allows. However sophisticated the computer or statistical techniques applied to such data it will yield spurious results until the emerging models are checked against the data from sedimentology, palaeogeography, and palaeoecology, evidence which accompanies the fossil cephalopod data. Already, a correlation between environmental perturbations and cephalopod evolution is apparent, both at high and low taxon level. The next stage is a wider integration of the data available.

References

Arkell, W. J., *et al.* (1957). *Treatise on Invertebrate Paleontology, Part L, Mollusca 4, Cephalopoda–Ammonoidea.* Geological Society of America and University of Kansas Press, New York and Lawrence, Kansas.

Bayer, U. and McGhee, J. R. (1984). Iterative evolution of Middle Jurassic ammonite faunas. *Lethaia* **17,** 1–16.

Brinkmann, R. (1929). Statisch-biostratigraphische Untersuchungen an mitteljurassischen Ammoniten über Artbegriff und Stammesentwicklung. *Abh. Ges. Wissensch. Göttingen, math.-Phys. Kl. N.F.* **13**(3), 1–249. Monographie der Gattung *Kosmoceras* ibid. (4), 1–123.

Callomon, J. C. (1985). The evolution of the Jurassic ammonite family Cardicoeratidae. *Spec. Pap. Palaeontol.* **33,** 49–90.

Crick, R. E. (1981). Diversity and evolutionary rates of Cambro-Ordovician nautiloids *Paleobiology* **7,** 216–29.

Donovan, D. T. (1964). Cephalopod phylogeny and classification. *Biol. Rev.* **39,** 259–87.

Donovan, D. T. (1977). Evolution of the dibranchiate Cephalopoda. *Symp. zool. Soc. London* **38,** 15–48.

Enay, R. and Mangold, C. (1982). Dynamique biogéographique et évolution des faunes d'Ammonites au Jurassique. *Bull. géol. Soc. France* **24,** 1025–46.

House, M. R. (1965). A study of the Tornoceratidae: Tornoceras and related genera in the North American devonian. *Phil. Trans. R. Soc.* **250B,** 79–130.

House, M. R. (1967). Fluctuations in the evolution of Palaeozoic invertebrates in *The Fossil Record* (ed. W. B. Harland *et al.*), pp. 41–54. Geological Society, London.

House, M. R. (1985*a*). Correlation of mid-Palaeozoic ammonoid evolutionary events with global sedimentary perturbations. *Nature, Lond.* **313,** 17–22.

House, M. R. (1985*b*). The ammonoid time scale and ammonoid evolution. In *Geochronology and the Geological Record* (ed. N. J. Snelling), pp. 273–83. Geological Society, London.

House, M. R. (1987). Major features of cephalopod evolution. (In press).

House, M. R. and Senior, J. R. (ed.) (1981). *The Ammonoidea.* Academic Press, London and New York.

Jeletzky, J. A. (1964). Comparative morphology, phylogeny and classification of fossil Coleoidea. *Univ. Kansas paleontol. Contr. Mollusca,* Article 7, 1–112.

Kennedy, W. J. and Cobban, W. A. (1976). Aspects of ammonite biology, biogeography, and biostratigraphy. *Spec. Pap. Palaeontol.* **17,** v + 1–94.

Raup, D. M. and Crick, R. E. (1981). Evolution of single characters in the Jurassic ammonite *Kosmoceras. Palaeobiology* **7,** 200–215.

Raup, D. M. and Crick, R. E. (1982). *Kosmoceras*: evolutionary jumps and the sedimentary record. *Paleobiology* **8,** 90–100.

Reeside, J. B. J. and Cobban, W. A. (1960). Studies of the Mowry Shale (Cretaceous) and contemporary formations in the United States and Canada. *Prof. Pap. U.S. geol. Surv.* **355,** 1–126.

Saunders, W. B. and Swan, A. R. H. (1984). Morphology and morphologic diversity of mid-Carboniferous (Namurian) ammonoids in time and space. *Paleobiology* **10,** 195–228.

Sepkoski Jr, J. J. (1982). A compendium of fossil marine families. *Milwaukee Publ. Mus. Contr. Biol. Geol.* **51,** 1–125.

Teichert, C. (1967). Major features of cephalopod evolution. *Spec. Publ. Dept. Geol. Univ. Kansas* **2,** 162–210.

Teichert, C., *et al.* (1964). *Treatise on Invertebrate Paleontology, Part K, Mollusca 3, Cephalopoda—general features, Endoceratoidea-Actinoceratoidea-Nautiloidea-Bactritoidea.* Geological Society of America and University of Kansas Press, Lawrence and New York.

Ward, P. D. and Signor III, P. W. (1983). Evolutionary tempo in Jurassic and Cretaceous ammonites. *Paleobiology* **9,** 183–98.

8. Extinction and survival in the echinoderms

C. R. C. PAUL

Department of Geological Sciences, Liverpool University, Liverpool, UK

Abstract

Rigorous analysis of extinction and survivorship requires estimates of the reliability of the fossil record, phylogenetic analysis of taxa at specific level, refined global stratigraphy, and a reliable timescale. Since no echinoderm data meet all these requirements only broad patterns of taxonomic turnover can be discussed. At generic level echinoderm diversity reveals three faunas: a Cambrian fauna dominated by filter-feeding 'eocrinoids', a Palaeozoic fauna dominated by filter-feeding crinoids, and a post-Palaeozoic fauna dominated by deposit-feeding and macrophagous eleutherozoans. Echinoderms suffered a major life crisis at the end of the Palaeozoic, but were scarcely affected by the Cretaceous–Tertiary boundary event.

Introduction

In this chapter I wish first to review the quality of data necessary to reveal real patterns of extinction and survivorship rather than artifacts of fossil collecting or taxonomic practice. I am forced to conclude that at present there are no really adequate data for echinoderms, although the best current example from the blastoids is briefly reviewed. These inadequacies mean that detailed analysis of periodicity in extinction peaks, for example, cannot be undertaken with echinoderms. Nevertheless, broad patterns of change are evident and can be compared with those seen in the fossil

Extinction and Survival in the Fossil Record (ed. G. P. Larwood), Systematics Association Special Volume No. 34, pp. 155–70. Clarendon Press, Oxford, 1988.

record as a whole. In particular, echinoderms seem to have followed the model of three main faunas (Cambrian, Palaeozoic, and post-Palaeozoic) advanced by Sepkoski (1979, 1981). The bulk of this chapter is therefore concerned with describing this model for echinoderms and interpreting the changes observed in terms of the palaeobiology of the groups involved. Much of the interpretation is speculative and likely to change as more detailed analysis of the fossil record is undertaken.

Requirements for analysis

Adequate analysis of survivorship requires five things: (1) estimates of the completeness of the fossil record both by taxon and by stratigraphic interval; (2) phylogenetic assignment of all taxa; (3) analysis at the lowest taxonomic level, i.e. of specific ranges; (4) sufficiently refined stratigraphy on a worldwide basis; and (5) an adequate timescale. The last two points are generally outside the control of the systematist, but the former three are not and are worth considering briefly.

1. Analysis of the record

Estimates of completeness of the fossil record can most easily be done by analysis of gaps (Paul 1980, 1982; Fig. 8.1). Gaps can be proved to occur where a taxon is known from below and above, but not actually within, a given stratigraphic interval. Figure 8.1 summarizes available data for families of several Palaeozoic echinoderm classes. Reading across the rows in Fig. 8.1 enables calculation for each taxon of the proportion of the total range actually represented by fossils. Although a crude estimator, it allows comparisons between taxa. It is inherently likely that taxa with very poor records probably extended well beyond their known ranges and will give poor data for survivorship analysis. Reading down the columns in Fig. 8.1 allows a similar comparison between geological periods. The Caradoc apparently had an excellent record, whereas that of the Llandovery is particularly poor. This gives a false impression of a life crisis at the Ordovician–Silurian boundary. Moreover, we may question whether at least some of the six families of cystoids which apparently became extinct at the end of the Ordovician might be found in the Lower Silurian were the record in the Llandovery better. It is conceivable, but I think unlikely, that the important extinction event at the end of the Ordovician is an artifact of the exceptionally poor fossil record in the Lower Silurian. Thus, analysis of gaps draws attention to potentially unreliable data and should routinely be undertaken for this reason alone.

	CAMBRIAN				ORDOVICIAN					SILURIAN				DEVONIAN							CARBONIFEROUS					PERMIAN						
	Comley	St David's	Merioneth	Tremadoc	Arenig	Llanvirn	Llandeilo	Caradoc	Ashgill	Llandovery	Wenlock	Ludlow	Downton	Gedinnian	Siegenian	Emsian	Eifelian	Givetian	Frasnian	Famennian	Tournaisian	Viséan	Namurian	Westphalian	Stephanian	Asselian	Sakmarian	Leonardian	Guadalupian	Dzhulfian	Total Range	Gaps
RHOMBIFERA																																
Macrocystellidae				▒	▒	G	▒																								4	1
Cheirocrinidae				▒	▒	▒	▒	▒	▒																						6	0
Pleurocystitidae					▒	▒	▒	▒	▒	G	G	G	G	G	▒	G	▒														13	6
Cystoblastidae							▒	▒																							2	0
Rhombiferidae								▒	▒																						2	0
Callocystitidae									▒	▒	▒	▒	▒	▒	G	G	G	▒	▒												11	3
Echinoencrinitidae					▒	▒	G	▒	▒																						5	1
Hemicosmitidae					▒	▒	▒	▒	▒																						5	0
Caryocrinitidae								▒	▒	G	▒	▒	G	▒																	7	2
Echinosphaeritidae					▒	▒	▒	▒	▒																						5	0
Caryocystitidae					▒	▒	▒	▒	▒																						5	0
DIPLOPORITA																																
Glyptosphaeritidae				▒	G	▒	▒	▒																							5	1
Protocrinitidae						▒	G	▒																							3	1
Dactylocystidae								▒	G	G	▒																				4	2
Gomphocystitidae						▒	G	▒	G	G	▒																				6	3
Sphaeronitidae				▒	▒	▒	▒	▒	▒	G	G	G	G	▒	G	▒	▒														14	5
Holocystitidae									▒	G	▒																				3	1
Aristocystitidae				▒	▒	▒	▒	▒	▒	G	▒																				8	1
CYCLOCYSTOIDEA																																
Cyclocystoididae							▒	▒	▒	▒	▒	▒	G	G	▒	▒	▒														11	2
EDRIOASTEROIDEA																																
Cyathocystidae						▒	▒	▒	▒																						4	0
				(G	G)																										(2	2)
Stromatocystitidae	▒	▒	▒																												3	0
Totiglobidae		▒	▒																												2	0
				(G	G	G	G	E)																							(5	4)
Lebetodiscidae							▒	▒	▒	▒	▒	G	▒	G	▒	G	▒														11	3
Carneyellidae							▒	▒	▒																						3	0
Isorophidae								▒	▒	G	▒																				4	1
												(G	G)																		(2	2)
Agelacrinitidae														▒	G	G	▒	▒	▒	▒	▒	▒	G	▒	▒						12	3
FISSICULATA																																
Phaenoschismatidae											▒	▒	▒	▒	G	▒	▒	▒	G	G	▒	▒	▒								13	3
Orophocrinidae																▒	▒	▒	G	G	▒	▒	▒	G	G	G	▒	G	▒		14	6
Nymphaeoblastidae																▒	G	G	G	G	▒	▒	G	G	G	G	▒				12	8
Neoschismatidae																					▒	▒	G	G	G	G	▒	▒	▒		9	4
Codasteridae																						▒	G	▒	▒	G	▒	▒	▒		8	2
Total Occurrences	1	2	2	7	12	14	18	21	19	11	12	8	8	8	7	9	9	5	5	4	5	6	6	5	5	4	4	3	3		223	
Gaps	0	0	0	2	3	2	4	0	2	8	2	4	5	3	4	4	2	1	3	3	0	0	4	3	3	4	0	1	0			67

Fig. 8.1. Ranges, including gaps, of families of some Palaeozoic echinoderms which are known from more than one stratigraphic interval. Known occurrences stippled, gaps indicated by the letter G. Gaps in parentheses are inferred from the phylogeny of the edrioasteroids. E = family Edrioasteridae. Total ranges and gaps for families are listed at the right, for stratigraphic intervals along the bottom. Average gap for these echinoderms is 30.04 per cent; the range for individual families is from 0 to 66.7 per cent and for stratigraphic intervals from 0 to 72.7 per cent.

2. Phylogenetic analysis

Assigning taxa to a phylogeny further enhances survivorship analysis in two ways. First, crude analysis of gaps cannot detect gaps beyond known ranges of taxa and hence, for example, cannot be applied to taxa known from only one stratigraphic interval, such as the family Edrioasteridae in Fig. 8.1. Arranging taxa in a phylogenetic tree can detect further gaps in the record. In Fig. 8.1 several gaps are recorded for edrioasteroids because they occur between families that are thought to be directly related, although the gaps cannot be assigned to any particular family. However, if Lower and Middle Cambrian edrioasteroids are really related to Ordovician and younger edrioasteroids, some examples must have continued through the intervals concerned. Secondly, and perhaps more importantly, phylogenetic analysis allows distinction between real extinction in the sense of termination of a clade and pseudoextinction. Figure 8.2 shows the stratigraphic occurrence in southern Britain and phylogenetic relationships of the chalk sea urchins *Infulaster* and *Hagenowia*. Although species apparently first appear earlier in Germany than in Britain, as far as we know in neither genus do two species occur together. Thus, for example, when *Hagenowia rostrata* disappeared *Hagenowia anterior* replaced it. There was no extinction event at the end of the range of *H. rostrata*, as there was at the top of the range of *Infulaster infulasteroides* when the entire *Infulaster* lineage became extinct. Clearly, in assessing extinction and survivorship we must distinguish between real extinction and pseudoextinction. Only monophyletic taxa have any real biological existence. What we are really interested in is extinction of monophyletic taxa (i.e. termination of clades) and only by assigning taxa to their correct places in a phylogeny can we detect monophyletic clades. All other cases are pseudoextinction and confuse the issue. In extreme cases pseudoextinction is no more than a taxonomic artifact.

3. Taxonomic level

Consideration of species rather than higher taxa is desirable for two reasons. First, there are numerous taxa which undergo major contractions well before the last member species died out. In the extreme case of 'living fossils' such as the coelacanth, *Latimeria*, no extinction can be recorded because the group is extant, but the fact remains that by far the vast majority of coelacanth species became extinct way back in the fossil record. Consideration of extinction at specific level is the only way to overcome this problem. Secondly, survival analysis at higher taxonomic levels than species tends to overestimate pseudoextinction and underestimate true extinction. While it is true that one family (for example) may give rise to another, if the daughter family is truly monophyletic it must have arisen

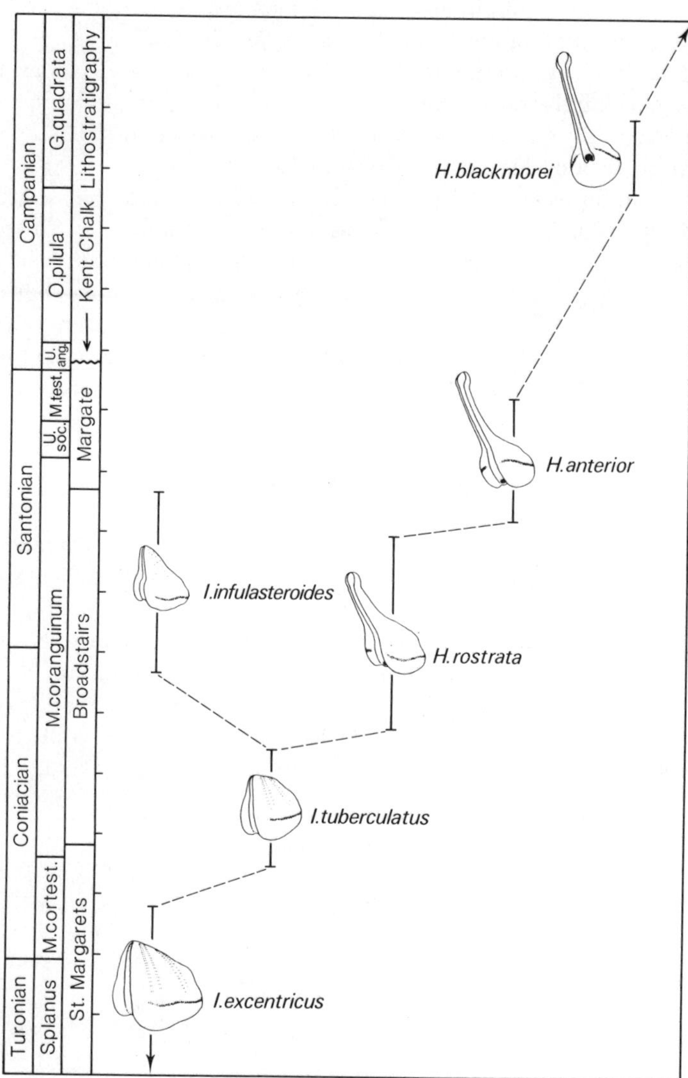

Fig. 8.2. Inferred phylogeny and known stratigraphic ranges in the chalk of southern England of the sea urchins *Infulaster* and *Hagenowia*. Columns on the left give standard stages and zones of the chalk and members of the Ramsgate Chalk Formation of Robinson (1986). Scale in 10-m intervals. Range of *H. blackmorei* from Mortimore (1986); ranges of the other species from Gale and Smith (1982). Note that in neither genus are species known to have coexisted. Redrawn from Smith (1984, fig. 5.4).

from only one species within the parent family. All the other genera and species in the parent family may have become extinct. Although the higher taxon (e.g. the superfamily) to which both families belong has not become extinct, all but one lineage within the parent family have. To argue that because one family may give rise to another the parent family suffered pseudoextinction ignores a great deal of real extinction. These problems can best be avoided by confining our analysis to species.

Consideration of species only has the major drawback that stratigraphy is rarely sufficiently refined (a) to estimate specific longevity, (b) to allow analysis of gaps for species, and (c) for world-wide correlation of specific extinction. It is because of this problem more than any other that familial and generic data have generally been used in survivorship analysis. Even so, most data have not been subjected to phylogenetic analysis so that what is really being investigated is taxonomic turnover not true survivorship.

Examples from echinoderms

To date the most detailed analysis of survivorship in the echinoderms is Horowitz *et al.*'s (1985) study of blastoids (Fig. 8.3). It is particularly good because it has been done at generic level and because all generic concepts were defined by Macurda alone from personal examination of material. Thus, it represents probably the most consistent taxonomic data set for any fossil group. However, the data were still not subjected to phylogenetic analysis and, as the authors themselves acknowledge, still represent taxonomic survivorship. Even so, their estimates that the average blastoid

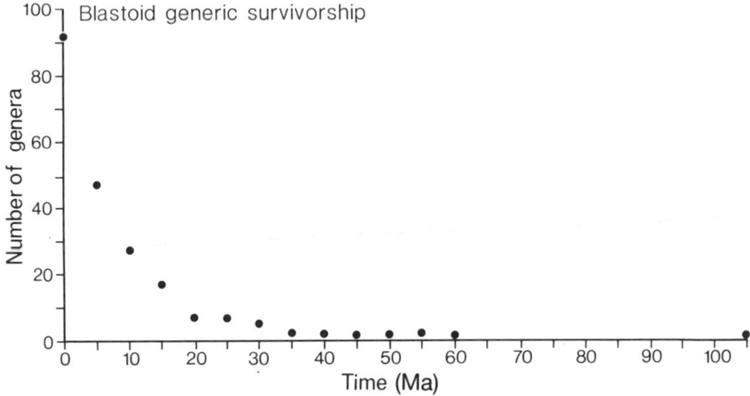

Fig. 8.3. Survivorship curve for blastoid genera plotted at 5 Ma intervals. Data from Horowitz *et al.* (1985), but corrected for the time-scale used in this symposium (Harland *et al.* 1982). Total time span for each genus includes only half the durations of the first and last known stratigraphic intervals.

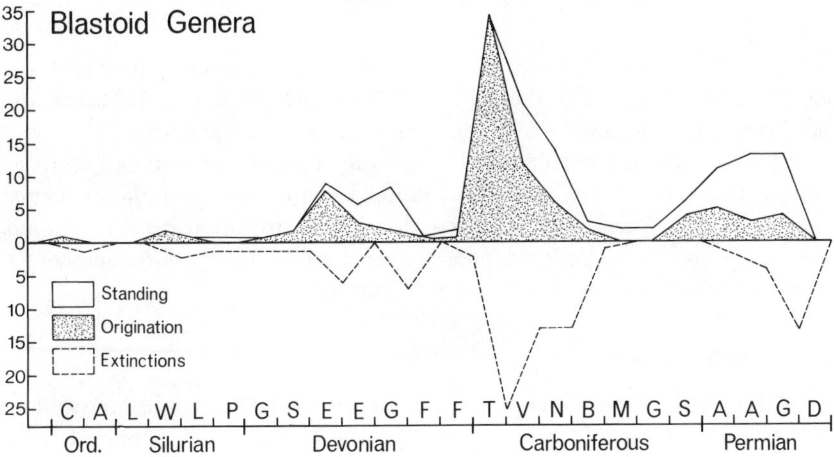

Fig. 8.4. Origination, standing crop, and extinction rates for blastoid genera. Note that extinction and origination rates mirror each other. Peaks for origination occur immediately below, and peaks for extinction immediately above, levels where blastoids are most diverse. They do not reflect radiations and mass extinctions, respectively, merely the relatively patchy fossil record of blastoids.

genus survived for 9.0 Ma and the average species for 5.1 Ma are probably reasonably accurate. However, even these figures are open to some question since the time-scales used (Harland *et al.* 1964; Izett *et al.* 1982) seem to allow for an excessively long Artinskian stage (25 Ma as compared with 5 Ma in Harland *et al.* 1982, the time scale used for this symposium volume). This results in an overly long Permian Period, but a shorter Visean (13 Ma versus 19 Ma). Using the latter time-scale I calculate the arithmetic mean duration of blastoid genera to be 10.4 Ma and the half-life to be 6.5 Ma.

Horowitz *et al.* did not analyse their data for major extinction episodes, so I present one here (Fig. 8.4) using their data (Horowitz *et al.* 1985, table 2), plus the Ordovician genus *Macurdablastus,* Broadhead, 1984. This analysis is again open to question as most blastoid genera are known from one horizon only, and, hence, peak extinction rates coincide with peak origination rates and peak standing crop. It is therefore doubtful whether any of the peak extinctions reflect more than the approximate stratigraphic location of blastoids (Fig. 8.4). In this context it may be significant that of the major groups depicted in Fig. 8.1 the fissiculate blastoids have the most incomplete fossil record with an average of 40.1 per cent gap. Treatise data for crinoids (Moore and Teichert 1978) are at too coarse a stratigraphic level (1/3 system) to compare major extinction events with Raup and Sepkoski's (1982) peaks for extinction. No other large data sets are available at present for echinoderms.

Broad patterns

Sepkoski (1979, 1981) suggested that Phanerozoic diversity patterns incorporated three main radiations (Fig. 8.5), which I shall refer to as the Cambrian, the Palaeozoic, and the post-Palaeozoic, respectively. The fossil record of echinoderms seems to be broadly compatible with these three faunas. The Cambrian echinoderm fauna is typified by unfamiliar forms; the rest of the Palaeozoic fauna is dominated by filter-feeding crinoids whereas post-Palaeozoic echinoderms are predominantly eleutherozoans. In this section I shall elaborate on this pattern.

1. The Cambrian fauna

Cambrian echinoderms include helicoplacoids and *Camptostroma* as well as the earliest cystoids s.l. (conventionally called 'eocrinoids', a paraphyletic group which should be abandoned), the first crinoid, *Echmatocrinus,* and the ancestor of the eleutherozoans, *Stromatocystites.* The fauna thus includes some groups confined to the Cambrian (or even the early Cambrian) as well as forerunners of groups which would dominate later faunas. If one considers the carpoids, these too show a similar pattern. The Cambrian carpoid fauna includes ctenocystoids and cinctans unique to the Cambrian, as well as the first solutes and cornutes which were characteristic of the Palaeozoic fauna.

2. The Palaeozoic fauna

As defined here, the Palaeozoic fauna is dominated by crinoids, the total generic diversity of which at its maximum in the Carboniferous, exceeded that of all other Palaeozoic groups combined (Paul and Smith 1984, fig. 14, p. 465). Crinoids were filter-feeders and this mode of feeding dominated in the Palaeozoic. Indeed, it is likely that the only significant modes of feeding among Palaeozoic echinoderms were filter- and deposit-feeding (see below). The success of the crinoids and their replacement of the cystoid groups after the Ordovician may well be related to their extensive arms which formed an extremely efficient filter. All groups which survive to the Upper Palaeozoic show a decline from the Carboniferous to the Permian. Even so, the Permo-Triassic extinction event was a major catastrophe for echinoderms, not only in the proportion of taxa that survived the event, but also in the resulting faunal turnover. The post-Palaeozoic fauna is dominated by groups quite different from those in the Palaeozoic fauna.

3. The post-Palaeozoic fauna

This fauna is characterized by eleutherozoans, particularly echinoids and asteroids, with crinoids surviving in reduced numbers and many in deeper

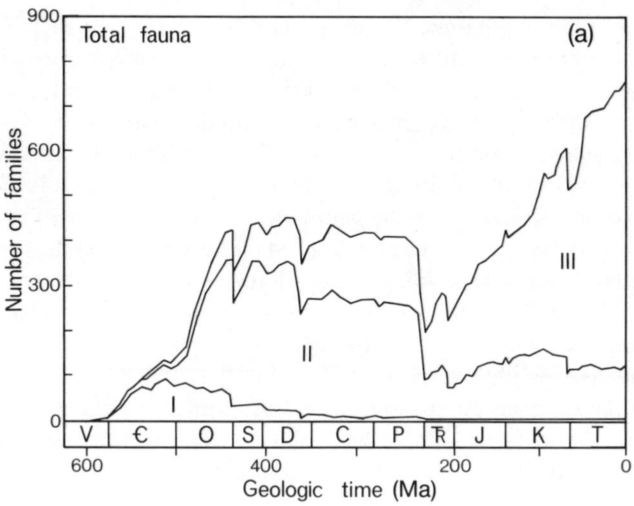

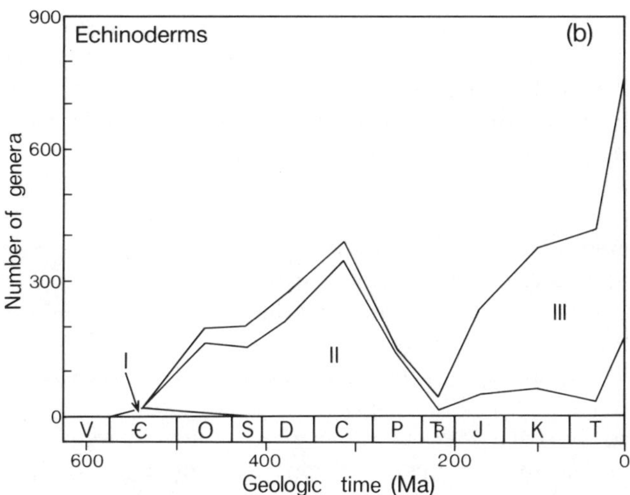

Fig. 8.5. (A) Relative familial diversity of the three main Phanerozoic marine faunas (Roman numerals). Redrawn from Sepkoski (1981, fig. 5). (B) Relative generic diversity of echinoderm faunas to show similarity to the pattern for all marine Phanerozoic families. The Cambrian faunas (I) includes all exclusively Cambrian echinoderms plus all 'eocrinoids'. The Palaeozoic fauna (II) includes all pelmatozoans other than 'eocrinoids' plus post-Cambrian edrioasteroids, while the post-Palaeozoic fauna includes all eleutherozoans except the edrioasteroids. Total diversity is plotted at the mid-points of geological systems with no allowance for variations in the duration of different systems. Note the major decline towards the Permo-Triassic boundary, but the minimal effect of the Cretaceous–Tertiary extinction event.

waters. The other pelmatozoan groups had all become extinct before the end of the Permian. Within echinoids it has long been accepted that there was a major life crisis at the Permo-Triassic boundary. Only one genus, *Miocidaris*, is known to have crossed this boundary, although Kier (1984, p. 3) has argued that two lineages survived into the Triassic and gave rise to the later radiation of echinoids. Coincidentally, miocidarids fixed the standard pattern of plating in the echinoid corona with pairs of columns of ambulacral and interambulacral plates. Among crinoids it has also been accepted that there was a major faunal turnover at the Permo-Triassic boundary. The Palaeozoic subclasses Camerata and Flexibilia did not survive the era, while the inadunates are represented by one genus, *Encrinus*, in the Middle Trias. All other post-Palaeozoic crinoids are referred to the subclass Articulata, although *Encrinus* is almost certainly not the ancestral form. Roux (1985) has recently described a living crinoid as an inadunate, but personally I am unconvinced by the published anatomical details.

In contrast to these two classes, published data on asteroids (Spencer and Wright 1966) infer that all modern orders can be traced well back into the Palaeozoic, although it has long been suggested that asteroids had two major radiations, the first in the Ordovician and the second in the Jurassic (Wright 1967; Blake 1982, 1986). Gale (1987) has recently re-evaluated the post-Palaeozoic asteroids and concluded that they form a monophyletic group sharing many synapomorphies and differing significantly from known Palaeozoic genera. Thus, it seems the asteroids also suffered a major crisis at the Permo-Triassic boundary. Given this reversal of interpretation of the asteroid fossil record, it would seem appropriate to re-examine that of ophiuroids before commenting further. Nevertheless, it does seem reasonably well established that the echinoderms as a whole suffered a major extinction event at the Permo-Triassic boundary which was accompanied by a major turnover in the composition of the echinoderm fauna.

Following the Palaeozoic all echinoderm groups underwent radiations, but the eleutherozoan groups were apparently more successful. This was due to major new morphological innovations (see below) which opened up new modes of life. In contrast to the Permo-Triassic boundary, the end of the Mesozoic Era marked very little change in faunal composition nor was it a major extinction event for echinoderms. Among the crinoids, the millericrinids did not survive the Cenomanian, lowest Upper Cretaceous, and the roveacrinids the Campanian. On present taxonomic assignment all other groups continued into the Tertiary.

Among the echinoids, although half a dozen families became extinct at the Cretaceous–Tertiary boundary, all the higher taxa recognized by Smith (1984, Fig. A.1) survived. Treatise data on asteroids and ophiuroids probably need re-evaluation in the light of Gale's (1987) findings. Even so, it appears that both classes suffered little extinction at the Cretaceous–Tertiary boundary and that most extinctions were at the specific level. Blake (1986) has recently commented on the persistence to the present day of several genera of Jurassic and Cretaceous asteroids. Since the end of the Mesozoic all groups appear to have increased their diversity right up to the present.

Palaeobiology

Modern echinoderms are characterized by (a) an extensive water vascular system, (b) a mesodermal calcitic skeleton with a meshwork structure called stereom, and (c) pentamery. Fossil evidence from the Lower Cambrian indicates that pentamery was the last to evolve (Paul and Smith 1984). Stephenson (1974, 1978) outlined a theory for the origin of pentamery which would only apply to a fixed filter-feeding organism that could not rotate to face into (or away from) currents. Having five arms is the most efficient method of filtering from currents which may approach from any direction. Since we now know that the earliest Cambrian echinoderms were fixed filter-feeders, the theory seems a very reasonable explanation. Direct fossil evidence for an extensive water vascular system with tube-feet is confined to true echinoderms, as opposed to carpoids, suggesting that stereom may have been acquired before the water vascular system. However, if the 'arms' of solutes or the food grooves of cinctans contained radial water vessels, it would seem likely that the water vascular system preceded stereom. In addition, solutes have a pore interpreted as a hydropore, suggesting a water vascular system was present in the earliest carpoids. Furthermore, a direct homologue of the water vascular system is present in hemichordates (the tentacular system), which strongly suggests that the water vascular system preceded stereom.

Either way, in the Lower Cambrian the most primitive true echinoderms, the helicoplacoids, already had a water vascular system with external radial water vessels lining the floors of the ambulacra and giving rise to lateral branches (tube-feet) externally and internal compensation sacs (Paul and Smith 1984, fig. 18). Such a system was admirably adapted (or pre-adapted) for respiratory gas exchange, although several Lower Cambrian genera (*Camptostroma, Stromatocystites, Lepidocystis, Kinzercystis*) possessed epispires on the oral surface which supplemented gas exchange via the tube-feet. The skeleton was flexible, probably leathery, with embedded ossicles and often involved imbricate plating. Thus, although it provided some protection and resistance to muscular contraction, it was not rigid. All Lower Cambrian echinoderms were filter-feeders; however, the solitary solute (carpoid) may have been a deposit-feeder. By the Middle Cambrian thecae with more or less rigid frames supporting flexible plated membranous covers had evolved in both true echinoderms (*Cambraster*) and carpoids (cinctans, ctenocystoids), and deposit feeding was established. Among filter-feeders progressive differentiation of stalk from theca can be seen in *Gogia* and the first true column appeared in *Eustypocystis* (Sprinkle 1973). Some stalks were undoubtedly used for raising the theca off the substrate and so enhancing filtering efficiency by taking advantage of natural currents. Thus, the Cambrian fauna established the basic pattern of echinoderm biology found throughout the Palaeozoic, but many of the genera were quite unlike later echinoderms. The first crinoid, *Echmatocrinus* from the Middle Cambrian, is unfamiliar when compared with later genera.

The Ordovician saw a tremendous proliferation of respiratory pore-structures in the cystoids (s.l). Even some crinoid genera, such as *Porocrinus, Palaeocrinus,* and *Cleiocrinus,* developed thecal pore-structures. I have always suspected that the abundance and diversity of respiratory pore-structures reflected low oxygen levels in the early Palaeozoic oceans. There was a major radiation of cystoids which reached their peak diversity in the Ordovician, as well as a reasonably diverse crinoid fauna. With the recent description of *Macurdablastus* (Broadhead 1984), we know that all currently accepted classes of echinoderms with the probable exception of the holothurians, had evolved before the end of the Ordovician. The fauna was dominated by filter-feeders and although the first starfish and echinoids were present, their morphology suggests that they were not predators, but probably deposit-feeders. During the rest of the Palaeozoic, the crinoids progressively outcompeted the cystoid groups, while the eleutherozoans remained a very minor element of the fauna. Total diversity rose to a peak in the Carboniferous, due almost entirely to the crinoids which dominated by virtue of their very extensive food gathering apparatus and small theca implying small food requirements. Diversity declined through the late Carboniferous and Permian, and dropped drastically at the end of the Palaeozoic. However, the blastoids, the last of the cystoid groups, had gone before the last stage of the Permian. The decline seems to have been gradual rather than sudden, but this may reflect the availability of exposures.

In the post-Palaeozoic the echinoderm fauna altered drastically. All crinoids except *Encrinus,* are assigned to the new subclass Articulata, among echinoids only *Miocidaris* and possibly one other line survived from the Palaeozoic, and it seems the asteroids suffered a similar decline. The most successful post-Palaeozoic crinoids, the isocrinids and comatulids, share the possession of prehensile cirri which they use for attachment. Prehensile cirri enable the comatulids to anchor themselves to promontories for feeding at night and to retire into crevices to avoid predators during the day. Even the stalked isocrinids lack a holdfast and attach themselves to the substrate with the cirri. Although they are not known to swim or to retire during the day, Messing (1985, p. 189) has recorded evidence of one crawling along the sea floor. Recently, Simms (1986) has argued that the pentacrinids were pseudopelagic and attached themselves to logs using the cirri. Undoubtedly, the possession of prehensile cirri was a major innovation for these groups. Motility seems to have been at a premium in post-Palaeozoic seas.

Extinctions of crinoids during the Upper Cretaceous primarily removed the pelagic crinoids, *Uintacrinus, Marsupites,* saccocomids and roveacrinids, and truly pelagic crinoids never evolved again. However, all these pelagic forms were extinct before the end of the Campanian at the latest and do not seem to have been victims of the Cretaceous–Tertiary boundary event.

The tremendous radiation of the echinoids seems to have involved first the evolution of rigidly sutured plates with interpenetrating stereom which enabled more efficient lantern apparatuses operated by powerful muscles to develop. These, in turn, allowed the echinoids to utilize a wider variety of foods, involving rasping hard surfaces as well as breaking up animal and vegetable debris. The development of compound ambulacral plating also allowed post-Palaeozoic echinoids to overcome one major limitation of the plate arrangement in the corona of *Miocidaris.* It provided better defence due to the development of large ambulacral spines and more tube-feet to cling to surfaces, particularly when rasping for food (see Kier 1974, for further details.)

Undoubtedly the major post-Palaeozoic innovation among echinoids was the appearance of the irregulars. This can be traced to adaptation to an infaunal mode of life and much of the subsequent evolution of irregulars is related to progressive adaptation to life in finer and finer sediments. The earliest infaunal echinoids, such as the pygasteroids and holectypoids probably just covered themselves with sedimentary particles and were confined to relatively coarse-grained substrates. Later galeropygoids developed a canopy of fine spines which kept sand-grade particles away from the test surface. The evolution of penicillate tube-feet in the disasteroids allowed them to live in and feed on, fine-grained sediments. It was a major adaptive breakthrough and from them the spatangoids evolved burrow lining techniques using the tube-feet of the anterior ambulacrum. Clypeasteroids developed auxiliary tube-feet which are used in feeding, while the fine canopy of spines sifts sediment for food. More detailed accounts of post-Palaeozoic evolution among the echinoids can be found in Kier (1974) and Smith (1984).

Among the asteroids the major adaptive innovations were the appearance of flexible arms, suckered tube-feet and the eversible stomach (Gale 1987). The former two allow modern starfish to inhabit hard substrates without being washed away, while the latter enables them to prey upon epifaunal organisms. The relative lack of flexibility in the arms of Palaeozoic starfish suggests that they were incapable of active predation and were also poorly adapted to burrowing. They seem to have been confined to soft substrates in relatively quiet waters.

With the changes outlined above the eleutherozoans came to dominate modern echinoderm faunas. The crinoids form a relatively minor constituent of the present day fauna, particularly when compared with their dominance in the Palaeozoic, but are somewhat more common and diverse in deeper waters.

Conclusions

(1) Rigorous analysis of extinction and survivorship patterns in the fossil record require estimates of the reliability of the fossil record, phylogenetic analysis of the taxa at the lowest taxonomic level, refined global stratigraphy, and a reliable time-scale. Even ignoring the last two points, no available data for echinoderms meet these criteria.

(2) The best data, for blastoids, suggest that on average genera survived for 10 Ma and species for 5 Ma. Generic survivorship halved every 6.5 Ma. Peaks of originations and extinctions mirror each other and reflect periods when blastoids are known, not major radiations and extinctions.

(3) The generic diversity of echinoderms as a whole seems to follow the pattern of three major faunas, a Cambrian, a Palaeozoic, and a post-Palaeozoic, recognized by Sepkoski (1979, 1981). The Cambrian fauna is small and dominated by 'eocrinoids'; the Palaeozoic is dominated by filter-feeding pelmatozoans, especially crinoids; the post-Palaeozoic by deposit-feeding and macrophagous eleutherozoans.

(4) The Palaeozoic fauna declined sharply in diversity before the Permian; nevertheless, the Permo-Trias boundary marked a life crisis for echinoderms. It also marked a major turnover in composition of echinoderm faunas from pelmatozoan (crinoid) dominated to eleutherozoan dominated.

(5) The Cretaceous–Tertiary boundary scarcely affected echinoderms as a whole. Several groups, particularly pelagic crinoids, became extinct before the end of the Campanian stage.

(6) The Cambrian fauna shows that stereom and the water vascular system were acquired before pentamery. Stereom allowed the development of a protective test; the water vascular system allowed filter-feeding and was pre-adapted for respiratory gas exchange. Pentamery developed as the most efficient filtering mechanism in a fixed organism unable to orientate itself to currents.

(7) Early Palaeozoic faunas may well have been dominated by cystoids due to low oxygen levels in the oceans. Later Palaeozoic faunas were dominated by crinoids possibly due to intense competition between filter-feeding organisms. Post-Palaeozoic faunas seem to be dominated by

mobile or cryptic echinoderms (pelmatozoan or eleutherozoan) perhaps due to intense predation pressures.

(8) The major post-Palaeozoic innovations were: among crinoids prehensile cirri; among asteroids flexible arms, suckered tube-feet, and the eversible stomach; among echinoids rigidly sutured coronal plates, ambulacral plate compounding, more efficient lanterns, and penicillate tube-feet.

Acknowledgements

Thanks are due to Dr A. S. Gale, City of London Polytechnic, and Dr A. B. Smith, British Museum (Natural History), for valuable comments on an earlier draft of the manuscript, and to Joe Lynch, Liverpool University, for help in drafting the figures.

References

Blake, D. B. (1982). Recognition of higher taxa and phylogeny of the Asteroidea. In *Echinoderms: Proceedings of the International Echinoderms Conference, Tampa Bay* (ed. J. M. Lawrence), pp. 105–7. Balkema, Rotterdam.

Blake, D. B. (1986). Some new post-Palaeozoic sea stars (Asteroidea: Echinodermata) and comments on taxon endurance. *J. Paleont.* **60,** 1103–19.

Broadhead, T. W. (1984). *Macurdablastus,* a Middle Ordovician blastoid from the southern Appalachians. *Paleont. Contr. Univ. Kansas, Pap.* **110,** 1–9.

Gale, A. S. (1987). Phylogeny and classification of the Asteroidea (Echinodermata). *Zool. J. linn. Soc.* **89,** 107–32.

Gale, A. S. and Smith, A. B. (1982). The palaeobiology of the Cretaceous irregular echinoids *Infulaster* and *Hagenowia. Palaeontology* **25,** 11–42.

Harland, W. B., Smith, A. G., and Wilcock, B. (1964). *The Phanerozoic Time-scale. A symposium dedicated to Professor Arthur Holmes. Q. J. geol. Soc. Lond.* **120S,** 1–458.

Harland, W. B., Cox, A. V., Llewellyn, P. G., Pickton, C. A. G., Smith, A. G., and Walters, R. (1982). *A Geologic Time Scale.* Cambridge University Press, Cambridge.

Horowitz, A. S., Blakely, R. F., and Macurda, D. B. Jr (1985). Taxonomic survivorship within the Blastoidea (Echinodermata). *J. Paleont.* **59,** 543–50.

Izett, G. A., *et al.* (1982). Major geochronologic and chronostratigraphic units. *Bull. isotopic Geochron.* **34,** inside front cover.

Kier, P. M. (1974). Evolutionary trends and their functional significance in the post-Palaeozoic echinoids. *Paleont. Soc. Mem.* **5,** 1–95.

Kier, P. M. (1984). Echinoids from the Triassic (St Cassian) of Italy, their lantern supports and a revised phylogeny of Triassic echinoids. *Smithson. Contr. Paleobiol.* **56,** 1–41.

Messing, C. G. (1985). Submersible observations of deep-water crinoid assemblages in the tropical western Atlantic Ocean. *Echinodermata. Proceedings of the Fifth*

International Echinoderms Conference (ed. B. F. Keegan and B. D. S. O'Connor), pp. 185–93. Balkema, Rotterdam.

Moore, R. C. and Teichert, C. (1978). *Treatise on Invertebrate Paleontology. Part T. Echinodermata* 2. Geological Society of America and University of Kansas Press, Lawrence, Kansas.

Mortimore, R. N. (1986). Stratigraphy of the Upper Cretaceous white chalk of Sussex. *Proc. Geol. Ass.* **97,** 97–139.

Paul, C. R. C. (1980). *The Natural History of Fossils.* Weidenfeld, London.

Paul, C. R. C. (1982). The adequacy of the fossil record. In *Problems of Phylogenetic Reconstruction,* Systematics Association Special Publication Volume No. 2, (ed. K. A. Joysey and A. E. Friday) pp. 75–117. Academic Press, London.

Paul, C. R. C. and Smith, A. B. (1984). The early radiation and phylogeny of the echinoderms. *Biol. Rev.* **59,** 443–81.

Raup, D. M. and Sepkoski, J. J., Jr (1982). Mass extinctions in the marine fossil record. *Science, NY* **215,** 1501–3.

Robinson, N. D. (1986). Lithostratigraphy of the Chalk Group of the North Downs, southeast England. *Proc. Geol. Ass.* **97,** 141–70.

Roux, M. (1985). Découverte d'un représentant actuel de crinoïdes pédunculés Paléozoiques Inadunata (echinodermes) dans l'étage bathyal de l'Isle de la Réunion (Ocean Indiés). *C. R. Acad. Sci., Paris* **301,** 503–6.

Sepkoski, J. J., Jr (1979). A kinetic model of Phanerozoic diversity II. Early Phanerozoic families and multiple equilibria. *Paleobiology* **5,** 222–51.

Sepkoski, J. J., Jr (1981). A factor analytic description of the Phanerozoic marine fossil record. *Paleobiology* **7,** 36–53.

Simms, M. J. (1986). Contrasting lifestyles in Lower Jurassic crinoids; a comparison of benthic and pseudopelagic Isocrinida. *Palaeontology* **29,** 475–93.

Smith, A. B. (1984). *Echinoid Paleobiology.* George Allen & Unwin, London.

Spencer, W. K. and Wright, C. A. (1966). Asterozoans. In *Treatise on Invertebrate Paleontology. Part U. Echinodermata* 3 (ed. R. C. Moore), pp. U1–89. Geological Society of America and University of Kansas Press, Lawrence, Kansas.

Sprinkle, J. (1973). *Morphology and evolution of blastozoan echinoderms.* Special Publication of the Museum of Comparative Zoology, Harvard.

Stephenson, D. G. (1974). Pentamerism and the ancestral echinoderm. *Nature, Lond.* **250,** 82–3.

Stephenson, D. G. (1978). The origin of the five-fold pattern of echinoderms. *Thalassia jugosl.* **12** (for 1976), 337–46.

Wright, C. W. (1967). Evolution and classification of Asterozoa. In *Echinoderm Biology* (ed. N. Millott), Symposium of the Zoological Society of London, Vol. 20, pp. 159–62. Academic Press, London

9. Extinction and the fossil record of the arthropods

D. E. G. BRIGGS

Department of Geology, University of Bristol, Bristol, UK.

R. A. FORTEY

Department of Palaeontology, British Museum (Natural History), London, UK.

E. N. K. CLARKSON

Grant Institute of Geology, University of Edinburgh, Edinburgh, UK.

Abstract

The history of selected arthropod groups is reviewed in relation to supposed events of mass extinction. A distinction is drawn between the true extinction of major clades and 'taxonomic pseudoextinction', where the 'termination' of a group is a reflection of taxonomic practice, such as the acceptance of paraphyletic groups. The fossil record of different arthropod groups varies considerably; trilobites have an excellent record, but other groups are so poorly known that they are of doubtful relevance to a discussion of major extinctions. The Cambrian–Ordovician boundary is not a major event at familial level for trilobites, but there is a high generic turnover among platform dwelling forms; taxonomic pseudoextinctions are a problem at this horizon. 'Biomere' boundaries in the Cambrian are probably similar, with comparable taxonomic problems. The Ordovician–Silurian event is marked by a major extinction of trilobite clades; this cannot be 'explained' simply by the late Ordovician glaciation. A drop in generic diversity immediately preceded both the Cambrian–Ordovician and Ordovician–Silurian events. The Frasnian–Famennian boundary

Extinction and Survival in the Fossil Record (ed. G. P. Larwood), Systematics Association Special Volume No. 34, pp. 171–209. Clarendon Press, Oxford, 1988.

appears to be the culmination of a decline in trilobite clades which began much earlier in the Devonian, and the boundary is distinguished mainly by the disappearance of some distinctive groups. The major decline in eurypterids occurs in the Pridoli-Gedinnian and their geological history does not reflect the periods of mass extinction in other groups (e.g. the Ordovician–Silurian boundary). The evidence does not support the hypothesis that the decline of eurypterids was related to competitive displacement during the 'rise' of the jawed fishes. The end-Cretaceous extinction appears to affect the brachyuran decapod families to a similar extent to other invertebrate groups. When the possibility of taxonomic pseudoextinction is identified, however, no unequivocal Maastrichtian decapod extinction remains. Generic extinctions among the cirripedes are higher in the Maastrichtian than in the stages prior to and after it, but the differences in extinction level do not indicate an exceptional event. The insect record shows no evidence of an end-Cretaceous event, but this may reflect the concentration of its effects in high latitudes where the insects, like the plants, may have been more resistant to short-term temperature fluctuations.

Introduction

The purpose of this paper is to examine the fossil record of arthropods in relation to some horizons which are widely considered to coincide with major extinctions. We have confined our attention, for the most part, to those arthropods with a high fossilization potential and with stratigraphic ranges appropriate to a consideration of major extinctions. Thus, we have considered extinctions in the trilobites, eurypterids, decapods, and cirripedes (and also, briefly, in the insects because of their potential significance as a terrestrial group in illuminating the nature of the end-Cretaceous extinction). The ostracods are beyond the scope of this paper due, not least, to limitations of space; as a consequence, our coverage neglects the Permo-Triassic extinctions. We are primarily concerned with presenting the facts about extinction within the arthropods, but a summary of likely causes is given in some cases. Our concern is with major, or possible major extinction events; arthropod evidence may not be adequate to discriminate the minor post-Permian events of Raup and Sepkoski (1984).

Extinction and taxonomy

In considering arthropod extinctions it is important to distinguish between two phenomena, both of which may have superficially similar taxonomic expression. Recent discussions of mass extinctions (e.g. Jablonski 1986) have considered most of the historical, or causal factors involved in mass extinction, but have not stressed the taxonomic basis on which all such observations are founded (see Fig. 9.1).

1. Major boundaries of faunal turnover are frequently also the boundaries between the fields of endeavour of different specialists. This may result in taxonomies which are separated by the same boundary — family names applied below the boundary may be different simply *because* they are below the boundary. Or primitive forms below a boundary may be 'lumped' within families based on largely plesiomorphic characters; such families may well include species in phylogenetic continuity with early taxa of different families above the boundary (Fig. 9.1b,c). This results both in an apparent extinction of the plesiomorphic family below the boundary, and a spurious novelty in the families above the boundary. The only 'extinction' in this case is a loss of primitive characters. We term this taxonomic pseudoextinction. The Cretaceous crab family Dynomenidae may be cited as an example; it originally included a variety of more or less primitive forms which are now considered likely to be related to more than one post-Cretaceous decapodan family. The resolution of such taxonomic problems both serves to extend ranges of post-Maastrichtian families, and remove an apparent reduction in diversity of the Dynomenidae at the end of the Cretaceous. Taxonomies grew up for all kinds of reasons other than the morphology of the animals themselves; early taxa may have seemed 'important enough' or separated by enough time from known descendants to somehow merit high level taxonomic recognition. This phenomenon has everything to do with scientific procedure, and nothing at all to do with the important matter of true extinction at a time of crisis.

This is somewhat different from what has been termed the 'Lazarus effect', whereby a taxon (or its very close relative) re-appears above an extinction event after a temporary absence. In groups like many of the arthropods the stratigraphic gap may be very large, and the taxonomic distance may be uncertain. It would be extremely naive to expect to be able to read off the extinction history of these groups within single sections spanning boundary intervals. Rather, the crucial question is a taxonomic — or rather phylogenetic — one: do taxa below the boundary belong within a close monophyletic group with those above the boundary? If they are relatively plesiomorphic forms forming a sister taxon to forms above the putative extinction horizon then the group must be judged to have passed through the event unscathed as far as the evidence allows. This is regardless of whether the sub-boundary form has been accorded generic or family status — and really should be regardless of the stratigraphic distance above and below the boundary from which fossils have been recovered.

2. True extinctions are interruptions of the phylogenetic continuum,

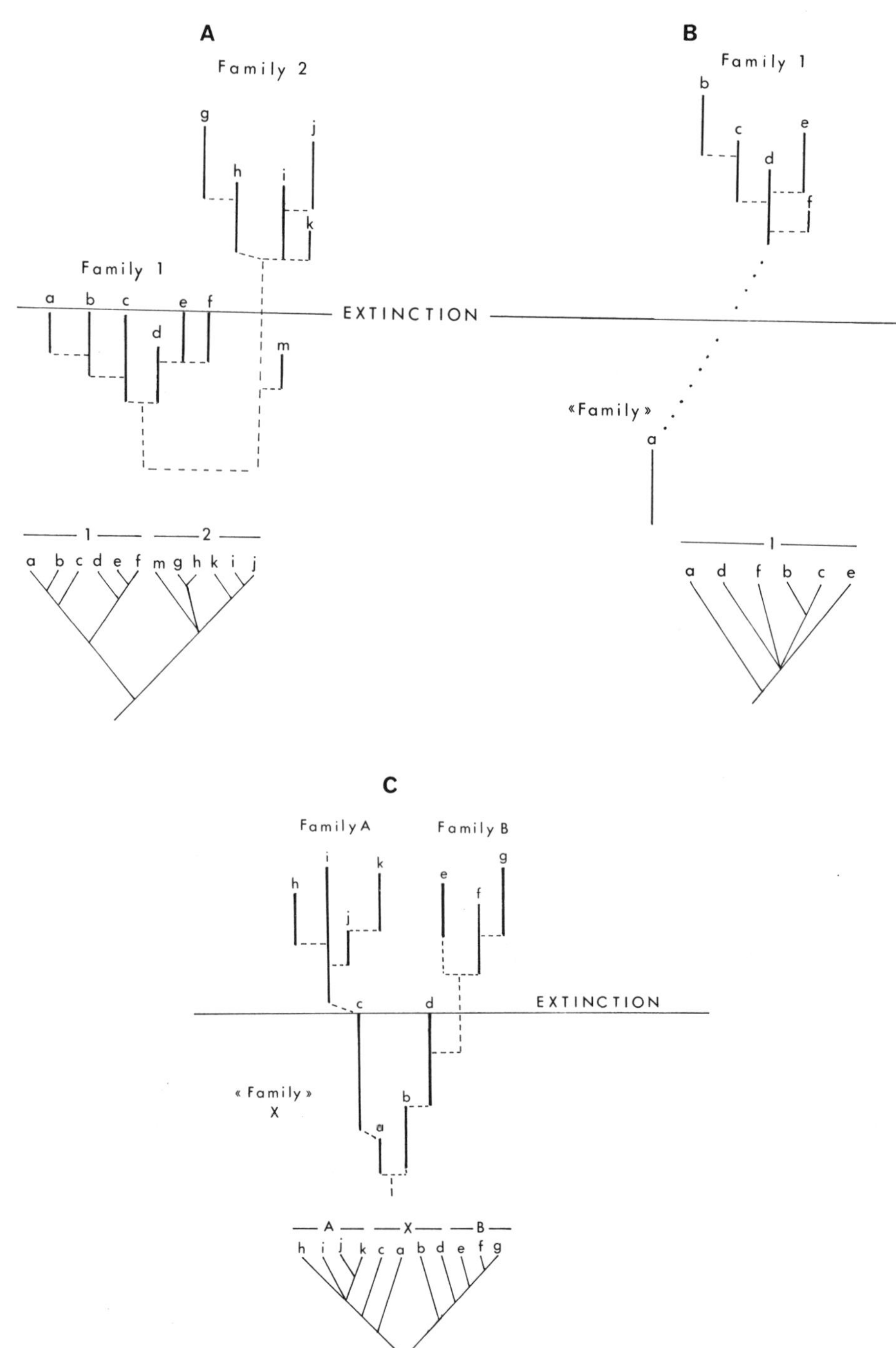
A
Family 2
g
h
i
j
k
Family 1
a
b
c
d
e
f
EXTINCTION
m
1
2
a b c d e f m g h k i j
B
Family 1
b
c
d
e
f
«Family»
a
1
a d f b c e
C
Family A
Family B
h
i
k
j
e
f
g
c
d
EXTINCTION
« Family »
X
b
a
A
X
B
h i j k c a b d e f g

with termination of a clade (Fig. 9.1a). The more distinctive the clade is (the autapomorphy characterizing the clade is therefore less likely to be derived polyphyletically) and the more rich in species, then the more confidently can the extinction event be recognized, and with less likelihood of confusion with 1 above. For fossil arthropods it is really only worth considering such terminations at generic level and higher, because recognition of palaeontological species is difficult to begin with, and the vagaries of the record — even with frequently-preserved fossils such as trilobites — make the disappearance of a species hard to distinguish from a taphonomic event (see Fig. 9.4).

Treatments of patterns of extinction based, for example, upon the *Treatise on Invertebrate Paleontology* are ambiguous because they do not take into account the major difference between the concepts explained above. It is feasible, indeed we think it probable, that particular extinction 'events' are dominated either by the first kind of taxonomic expression or the second. Only the latter will qualify for consideration as the kind of catastrophic event to which so much attention has been paid recently.

For this reason we attempt to distinguish between true and pseudoextinctions. Termination of clades are shown as triangles on diagrams; pseudoextinctions involving taxa which are in probable phylogenetic continuity (regardless of whether they happen to be recognized as a 'family' confined below a horizon under consideration) are shown as asterisks on diagrams.

Stratigraphic definitions of boundaries are frequently problematic, and such problems are discussed where appropriate below.

Cambrian–Ordovician boundary

1. The boundary and patterns of diversity across it

This boundary approximates to the base of the Tremadoc Series; it has been claimed as an important extinction event. Around this horizon appear

---

Fig. 9.1. Diagram to illustrate the differences between extinction of a clade and taxonomic pseudoextinctions. A phylogenetic tree in relation to an extinction horizon is shown in the upper diagram, and a cladogram of relationships in the lower diagram. (A) True extinction; a phylogenetically valid taxon (1) is extinguished at the boundary. (B) Taxonomic pseudoextinction because of poor record below a boundary. Here the sub-boundary taxon may have been accorded family status because of its stratigraphic position, or because it differed in being relatively plesiomorphic compared with the compact taxon *b* to *f*. However, *a* belongs within clade 1 and its disappearance cannot be taken as evidence of extinction. Arthropods with sporadic records approximate to this case. (C) Taxonomic pseudoextinction because of paraphyletic taxa. 'Family' X below the boundary is recognized on the basis of plesiomorphic characters. Its 'disappearance' may represent loss of those characters, but the true clades (A and B) cannot be said to have become extinct at the horizon.

many of the groups which survive today — and the appearance of what Sepkoski (1981) termed the 'Paleozoic ecological fauna'. Among the arthropods most data relating to the boundary are from trilobites.

In detail, there are considerable problems about deciding exactly where such a boundary should be taken, as with its international correlation. There is an International Working Group at present studying these problems, and the unpublished papers arising from this work run to hundreds of pages. Some of the results were summarized recently in a collection of published papers (Bassett and Dean 1982). This is obviously no occasion to review this welter of evidence. A world-wide eustatic event, possibly a polyphase regression, has been claimed at about this stratigraphic level. The final and arbitrary decision on the boundary may place it slightly above the climax of these eustatic events, which introduce disconformities into platform successions. For the purposes of this paper, however, we are obliged to accept an horizon against which to measure extinction events. The traditional base of the Tremadoc series based on the appearance of graptolites of the '*Dictyonema*' *flabelliforme* type is no longer adequate for this purpose, probably being slightly later than the faunal change in other groups. Conodonts are increasingly used as an international base for correlation — although not extending into clastic facies at high palaeolatitudes. The principal change in the trilobite faunas of North America has been recognized between the *Saukia* Zone of the Trempealeauan and the *Missisquoia* Zone of the Ibexian (= Canadian), often used as the Cambrian–Ordovician boundary on that continent. This horizon apparently lies just above the base of the *Cordylodus proavus* conodont zone and, on the most recent opinion of Miller (1986), within the basal part of the *Fryxellodontus inornatus* subzone. Even with this horizon there are possible problems of heterochrony.

Because this (*Saukia–Missisquoia*) boundary appears to correspond best with reported trilobite extinctions, we have attempted to correlate it worldwide, showing numbers of genera in various trilobite families across the boundary; we include also the *Cambroidstodus* conodont Zone and its equivalents below, and the *C. intermedius/lindstromi* conodont Zone (early Tremadoc) above. Because this interval has been much studied recently the data base includes faunas from USA, Canada, China, Britain, Scandinavia (Norway), Mexico, Australia, and Kazakhstan. Some unpublished data from Working Group papers is used; recent compendia of Chinese sections have been published (Lu *et al.* 1984*a*; Chen *et al.* 1985); many of the other sources are in Bassett and Dean (1982). Details may change once the many correlation problems are refined. An important addition, not apparent from faunal lists, is the inclusion of genera subject to the 'Lazarus effect' — those which have not been reported from one of the horizons under consideration, but which are known to reappear above. They must obviously be added to range bars; some were listed by Fortey (1983). Examples are the olenid *Plicatolina* and the catillicephalid *Buttsia*.

A few systematic notes are essential to explain Fig. 9.2. We have not included Upper Cambrian families which are believed to be polyphyletic, and therefore are not natural clades; this includes especially the highly effaced families Plethopeltidae and Tsinaniidae; also Raymondinidae. A number of generalized ptychopariids (*Hystricurus, Ketyna,* etc.) have been grouped together in the category 'Hystricurinae'. Incertae sedis are not included; nor are families reputedly confined to one or two zones (e.g. Missisquoiidae). Conservative taxonomy is adopted where we feel genera have been split too finely. The record of the continuation of the Ptychaspididae into the Ordovician is based on the assignment of the genus *Curiaspis* Sdzuy, 1955, from the Leimitzschiefer, to this family; continuation of the Saukiidae is based on the Tinu Formation in Mexico. We accept the phyletic connection previously proposed between Loganopeltidae and Harpidae, and suggest it between Lonchocephalidae and Calymenidae, but the status of the Cambrian families as 'pseudoextinctions' is indicated by asterisks on the diagram in the manner explained above. A supposed extinction of the Lecanopygidae at the C–O boundary is not accepted, as the younger Ordovician genus *Benthamaspis* is placed in this family. Similarly, the range of the family Bynumiidae (hitherto considered confined to the Cambrian) is continued into the Ordovician because the Ordovician genus *Clelandia* is regarded as synonymous with *Bynumiella. Macropyge* and its relatives are included in the Ceratopygidae. The continuation of Odontopleuridae into the Cambrian is because we recognize one of the species (*Acidaspides praecurrens*) figured by Bruton (1983) as a true odontopleurid. Clearly, the introduction of these examples of taxonomic pseudoextinction makes a good deal of difference to the C–O boundary as an horizon both for disappearance and appearance of major clades.

A second diagram (Fig. 9.3) treats the data at a lower taxonomic level, showing the first and last appearance of genera within the boundary interval zone-by-zone. In this case we can include genera belonging to families which are not natural phylogenetic units, but obviously we have to exclude genera with ranges within a single Zone. This diagram also excludes the genera which pass unscathed through the interval, i.e. it simply ignores continuity.

2. *Discussion of the Cambrian–Ordovician interval*

Although it is hardly possible to deny the radical change in faunas between the Upper Cambrian and, say, the end of the Tremadoc, the nature of the boundary interval itself is far more equivocal. Figure 9.2 shows that at family level, particularly including 'Lazarus effects' and taxonomic pseudo-extinctions, the boundary or possible boundaries are not clearly marked by the simultaneous extinctions of major clades. A boundary taken between the *C. apopsis* and *Missisquoia* trilobite Zones sees the extinction of two

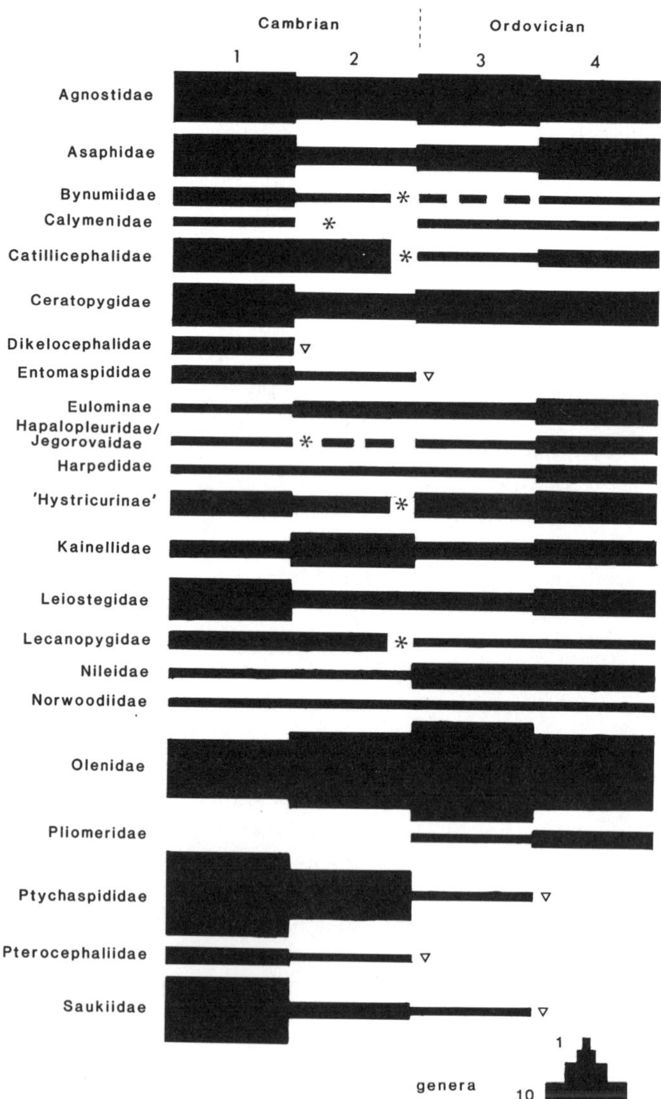

Fig. 9.2. Generic abundances worldwide in family clades of trilobites across the Cambrian–Ordovician boundary, represented by four successive zones (1–4) spanning the interval. Triangles show demise of true clades, asterisks represent cases where taxonomic pseudoextinction is surmised. Polyphyletic taxa omitted. 'Lazarus' taxa included. For full explanation see text.

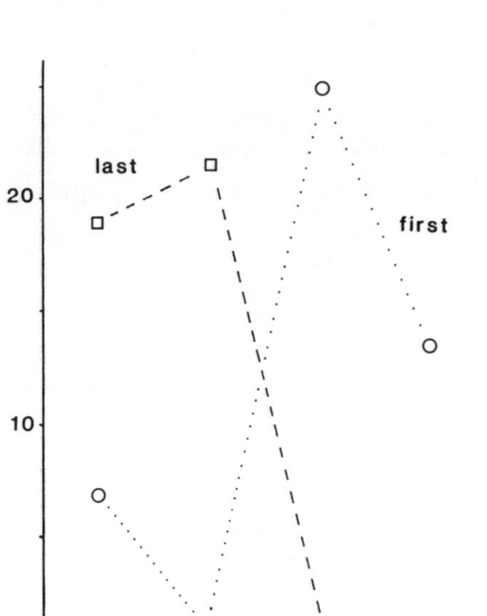

Fig. 9.3. Trilobite genera making their first and last appearances within or at the end of the four zones across the Cambrian–Ordovician boundary. This shows maximum last appearances, and low originations at the end of the Cambrian, and very high first appearances early in the Ordovician. Genera from polyphyletic families are included here, and effects of possible taxonomic pseudoextinctions are ignored. This is heavily biased towards platform taxa.

trilobite families; the number is greater only if the concept of pseudoextinction is not accepted, or polyphyletic taxa are allowed. The same horizon shows the decline, but not the termination of the saukiids and ptychaspidids. However, it is also possible to point to a decline in these and a few other families somewhat earlier, at or about the top of the *Proconodontus* Zone (*S. serotina* subzone). This 'double' effect — which may in part be a stratigraphic artefact as the result of the necessary lumping of the data into zones — is not unlike that described for the Ordovician–Silurian boundary below. As in that case, the interval embraced by the decline in these families is a short one, corresponding to a regression on the platform. Certainly, those families most affected were typical inhabitants of inshore sites.

Many of those trilobites least affected through the interval tended to be found in off-platform, slope or deep-water environments (Fortey 1983) — most Olenidae, Agnostida, Ceratopygidae, and some of the longer-ranging asaphids, for example. The differences in response of platform and more marginal faunas is reflected in the individual species-ranges through

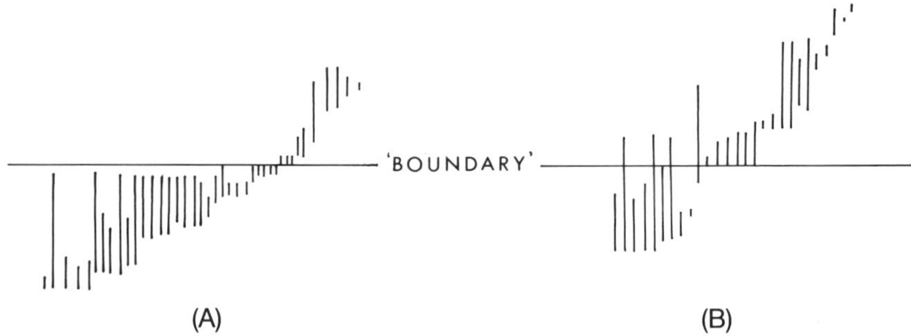

Fig. 9.4. Trilobite species ranges across Cambrian–Ordovician boundary sections, contrasting platform, and relatively off-platform sites. (A) After publications of Stitt on platform Oklahoma, showing sharp truncation of all species at a boundary horizon which may be a regressive maximum. (B) After Ludvigsen (1982) on north-west Canada, showing more overlapping species ranges in a more marginal palaeogeographic site.

sections (Fig. 9.4). Compare, for example, the 'biomere' patterns shown by Stitt (1977) from carbonate platform rocks in the USA, in which there is a complete taxonomic break at the boundary, with the patterns from marginal sites on Kazakhstan (Chugaeva and Apollonov 1982) or Zhejiang, China (Lu *et al.* 1984*b*), with staggered and overlapping ranges. If the correlation accepted here is correct even the saukiids did not become extinct at either horizon, but persisted into 'Ordovician' time in a few places.

At generic level (Fig. 9.3) the picture is different, bearing in mind that this plot removes the effect of continuity of other taxa. There are high last appearance rates in the last two Cambrian zones — which fall off greatly in the first two zones of the Ordovician. First appearances of genera are extremely low at the end of the Cambrian, and very high in the basal Ordovician, hence, on both criteria, the time of crisis is the latest Cambrian (*C. apopsis* = earliest *proavus* zone) with new appearances at a minimum and last appearances at a maximum. Even allowing that the curves may be exaggerated as a result of pseudoextinction there seems no reason to doubt the change at this horizon.

Figures 9.2 and 9.3 are not incompatible. High last appearances of genera in the latest Cambrian is a reflection of the contraction in numbers of genera of platform groups at or near the base of the *C. apopsis* interval, and again at its end; high appearance rate coincides with the repopulation of the platform in the Ordovician (taxonomic problems are likely to be most important here). Off-platform taxa are less affected and this is responsible for broad continuity through the interval. Not all platform taxa are extinguished at the boundary — some reappear in the Ordovician after some time lapse. However, a changeover in platform taxa at generic level is well-established, even allowing for possible pseudoextinctions.

3. Causes of the extinctions

The Cambrian–Ordovician 'event' is doubtfully considered a major extinction in terms of the simultaneous termination of major trilobite clades. By the time taxonomic pseudoextinctions are allowed for, extinctions coinciding with the boundary are relatively few, and many families maintain a nearly constant diversity through the interval. However, there was a pronounced extinction and replacement of platform taxa, even allowing for possible pseudoextinctions. The same interval has been equated with a eustatic regression and it is reasonable to link the taxonomic with the eustatic change. As with the Ordovician–Silurian boundary the eustatic event may exaggerate the apparent simultaneity of the extinction events for taphonomic reasons.

The Cambrian–Ordovician boundary has been regarded as the last of the biomere 'events' (Stitt 1975). Complex migration of the biofacies at the boundary has been described by Ludvigsen and Westrop (1983). The extinction event on the platform has been attributed to the incursion shelfwards of cool water (possibly accompanied by displacement of the thermocline) and the accompanying biofacies at the boundary between Cambrian and Ordovician (e.g. Ludvigsen 1982, fig. 18), thereby restricting the area available for shelf-adapted forms. Associated with transgressions of biofacies is the sudden appearance of normally more oceanic taxa such as the olenids *Jujuyaspis* and *Bienvillia* in platform sequences at the base of the Ordovician. This explanation essentially places the extinction 'event' on the horizon between Cambrian and Ordovician. This is not altogether adequate, because, as we have seen, there was an interval preceding this event where appearance of new genera was at a minimum, and last appearances at a maximum. Hence, it is probably necessary to take the regressive event and the ensuing transgressive or onstep event *in tandem* to explain the changes at the boundary.

Ordovician–Silurian boundary

1. The boundary and patterns of diversity across it

For some years the Ordovician/Silurian boundary has been recognized as what Sheehan (1975) termed 'a time of crisis'. The recognition of a widespread late Ordovician glaciation has reasonably been linked with a period of rapid faunal turnover and, in some cases, extinction. Stratigraphic research on this boundary has been actively pursued by a Working

Group of the International Geological Correlation Programme, and recently a decision has been reached on the definition of the boundary at the base of the *Akidograptus acuminatus* graptolite biozone (Cocks 1985), lying above the Ashgill Series of the Ordovician and at the base of the Llandovery Series of the early Silurian.

As far as arthropods are concerned the most detailed evidence on the effect of this boundary upon the phylum comes from trilobites, which have been extensively studied from this interval for more than 50 years. In our discussion of the Ordovician/Silurian history of the group we have been helped by papers reviewing aspects of the subject by Owen (1987) and Lespérance (1987) who generously allowed us access to their information prior to publication. Stratigraphic division of the late Ordovician is not fully agreed on the international scale; for example, the term 'Gamachian' is used for late Ordovician strata on Anticosti Island, where the time interval is most fully represented in North America. For the purposes of this paper we have calibrated extinction events against three stages: Rawtheyan, Hirnantian, and Rhuddanian, the last two of the Ordovician and the basal stage of the Llandovery, respectively. Because the Hirnantian is recognized primarily by the characteristic *Hirnantia* brachiopod fauna, which itself is diachronous (Rong 1984), recognition of the Hirnantian is not without difficulties; here we have accepted the recent opinions of Owen (1987). In any case, all authors seem to agree that the Hirnantian was brief compared with a typical stage of 5–6 Ma, and doubts about its use as a term seem to stem from this, rather than doubts about its equivalence to a worldwide 'event'.

Data on Fig. 9.5 are compiled for numbers of genera within trilobite families from all available literature sources we could gather from Scandinavia, Ireland, Poland, North America, Scotland, USSR (Kazakhstan), China, and Wales. Space does not permit listing of all sources: most are given in Thomas *et al.* (1984), and Owen (1987). Rawtheyan is our own compilation; Hirnantian is largely taken from Owen and early Silurian from Lespérance (1987). However, we have added to range bars genera exhibiting the 'Lazarus effect'. To give one example, the distinctive aulacopleurid *Scharyia* is known from Caradoc rocks and again from the mid-Silurian; despite its lack of record through our interval, it must have existed. Figure 9.5 shows three main points:

1. A considerable diminution in numbers of genera between Rawtheyan and Hirnantian for nearly all families (see also Brenchley and Newell 1984).

2. However, almost all families survive into the Hirnantian, only to become extinct at the end of the Ordovician. All terminations here are extinctions of well-defined clades, i.e. extinctions are real ones of well-documented groups with long histories.

3. The early Silurian does not apparently represent an explosive

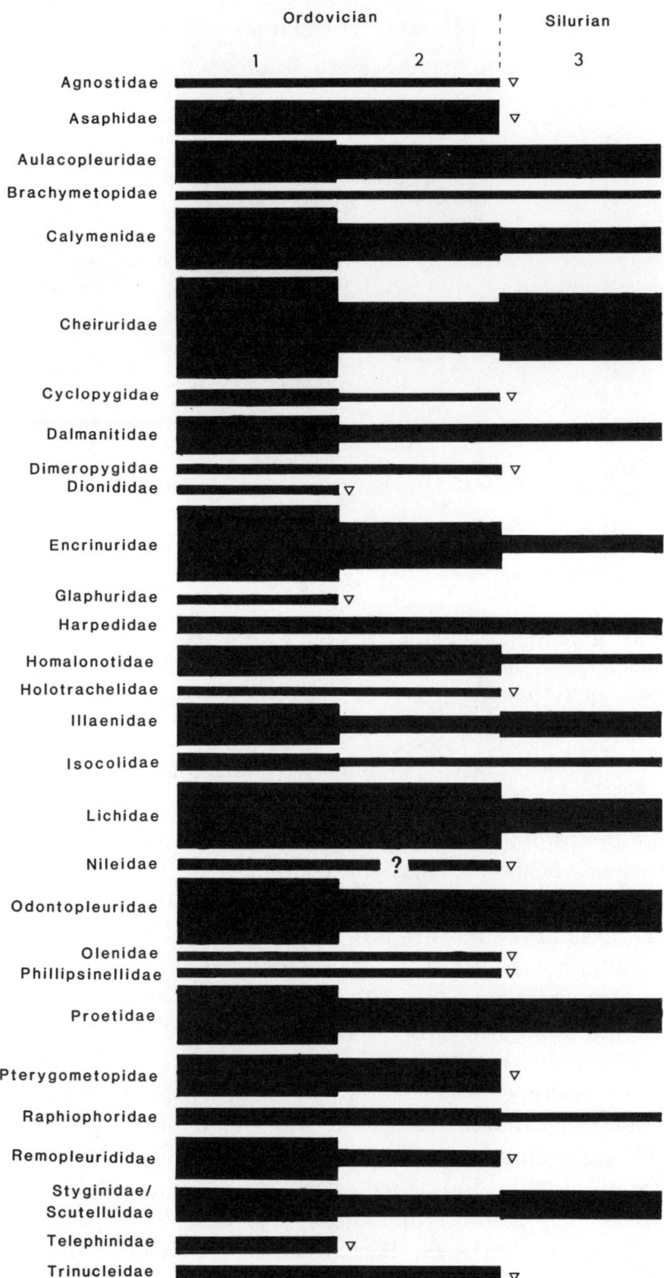

Fig. 9.5. Generic abundances worldwide in family clades of trilobites across the Ordovician–Silurian boundary, late Rawtheyan (1), Hirnantian (2) and early Rhuddanian, basal Llandovery (3), showing concentration of extinction at the end of, or within the Hirnantian. Scale as Fig. 9.2.

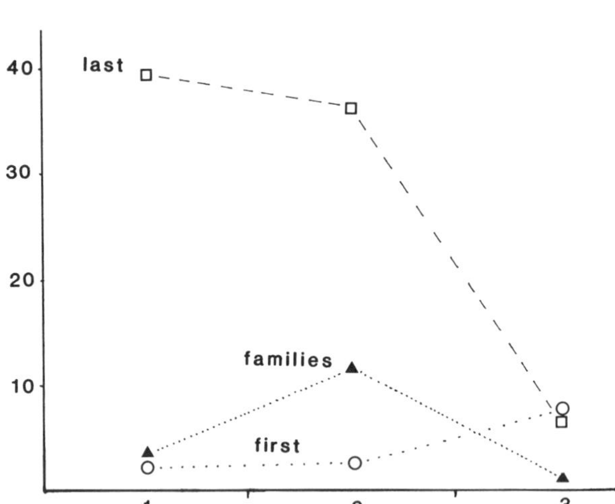

Fig. 9.6. Last appearances of trilobite genera (squares) across the Ordovician–Silurian boundary, at the end of or within time divisions 1–3; circles show low generic origination throughout the interval (compare Cambrian–Ordovician). Triangles show peak in last family appearances within or at the end of the Hirnantian. 'Lumping' of species into stratigraphic intervals for correlation purposes may obscure staggered disappearances.

radiation of the trilobites — low diversity continues (Fig. 9.6). There are a few new generic taxa here, and the continuation of a number of 'relict' genera. However, Rhuddanian trilobite faunas are rare.

It is not likely that the low Hirnantian generic count is entirely an artefact: faunas are known from more than 30 successions and generic diversity within single successions (e.g. Girvan District) mirrors that of the whole. On the other hand, the regressive phase of the Hirnantian has biased against the recovery of deep water facies, and the apparent disappearance of deeper water families below the Hirnantian (e.g. Dionididae, Telephinidae) should be viewed with some circumspection. Deeper water facies are again known from the Llandovery, and the disappearance of such forms by then is assured. Problems of stratigraphic refinement remain, which may prove extinction of taxa to be non-synchronous, but given the short duration of the Hirnantian the available evidence shows a sudden termination of family-level taxa.

2. *Discussion of the Ordovician–Silurian event*

The trilobite data suggest a double event — diminution in numbers of generic taxa, with (possibly an artefact of the record) extinction of some clades at or near the end of the Rawtheyan; and complete extinction of a number of important groups with long histories probably at the end of the Hirnantian. Which of these is regarded as the major event depends upon your point of view: if overall generic diversity is the prime criterion, then it is the former event, if the extinction of major clades is the criterion then the end-Ordovician event is the more significant. The latter is our view here, while recognizing that the prior diminution in diversity may have rendered clades more vulnerable to final extinction. Some of the taxa which are terminated at the boundary (Asaphidae, Olenidae, Pseudagnostidae, Remopleurididae) had been abundant since the Cambrian; others such as the benthic trinucleids and pelagic cyclopygids were successful and diverse members of the typical Ordovician fauna. Cyclopygids comprised a probable mesopelagic community at the periphery of Ordovician palaeocontinents, with 10 or so genera which remained almost unchanged from the Arenig to the Ashgill. Apparently synchronous disappearance of such groups surely indicates an extinction event of some importance.

The coincidence of the diminution in generic numbers with the late Ordovician glaciation is well-known. Owen (1987) points out that climatic deterioration probably began in the later Rawtheyan, with a climax in the later Hirnantian (Brenchley and Newell 1984). There is a marked glacio-eustatic regression during the Hirnantian. The associated reduction in trilobite diversity has been attributed to the contraction of climatic belts, with an extension of cold water regions at the expense of warmer ones, which were broad in the earlier part of the Ordovician. During the regressive phase, reduction in the area of peri-continental seas reduced the number of niches available (although the 'species area' effect is not sufficient to account for mass extinction, according to Jablonski 1986). A concomitant increase in clastic facies also probably influenced which genera did and did not survive. The spread of cool water faunas of low diversity is shown by the exceedingly widespread occurrence of the *Mucronaspis* trilobite fauna (Jaanusson 1979), a fauna with constituent genera having prior distributions only in the relatively high latitude peri-Gondwanan region, but which extends at this time even into the North American continent. With such an important glacial event recognized from independent evidence there seems to be no good reason to look further for the cause of the Hirnantian decline.

However, it is not obvious that the same cause was necessarily responsible for the extinction of families at the *end* of the Hirnantian. We see no reason in principle why, say, agnostids or cyclopygids, which survived into the Hirnantian, should not have passed into the Silurian and diversified subsequently like encrinurids or cheirurids. Post-glacial colonization was presumably possible from relatively small populations, indeed some of the surviving genera on Owen's (1987) list are known from a single locality. If the cyclopygids were pelagic as has been claimed there must have been a change in the oceanic habitat; the same phenomenon may have been responsible for the drastic decline in graptolites during the Hirnantian. Agnostids, olenids, remopleurids, and trinucleids were all found in deeper shelf habitats earlier in the Ordovician. If an extension of the glacial explanation is sought, it is possible that widespread anoxia in the oceans at the transgression following the glaciation may have been responsible for the selective extinctions, such as have been claimed elsewhere in the geological column. If any extraneous cause is to be looked for, evidence for it should be sought at or close to the Ordovician–Silurian boundary.

Some of the surviving genera, and those with the longest geological histories (*Scotoharpes, Decoroproetus, Sphaerexochus,* illaenids) were inhabitants of bioherms. It seems reasonable to invoke the continuation of such bioherms through the crisis period as a refuge for their survival. A few of these genera are the longest-lived of any trilobite (Fortey 1980). In spite of the contraction of the climatic belts there is evidence of the continuation of small bioherms at near palaeoequatorial latitudes, e.g. on Anticosti Island (Petryk 1981). However, it is not clear at the moment where the Llandoverian equivalents are to be found. Other survivors, dalmanitaceans and lichids, were familiar components of Gondwanan clastic facies, and may be regarded as having been preadapted to weather the crisis.

3. *Summary*

The Ordovician–Silurian boundary included a major extinction event affecting the trilobites. The available evidence on stratigraphic distribution of genera (not yet highly refined) suggests that the extinction was a double event: a reduction in numbers of genera close to the Rawtheyan–Hiranantian boundary, and the final extinction of a number of important families at our very near the boundary between the Systems. Outer shelf and 'oceanic' taxa were affected; some biohermal and previously cooler water forms survived. The reduction in generic number is coincident with an important and well-documented glacial event, and a concomitant glacio-eustatic regression; the final extinction is not so obviously related to that event, but coincides with the eustatic rebound.

Minor Cambrian–Ordovician extinction events

In the light of the account of the Cambrian–Ordovician boundary given above it is necessary to briefly consider some other, possibly comparable events in the Lower Palaeozoic.

1. *Biomeres*

Biomeres (Palmer 1984 and references therein) are more or less cyclical packages of trilobite diversification and extinction recognized in the Upper

Cambrian of the North American craton; some of them correspond to broad sedimentary cycles. The Cambro–Ordovician boundary has been claimed as the last biomere event. Certain generic taxa, even families, are confined within particular biomeres, with supposed extinction events at their boundaries. These boundaries are subtly expressed in rock sections; Palmer (1984) argued against regressions and the like as their control, even though the boundaries are traceable continent-wide. However, he does state that 'the event is followed at many localities by evidence of relative deepening', but 'that the extinction event takes place before the record of any significant change in depositional environments'.

This is so similar to the Cambrian–Ordovician boundary that a common cause is likely. Palmer's preferred scenario is of a displacement shelfwards of cool (possibly deoxygenated) water, but without accompanying transgression (his fig. 14). It seems to us that a relative sea level rise is reasonably invoked to explain biofacies shifts — for example, to introduce the pelagic trilobite *Irvingella* on to the platform at the base of the ptychaspid biomere A further problem with biomeres, and in our view a more important one, it concerned with trilobite taxonomy. Palmer and his co-workers have emphasized 'iterative evolution' in the biomere concept: closely similar trilobites, usually ptychopariοids, are found either side of a biomere boundary. Their resemblance is attributed to convergence, and this is reflected taxonomically — they are placed in different genera or even families. A prime example is the resemblance between *Elrathia* (Marjumiid biomere), *Aphelaspis* (Pterocephaliid biomere), and *Parabolinoides* (Ptychaspid biomere). Such iterative explanations quickly become circular arguments: genera are unrelated just *because* they are either side of a biomere boundary; naturally their non-appearance in between is attributed to extinction. Since many of these taxa are plesiomorphic forms it is impossible to justify such assumptions morphologically. A phylogenetic classification based on morphology would certainly 'lump' many together, and with the 'lumping' would go the extinction event, to become another example of taxonomic pseudoextinction. Quite how important such pseudo-extinctions are in the biomere concept is beyond the present review, but they may well reduce the distinctiveness of each biomere. However, there are certainly some distinctive endemics characterizing platform sequences that do become extinct at biomere boundaries. It is probably wisest here to follow Jablonski (1986) in regarding biomeres as interesting, but relatively minor extinction episodes.

2. *Ordovician–Silurian cycles*

The Series into which the Ordovician and Silurian periods are divided were originally recognized by nineteenth-century geologists on the basis of faunal differences. The same Series boundaries relate to eustatic events (Fortey 1984). It may prove possible to repeat the kind of analysis for the C–O boundary at each Series boundary. However, although they may prove comparable with biomere boundaries, none qualifies as a major event.

Frasnian–Famennian boundary

1. The boundary and patterns of diversity across it

The boundary between the Frasnian and Famennian stages in the Devonian has been claimed as an horizon of major extinction (Kellwasser event), and an extra-terrestrial cause has been invoked to explain it (McLaren 1970, 1982). The trilobites again provide sufficient evidence to examine this claim. There is clearly a general decline in the group between Devonian and Carboniferous — by which time only the Order Proetida remains. However, it is a different question whether this decline was relatively gradual or achieved in one or more major pulses of extinction. For this reason we have taken a broader time spread than was the case with the Cambrian–Ordovician or Ordovician–Silurian extinction events. A major event is again taken as the simultaneous extinction of several major clades, of a greater magnitude than extinctions at stages to either side. Figure 9.7 takes the subfamily as the clade unit for plotting diversities from the Gedinnian to the Famennian. Data is derived especially from Alberti (1979) with updates from ENKC and R. Feist.

We do not express opinions on whether or not any of these subfamilies are paraphyletic groups, largely because we do not have personal expertise in the detailed taxonomy of the groups chosen. Hence, there may be taxonomic pseudoextinctions concealed in the diagram as it stands at present. However, most of the groups are true clades, and in some cases represent the last appearances of major families (e.g. Scutelluidae); taxonomic pseudoextinctions would relate to proetide groups with a post-Famennian history. Dr R. M. Owens informs us that Proetidellinae, Decoroproetinae, Astycoryphinae, and Denemarkiinae may all be included within the clade Tropidocoryphinae and, if this is so, the level of subfamily extinctions before the end of the Frasnian would be correspondingly reduced. The Pseudotrinodinae is an unsatisfactory taxon, and may be omitted when its affinities are resolved. Stratigraphic refinement is a problem: most data are 'clumped' at stage level. However, some groups extending into the Frasnian disappear at different times within the stage and this is recorded on the figure.

Extinctions of major clades coinciding exactly with the end of the Frasnian are relatively few: Pteropariinae, Cornuproetinae, Dechenellinae, Aulacopleurinae, and Scutelluidae (family). The last is an especially important group. Another eight groups are added *within* the Frasnian. It is of course possible that taphonomic bias has simply not allowed us to see

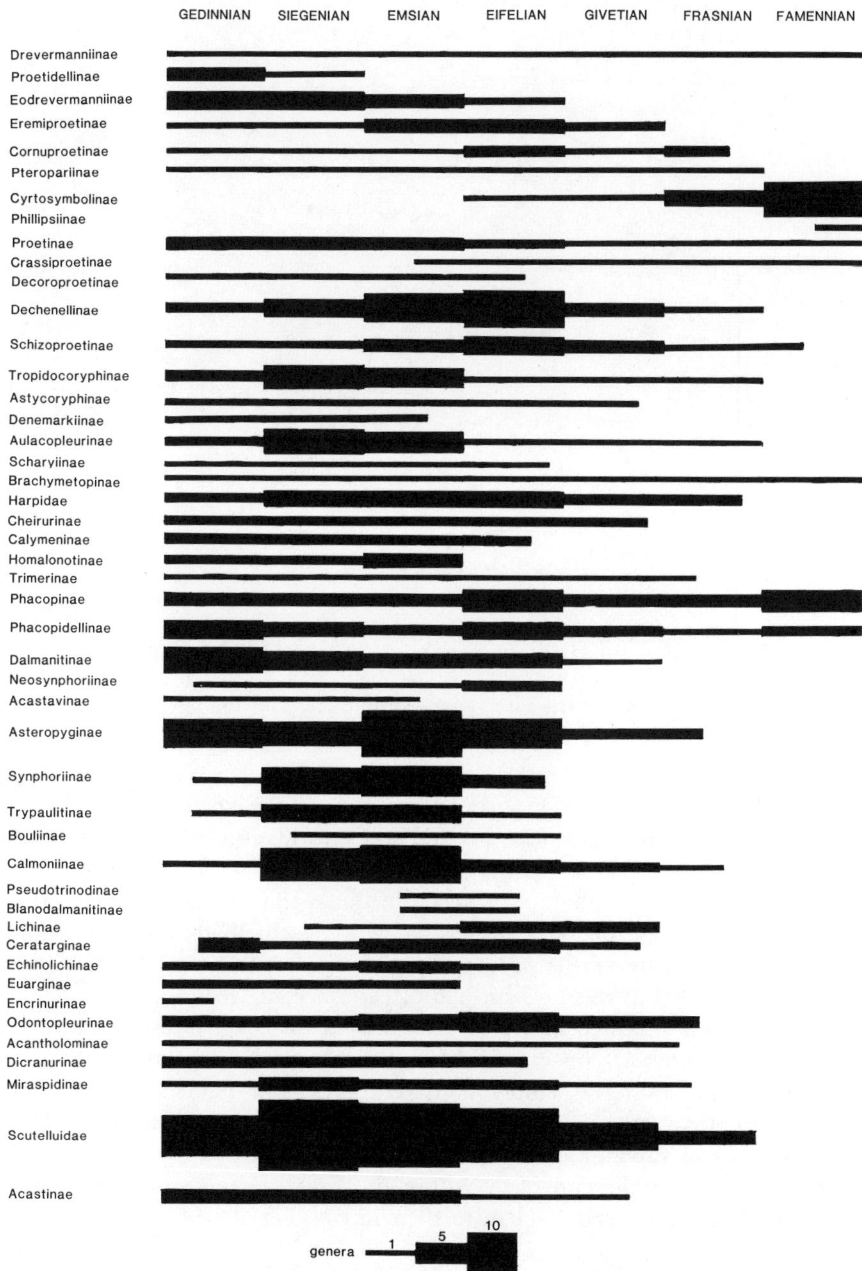

Fig. 9.7. Generic abundances within subfamily clades of trilobites from Gedinnian to Famennian, showing decline in most clades, and disappearance of many, before the Frasnian–Famennian boundary.

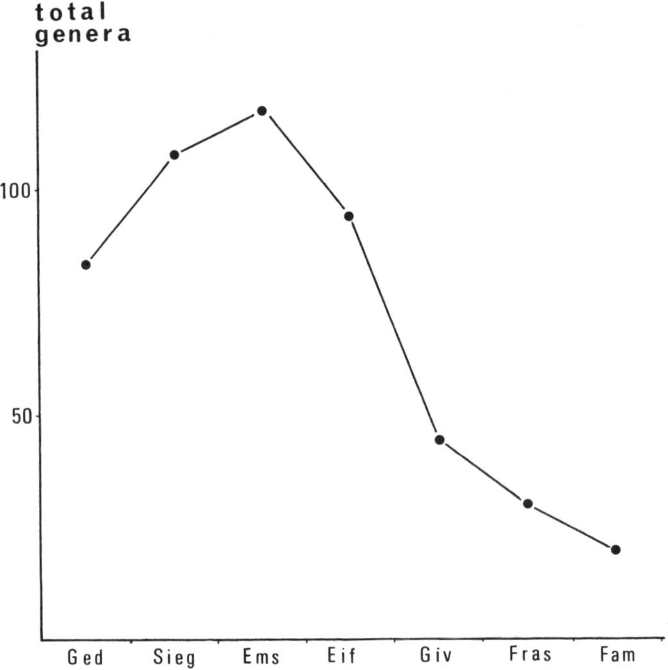

Fig. 9.8. Total number of trilobite genera within stages in the interval Gedinnian to Famennian. Steady reduction in numbers after the Emsian.

the continuation of these groups to the end of the stage, although the extensive work on sequences around the world through this interval inclines us against this possibility. Also, for virtually all of the groups with more than one genus there is a clear decline within the group since the Emsian. This is particularly noticeable within scutelluids and dechenellines. Hence, whatever the nature of the Frasnian–Famennian event, the clades terminated at this horizon had already become reduced.

This trend is reflected in a plot (Fig. 9.8) of the total number of genera through the interval concerned. After a peak in the Emsian there is a steady decline into the Famennian. The Frasnian–Famennian drop is much less than that between Eifelian and Givetian. In terms of last appearances of subfamilies almost as many finish at the end of or within the Eifelian as at the end of or within the Frasnian, and extinctions are high throughout the end Eifelian to Famennian (Fig. 9.9).

2. Discussion of the Frasnian–Famennian event

The interpretation of the data is not unequivocal (House 1975, 1985). The trilobite evidence does not necessarily support a catastrophic extinction at

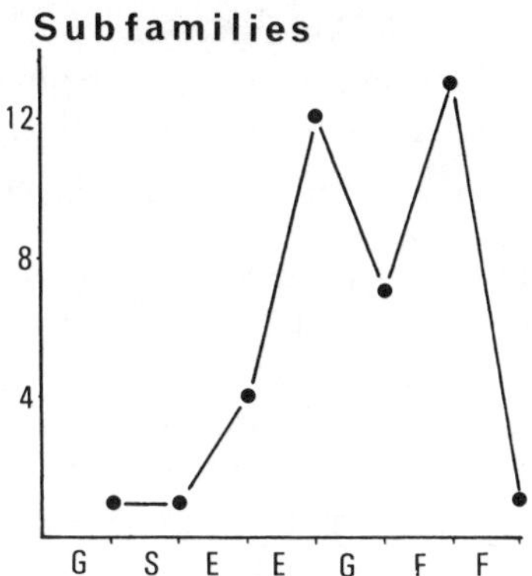

Fig. 9.9. Number of trilobite subfamilies becoming extinct within, or at the end of Devonian stages, showing only slight prominence of Frasnian event.

this horizon. If the disappearance of clades is permitted as the measure of extinction (and taken with total generic number) there appears to have been a long decline prior to the Frasnian–Famennian boundary. The measure of the particular importance of this boundary then becomes the disappearance of certain trilobite groups — harpids, scutelluids, asteropygids, and odontopleurids — which give the earlier faunas their particular morphological richness, while their disappearance appears to impoverish Famennian and later faunas. It is unclear whether harpids, asteropygids, and odontopleurids actually extended to the boundary: present evidence suggests they did not. This assessment of an extinction boundary is obviously subjective. The removal of certain 'key' families invites particular explanations as to why the harpids or whatever were unsuited to conditions pertaining in the Famennian.

The destruction of the coral-stromatoporoid reef habitat (possibly caused by global cooling) is one such explanation, perhaps operating over as much as seven million years (McGhee 1982). The disappearance of trilobite taxa is then related to the disappearance of the habitat to which they were precisely adapted. If this is true, one might envisage a slow contraction of available habitats through the long Devonian decline. Although the harpids were perhaps obligate reef-dwellers, this has not, to our knowledge, been demonstrated for the odontopleurids, which in their earlier history are found in a variety of former habitats, including deep-water ones.

There were also changes connected with provinciality. Alberti (1979) has shown that there were major differences between cooler water Malvinokaffric province trilobites and those elsewhere prior to the Kellwasser event. In the Famennian, Proetidae, Brachymetopidae, and the last Phacopida are present in Europe (Struve 1976), but trilobites of this age are virtually absent in North America. The implication might be that survivorship across the Frasnian–Famennian boundary was enhanced for members of high latitude faunas (cf. Copper 1977). This, in turn, may be taken as evidence that the extinction event, or rather decline, was temperature related.

In summary, trilobite evidence suggests a protracted decline in diversity, in both numbers of genera and clades, throughout a long period of the Devonian. The Frasnian–Famennian 'event' is distinguished primarily by the disappearance of a few clades distinctive enough to invite speculation about special causes, but on the data alone it is not marked out as different in kind from earlier events. Progress in evaluating details across the boundary is likely to be made by more refined zonal division derived from the study of continuous sections, of which those in the Montagne Noire, SW France (Feist 1970, 1976; Feist and Klapper 1985), look promising.

Eurypterid extinctions

Apart from the trilobites and ostracodes (the latter beyond the scope of this review) the eurypterids are the only group of arthropods with a sufficiently high fossilization potential and diversity to merit attention in a discussion of major Palaeozoic extinction events. The occurrence and systematics of eurypterids has been reviewed recently by Plotnick (1983, 1987) who has kindly made his compilation and interpretation of eurypterid stratigraphic and environmental ranges available to us. Eurypterids, like other chelicerates, lacked a mineralized component in the exoskeleton which was therefore subject to normal organic decay processes. For this reason eurypterids are found mainly in fine-grained carbonates and clastics, and as their preservation relies on exceptional conditions, their fossil record must be regarded as relatively incomplete.

The radiation of the eurypterids in the Ordovician shown in Fig. 9.10A is based largely on the diverse faunas of New York State, which have yielded over 30 species. Plotnick (1983, 1987) regards the vast majority of these supposed eurypterids as either random clumps of organic material or inorganic fine-grained infilling of the bedding surface. The first unequivocal eurypterids, however, which occur in the Caradoc, include *Megalograptus* (Caster and Kjellesvig-Waering 1964), one of the most derived eurypterids known (Plotnick 1983). Thus, it is probable that a radiation of more primitive eurypterids did take place in the middle Ordovician or earlier. The eurypterids increased in diversity from mid-Ordovician to Pridoli (Fig. 9.10A), after which they declined sharply and then more gradually from mid-Devonian times, finally becoming extinct in the late Permian. They occupied mainly shallow marine environments until the mid-Devonian. The first non-marine forms occur in the Wenlock. The

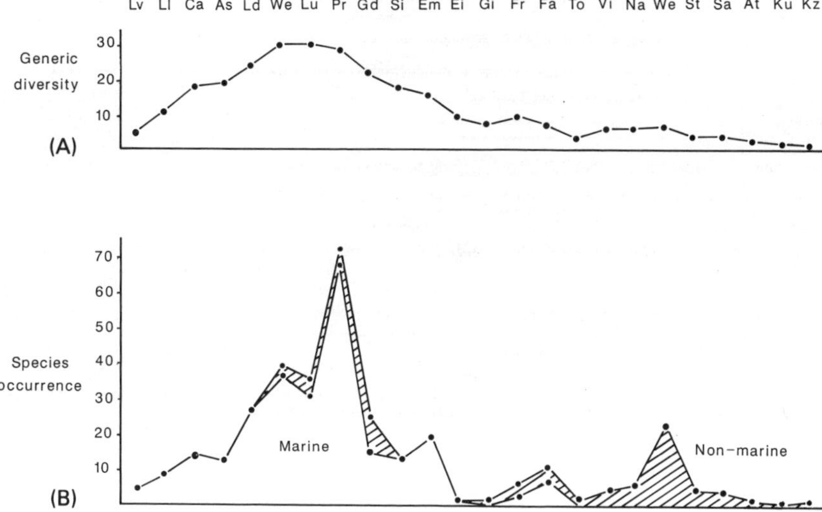

Fig. 9.10. Diversity of eurypterids worldwide through their known stratigraphic range. (A) Generic diversity based on ranges. (B) species diversity based on occurrences rather than ranges. Stratigraphic divisions are series except for Devonian and Permian which are stages. Data from Plotnick (1983).

Carboniferous–Permian eurypterids were all non-marine (Plotnick 1983, 1987, 1988).

Plotnick's (1983) detailed analysis of the characters of eurypterid genera led him to tentatively propose a new familial classification. This is the basis for the plots of generic diversity of clades in Fig. 9.11. Plotnick's classification is based on a clearly argued cladistic analysis, and if this is accepted it is clear that the taxa are true clades and their extinctions are therefore real. The most significant extinctions of eurypterids did not coincide with the periods of mass extinction. The eurypterids were essentially unaffected by the end-Ordovician and Frasnian–Famennian events, for example. Their diversity is too low by the late Permian for the timing of their final disappearance to have any significance. Thirty-three per cent of families (three out of nine) became extinct in the Pridoli and the major generic extinctions occurred at this time also. Eleven genera (39 per cent of standing diversity) became extinct in the Pridoli and eight (42 per cent of standing diversity) in the following stage, the Gedinnian. (This interval coincides with maximum generic diversity, and the divisions are as short as any used in the compilation of Fig. 9.11 except for the Permian stages (durations from Harland *et al.* 1982). Thus, normalization would not materially affect the identification of this peak of extinction.) A plot of the diversity of eurypterid genera (Fig. 9.10A) shows no strikingly high rate of

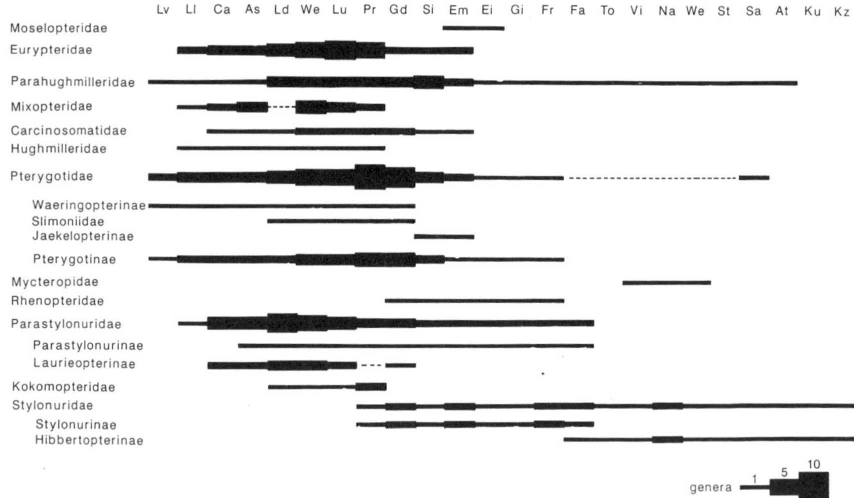

Fig. 9.11. Generic abundances world-wide in clades of eurypterids through their stratigraphic range. Subfamilies are plotted *in addition* to the families which include them. Stratigraphic divisions as for Fig. 9.10. Classification and ranges from Plotnick (1983).

extinction at any horizon. A plot of species diversity (Fig. 9.10B), however, shows more rapid fluctuations, including a dramatic drop from 71 to 25 species from Pridoli to Gedinnian. This plot is based on occurrences rather than ranges and so is more subject to taphonomic effects than that of generic diversity (both plots based on data from Plotnick 1983). The Pridoli-Gedinnian reduction may be a real effect, however; the diversity of eurypterids in localities representing shallow shelf/marginal marine environments is much greater for the Pridoli than Gedinnian (74 species occurrences at 22 Pridoli localities, 10 at 8 Gedinnian localities).

The eurypterids disappeared from marine environments by the end of the Devonian (Plotnick 1983, 1987) and the slight radiation of non-marine forms during the early Carboniferous made little impact on the overall trend in declining diversity (Fig. 9.10B). The decline and extinction of the eurypterids has long been considered to be linked to the diversification of the fishes. Simpson (1953) considered that 'the eurypterids and early fishes covered broadly similar adaptive ranges and were apparently incompatible, for the eurypterids declined and became extinct, as adaptively comparable fish became more abundant'. The early eurypterids rapidly evolved to include substantial predators (e.g. *Megalograptus* in the inner shelf/marginal marine environment by the late Ordovician, *Mixopterus* in non-marine environments by late Silurian), but large predaceous fish did not appear until the Devonian. After this initial rapid

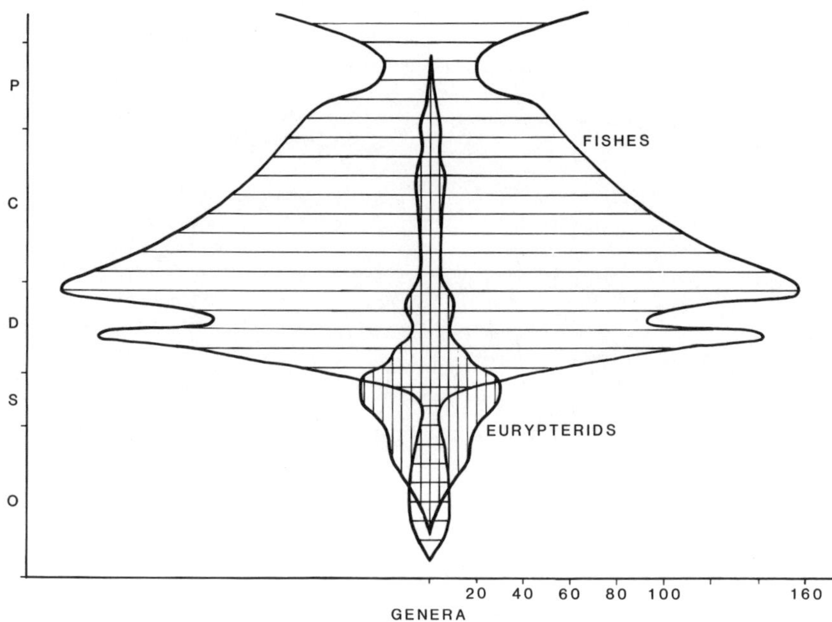

Fig. 9.12. Generic diversity of eurypterids and fishes plotted on an absolute scale. Eurypterid diversities plotted at the same stratigraphic intervals as in Fig. 9.10; fish diversities plotted at coarser intervals. Data from Plotnick (1983) and Thomson (1977).

ecological diversification of the arthropods, the expansion of the fishes appears to be complemented by a decline in the eurypterids.

The hypothesis that the patterns of diversity of eurypterids and fishes are somehow related is very difficult to test with data from the fossil record. There is little evidence for direct interaction (predation by fish on eurypterids, for example); only 24 per cent of 169 eurypterid localities yield vertebrates (Plotnick 1983). MacArthur (1972) has shown that a predator-prey relationship is unlikely to produce extinction (except in island conditions) and taxa that compete directly (as suggested in this case) do not tend to occur together. When the clades are plotted on an absolute scale (Fig. 9.12; Simpson 1953, fig. 18), however, they do appear complementary (the eurypterids decline as the fishes diversify). Such a comparison is misleading, however, as the fish clade is significantly more diverse. Plotted logarithmatically (Fig. 9.13), it becomes clear that both groups decline in the mid-Devonian, peak again in late Devonian and then decline to the end of the Palaeozoic. If the gnathostomes, which are more likely to have been 'adaptively comparable' to the predaceous eurypterids, are considered alone, the onset of eurypterid decline appears to coincide with their appearance in the late Silurian–early Devonian. The early

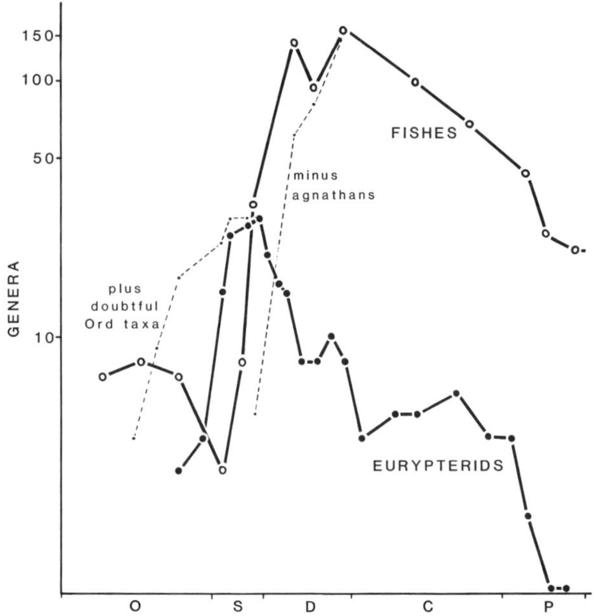

Fig. 9.13. Generic diversity of eurypterids and fishes plotted on a logarithmic scale (cf. Fig. 9.12). Dotted lines show the result of including doubtful Ordovician eurypterids, and of removing agnathan fishes. Diversities plotted at the same stratigraphic intervals as Fig. 9.12. Data from Plotnick (1983) and Thomson (1977).

record of the gnathostomes is problematic, however. The placoderms, for example, appear in the Lower Devonian with an 'instant' diversity of some 40 genera, mainly marine, implying that they may have undergone an earlier radiation. Although there may have been some overlap in marine environments during the Devonian, the eurypterids began to decline before the gnathostomes became well established. More importantly, the decline in eurypterids is not complemented by an increase in fishes, but paralleled by a similar decline (Briggs 1985).

The post-Devonian eurypterids were non-marine, some becoming amphibious (Rolfe 1980; Plotnick 1983; Selden 1985). The non-marine (freshwater and euryhaline) families of fish were less severely affected by the late Permian extinction (Pitrat 1973) than the marine families, but they nonetheless show a decline from the late Carboniferous to early Permian. Although it is clear that Pitrat's compilation includes a high degree of taxonomic pseudoextinction among fish families (Patterson and Smith 1987) there is no evidence of changes in diversity in response to the decline of the eurypterids. Both the number of taxa and the preservation potential of eurypterids, in particular, are inadequate to allow reliable statistical inferences to be drawn (cf. Gould and Calloway 1980 on the geometry of bivalved mollusc and brachiopod clades). The data presented here, however, do not favour competitive displacement of eurypterids by fish. Both taxa may have been influenced by similar global controls on diversity, but the smaller eurypterid clade failed to survive and radiate after the Permian extinction event.

Cretaceous–Tertiary boundary

1. Introduction

The terminal Cretaceous event has attracted more attention than any other mass extinction in recent years since the demonstration of an iridium anomaly in association with boundary strata (Alvarez *et al.* 1980). This has focused attention on the possibility of a massive asteroid impact at this horizon, providing an extraterrestrial cause for this extinction. It is now widely recognized, however, even by the major proponents of an extraterrestrial agent, that the supposed impact was preceded by a period of slow decline in many taxa for several million years (Alvarez *et al.* 1984).

The recognition of mass extinctions has relied mainly on calibrations of the diversity of taxa through time. Pragmatic considerations, in addition to the reduction in completeness of the record of taxa at successively lower ranks in the taxonomic hierarchy, have dictated that such calibration should be carried out initially at the level of families (Sepkoski 1979, 1982*a*,*b*). Compilations have largely concentrated on total diversities of large numbers of marine invertebrate taxa on a global scale. Such gross representations can be deceptive. The effects of extinction on susceptible groups may conceal much lower rates in others. As Jablonski (1986) pointed out, even if rates of extinction in most invertebrate groups were doubled they would not approach the 'breakneck background rates typical of many ammonoid clades'. Thus, comparisons between background and elevated rates in specific higher taxa will reveal their response to major extinction events and may illuminate the nature of the event itself.

Only a small number of arthropod groups have a sufficiently representative fossil record to warrant their use in a consideration of the end Cretaceous event. Of marine forms only the ostracodes, decapods, and cirripedes have a calcified skeleton, and are therefore represented in other than exceptional preservations. (The evolution of calcified skeletons in the crustaceans, with the exception of the ostracodes, it an essentially Mesozoic phenomenon.) The ostracodes are beyond the scope of this paper; the effect of end Cretaceous events on the deep sea forms has been considered by Benson *et al.* (1984). Vagaries of preservation ensure that the record of the insects is grossly underrepresented. The effects of the end-Cretaceous event were much more severe on terrestrial than on marine taxa, however, so we briefly consider the insect record here.

The effect of the terminal Cretaceous extinction on marine taxa was much less pronounced than that of the Permian. About 15 per cent of marine animal families (i.e. about 97) became extinct, as opposed to over 50 per cent in the Permian (Sepkoski 1982*b*). Most severly affected were cephalopods (including all the remaining ammonoids), sponges, bivalves, gastropods, echinoids, and osteichthyan fishes.

2. *Decapods*

The decapods appear in the late Devonian (Schram *et al.* 1978), but the Palaeozoic record is fragmentary and consists only of carapaces (Schram 1986, p. 306). Figures 9.14 and 9.15 show the diversity of families of non-brachyuran and brachyuran decapods in the late Cretaceous and early Tertiary (classification based essentially on Schram 1986). Although all these taxa have calcified exoskeletons, they are nonetheless relatively rare and their fossil record must be viewed in this light. Feldmann (1981)

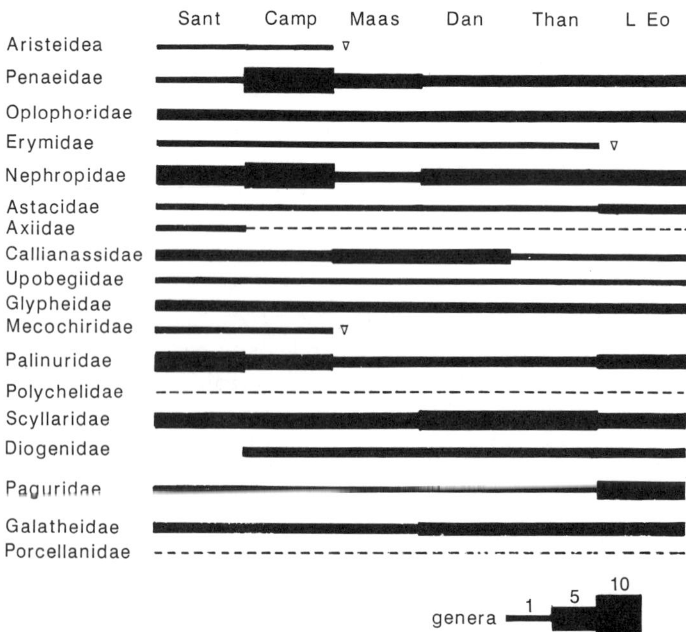

Fig. 9.14. Generic abundances world-wide in family clades of non-brachyuran decapods across the Cretaceous–Tertiary boundary. Classification from Schram (1986); data mainly from Glaessner (1969).

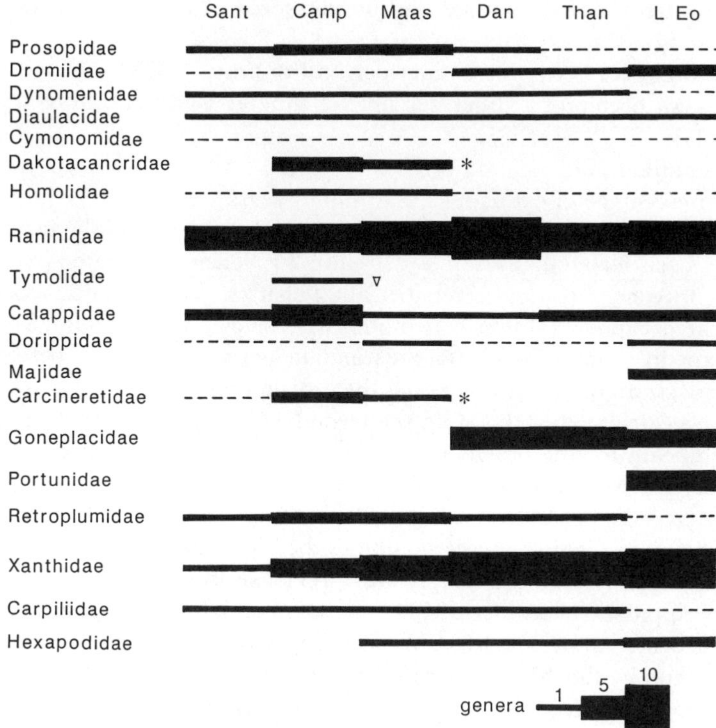

Fig. 9.15. Generic abundances world-wide in family clades of brachyuran decapods across the Cretaceous–Tertiary boundary. Classification from Schram (1986) with emendations from Wright and Collins (1972); data from Glaessner (1969) updated from various sources.

recognized 98 species in the North American Penaeidea, Astacidea, and Palinura (infraorders). Of these about 31 per cent 'are based on bits and pieces of material which are of poor enough quality that ... the identification (to infraorder) is questionable' (p. 455). Bishop (1986) has shown that 40 (42 per cent) of 96 species of North American Cretaceous crabs are based on a single specimen, and only 25 per cent of species are based on more than eight specimens. Plotnick *et al.* (1987) have shown that Recent crabs are highly susceptible to the attentions of scavengers, even though undisturbed carcases have a high preservation potential. Further evidence of the fragmentary record of fossil decapods is provided by the 'Lazarus effect' evident even at family level. Thus, the Polychelidae (Fig. 9.14), for example, is represented by three genera in the Jurassic and one in the Recent, but lacks any record (so far as we are aware) in between. The Homolidae and Dorippidae (Fig. 9.15) might appear to become extinct in the Maastrichtian, but the former have Recent representatives, and the latter reappear in the Eocene.

The intensity of extinctions can be measured in four main ways (Raup 1986):

(1) total taxa becoming extinct per unit of time;

(2) taxa becoming extinct per unit of time as a percentage of standing diversity;

(3) number; and

(4) percentage of extinctions per million years.

The calibration of ranges of decapod taxa is not sufficiently accurate to form a basis for methods 3 or 4. Measure 2 is considered more appropriate than 1 in view of the low diversities and vagaries of preservation associated with the decapods. (Unless stated otherwise below, we have not normalized the data in terms of percentage extinction per million years; percentages refer to stratigraphic divisions in the text-figures.) When data at family level are calculated in this way, the terminal Cretaceous event does emerge as significant for the brachyurans. Two extinctions occur during or at the end of the Maastrichtian, giving an extinction of 13 per cent which approaches the percentage extinction for all marine families at this level (15 per cent). The Campanian extinction of the Tymolidae gives a rate of 7 per cent for this interval; this family was previously considered a subfamily of the Dorippidae (Glaessner 1969). The Campanian is the most significant interval when all decapod families are considered, 9 per cent as opposed to 6 per cent in the Maastrichtian, and 3 per cent in the Thanetian (the sample is arguably too small to have any significance, however; three family extinctions in the Campanian, as opposed to two in the Maastrichtian, and one in the Thanetian).

The two families which became extinct during the Maastrichtian, the Dakotacancridae and Carcineretidae, are both low diversity (three genera) taxa of problematic affinity. Schram (1986, p. 306) commented that 'the classification of the Reptantia is a morass. A single overview using character analysis on all groups is badly needed'. This applies particularly to determining the position of the extinct groups. Both Maastrichtian extinctions may be examples of taxonomic pseudoextinction.

Glaessner (1969) assigned the Dakotacancridae to a superfamily within the Dromiacea. Guinot (1978, pp. 230 and 260) demonstrated that they are not dromiaceans and concluded that their affinities remain uncertain. Schram (1986) included the Dakotacancridae within the Archaeobrachyura, but this is not a natural taxon including 'à la fois des espèces primitives, qui sont sans doute a l'origine des autres sections (les 'vrais Crabes'), et des espèces apomorphes' (Guinot 1978, p. 232). Thus, the view of C. W. Wright (1980, unpublished diagram circulated at the I.G.C., Paris), that the Dakotacancridae are ancestral to the Cancridae (first recorded in the middle Eocene), deserves consideration and critical evaluation as part of a wider study in the future. If he is correct the loss of the dakotacancrids represents a taxonomic pseudoextinction. In any event

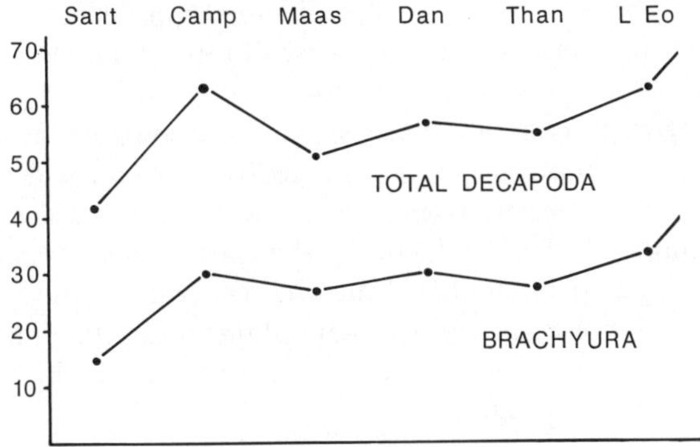

Fig. 9.16. Generic diversity of decapods across the Cretaceous–Tertiary boundary based on the compilations in Fig. 9.14 and 9.15.

their latest appearance is early to mid rather than late Maastrichtian (Bishop 1983). The families Carcineretidae and Portunidae together make up the Superfamily Portunoidca (Glaessner 1969). Lacking Recent representatives, the Carcineretidae are a less completely known primitive sister group of the Portunidae and Wright (1980, unpublished diagram) shows one evolving into the other. Thus, the disappearance of the Carcineretidae in the Maastrichtian may be interpreted as an example of a taxonomic pseudoextinction. Guinot (1978, p. 263), however, emphasizes that in the absence of data on the 'sternum thoracique' the affinities of the Carcineretidae cannot be established with confidence.

There are no unequivocal Maastrichtian extinctions of brachyuran families. The pattern of generic level extinctions is hidden in our compilation, just as it is in those of Sepkoski (1979) and Raup and Sepkoski (1982). Plots of generic diversity of brachyurans, and of all decapods, from late Cretaceous to Lower Eocene (Fig. 9.16) include both the end of the Cretaceous radiation (of mainly eubrachyuran forms) and the beginning of the modern radiation in the Eocene. The major fluctuations are in crab diversity, the number of non-brachyuran taxa remaining relatively constant. The plot shows a 20 per cent extinction of Maastrichtian decapod genera, compared to 31 per cent in the Campanian, and 26 per cent in the Thanetian (the sample is small, although not as small as for families; the difference of 10 per cent between the Campanian and Maastrichtian represents nine fewer generic extinctions in the latter). Normalization yields figures of 3.0 per cent per million years in the Campanian, 2.6 per cent in the Maastrichtian, and 3.0 per cent in the Paleocene taken as a whole (durations from Harland *et al.* 1982). Thus, where genera are concerned, the Campanian appears to have been a more significant time for decapod extinction than the Maastrichtian. The record is such, however, that this could be the result of the Signor–Lipps effect — that is, the observed time ranges are cut short because of taphonomic loss. Apart from general taphonomic considerations there is good evidence from the occurrence of Lazarus taxa in the Maastrichtian to support the possibility of truncated ranges. About 55 per cent of decapod genera which must have had Maastrichtian representatives have yet to be discovered in Maastrichtian strata, compared to 17 per cent for the Campanian. A combination of the nature of decapod taphonomy, the lack of stratigraphic refinement in the recorded ranges of taxa, and the state of our understanding of decapod taxonomy, ensures that few reliable deductions can be drawn regarding their extinction and survival.

3. *Cirripedes*

The cirripedes are represented by a small number of Palaeozoic examples, but the radiation of forms with calcareous skeletons did not get underway until the Triassic. The barnacle record is essentially that of the Thoracica, which bear calcareous plates. A few acrothoracican trace fossils, found primarily as burrows into calcareous substrates including shells, are sufficiently distinct to allow them to be treated as separate genera (Lithoglyptidae on Fig. 9.17). Tomlinson (1969) considered the similarities between these traces and the burrows made by Recent Lithoglyptidae to be insufficient to warrant assignment of the trace fossils to new families (cf.

Fig. 9.17. Generic abundances worldwide in family clades of cirripedes across the Cretaceous–Tertiary boundary. Classification from Newman (1988); data from various sources including Newman *et al.* (1969), Newman (1988), Schram (1986), Buckeridge (1983), and Tomlinson (1969).

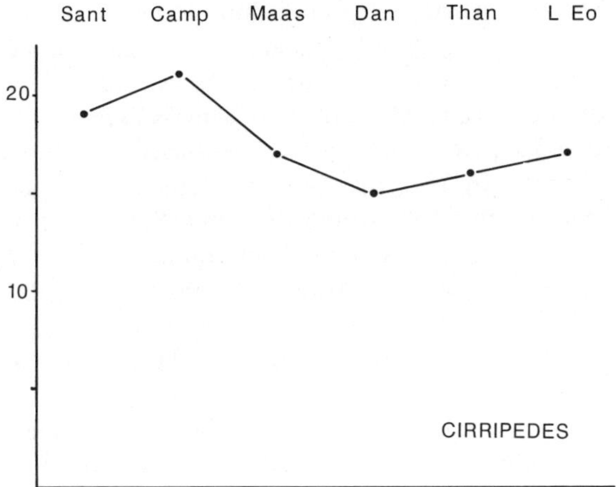

Fig. 9.18. Generic diversity of cirripedes across the Cretaceous–Tertiary boundary based on the compilation in Fig. 9.17.

Newman *et al.* 1969; W. A. Newman, *pers. comm.*, now agrees with the assignment to Lithoglyptidae). Burrows made by Ascothoracica, which are parasites of cnidarians and echinoderms, have been recorded from the Cretaceous, but they are omitted from consideration here due to their very sporadic record and uncertainty regarding their relationship to the living ascothoracicans (Schram 1986, pp. 485 and 515, provides a brief review of Acrothoracica and Ascothoracica.) Shifts in generic diversity in the late Cretaceous–early Tertiary interval under consideration are not pronounced. Diversity (Fig. 9.18) reaches a peak at about 21 genera in the Campanian and then declines to about 75 per cent of that number in the Danian, before the Tertiary radiation of the sessile barnacles.

The classification adopted here is that of Newman (1988). He elevates the Scalpellidae s.l. to a suborder Scalpellomorpha (Superfamily Scalpelloidea) comprising the families Eolepadidae and Scalpellidac. The Eolepadidae include the Stramentinae (= Stramentidae s.l.) and related subfamilies; the redefined Scalpellidae is consequently restricted. The Santonian to Eocene interval is dominated by the Eolepadidae. Newman's incorporation of the Stramentidae into the Eolepadidae downgrades the only extinction of a cirripede family in the late Cretaceous (cf. Sepkoski 1982*b*) to that of a subfamily Stramentinae (thus eliminating a taxonomic pseudoextinction). This subfamily disappeared before the end of the Cretaceous; no examples are known after the end of the Campanian. Generic extinctions are, however, greater during the Maastrichtian when four genera (24 per cent of the standing diversity) disappear, in a continuation of the reduction in diversity started with the loss of four genera (19 per cent) during the preceding stage (normalization yields figures of 1.9 per cent per million years for the Campanian, 3.0 per cent for the Maastrichtian, and 0.6 per cent for the Palaeocene as a whole). These levels can hardly be considered evidence of a mass extinction.

A pollicipedine scalpellomorph with a peduncle clad with heavy scales gave rise (?during the Jurassic) to the sessile Brachylepadomorpha, which in turn gave rise to the Verrucomorpha and Balanomorpha (Newman 1987). These sessile forms lack a peduncle and, as such, are much less susceptible to predators such as true crabs than the stalked barnacles. Today most scalpellids occupy deep water, where true crabs are rare. The evolutionary history of the Tertiary barnacles appears to be, at least in part, a reflection of the ease with which they could be prized off the substrate. The brachylepadomorphs are already in decline in the early Tertiary and became extinct in the Miocene; the verrucomorphs also decline from the Cretaceous and most species of the single extant genus are found in deep water; it is essentially the balanomorphs which have radiated into and dominate shallow water habitats today. [The Brachylepadomorpha was treated as a suborder of the Sessilia (as were the Verrucomorpha and Balanomorpha) by Newman (1987), not as a superfamily of the Balanomorpha as implied and so classified by Schram (1986, pp. 504–5).]

4. *Insects*

None of the terrestrial arthropods, with the exception of some diplopod myriapods, has a mineralized exoskeleton. Of the groups which traverse the Cretaceous–Tertiary boundary, the insects are by far the most diverse. Their fossil record is relatively impoverished, however. As Carpenter and Burnham (1985) emphasize, the insect exoskeleton is 'rapidly destroyed by a variety of micro-organisms, and insects must generally undergo immediate burial under anaerobic conditions for fossilization to take place'. In spite of this, the fossil insects, like their living counterparts, are more diverse than any other group of animals. No orders of insect become extinct after the early Jurassic (Harland *et al.* 1967). A detailed analysis of patterns of insect diversity must await publication of Carpenter's *Treatise on Invertebrate Paleontology* volumes. In the meantime we are grateful to Dr P. E. S. Whalley for information on Cretaceous–Tertiary insects. The major changes at family level occurred before the Early Cretaceous, so that while the Early Jurassic was dominated by families that are extinct, the vast majority of the Early Cretaceous families have living representatives. These include a range of dipterans (Chironomidae, Ceratopogonidae, Cecidomyiidae, Sciaridae, Tipulidae, Empididae, Psychodidae, Platypezidae, Phoridae, Bibionidae) and lepidopterans (Micropterigidae) many of which are highly temperature sensitive. Whalley (*pers. comm.*) argues that the survival of such taxa is incompatible with an extended period of cold and darkness, such as would follow a major asteroid impact. (Emiliani *et al.* 1981 have postulated an initial, transient rise in temperature which may have been responsible for causing extinctions.) Insects inhabiting the colder parts of the world would have been less susceptible than those in warmer latitudes. There is no evidence, however, of a major insect extinction at the end of the Cretaceous, nor of a period of rediversification/migration to lower latitudes in the Early Tertiary.

Detailed studies of the plant record are now being undertaken which will indicate more precisely the effects of the Cretaceous–Tertiary event on the terrestrial biota. There is clear evidence of a major vegetational trauma, spore-pollen ratios indicating a temporary attenuation of angiosperms and their replacement by ferns, which are very successful primary colonizers (e.g. Spicer *et al.* 1985). The extinctions appear to have been concentrated in mid-latitudes in the northern hemisphere, particularly affecting evergreens, with subsequent reintroduction mainly from further north. There is a marked selection amongst angiosperms for deciduousness and dormancy across the boundary and this is consistent with a brief low temperature dark event (Wolfe and Upchurch 1986). In spite of the trauma affecting vegetation during the boundary event, most plant lineages (including evergreens) passed through the boundary (Boulter *et al.*, this volume). Clearly, low preservation potential prevents the sampling of the insect record through boundary sections in the same manner as plants. However, like deciduous plants higher latitude insects presumably would have been resistant to seasonal extremes of climate and this would have minimized the effects of a low temperature event. Plant extinctions were apparently fewer in lower latitudes (<45 N) and in the southern hemisphere implying that the lower temperature event was limited in geographical extent (the pattern may include an element of sampling bias). The concentration of the effects in higher latitudes, where insects are likely to have been more resistant to temperature fluctuation, may explain why this group shows no evidence of a major extinction at the Cretaceous–Tertiary boundary.

Acknowledgements

We are very grateful to the following for very generously providing data for, and advice on particular parts of this paper: P. J. Lespérance, S. F. Morris, W. E. Newman, A. W. Owen, R. M. Owens, R. E. Plotnick, R. A. Spicer, P. E. S. Whalley, and C. W. Wright. Mrs E. Richards typed part of the text.

References

Alberti, H. (1979). Devonian trilobite biostratigraphy. *Spec. Pap. Palaeont.* **23,** 313–24.

Alvarez, L. W., Alvarez, W., Asaro, F., and Michel, H. V. (1980). Extraterrestrial cause for the Cretaceous–Tertiary extinction. *Science* (*Wash.*) **208,** 1095–108.

Alvarez, W., Kauffman, E. G., Surlyk, F., Alvarez, L. W., Asaro, F., and Michel, H. V. (1984). Impact theory of mass extinctions and the invertebrate fossil record. *Science* (*Wash.*) **223,** 1135–41.

Bassett, M. G. and Dean, W. T. (ed.) (1982). *The Cambrian–Ordovician Boundary: Sections, Fossil Distributions and Correlations.* The National Museum of Wales, Cardiff.

Benson, R. H., Chapman, R. E., and Deck, L. T. (1984). Paleoceanographic events and deep-sea ostracodes. *Science* (*Wash.*) **224,** 1334–6.

Bishop, G. A. (1983). Fossil decapod Crustacea from the Late Cretaceous Coon Creek Formation, Union County, Mississippi. *J. Crust. Biol.* **3,** 417–30.

Bishop, G. A. (1986). Taphonomy of North American decapods. *J. Crust. Biol.* **6,** 326–55.

Brenchley, P. J. and Newell, G. (1984). Late Ordovician environmental changes and their effect on faunas. In *Aspects of the Ordovician System* (ed. D. L. Bruton), pp. 65–80. Universitetsforlaget, Oslo.

Briggs, D. E. G. (1985). Gigantism in Palaeozoic arthropods. *Spec. Pap. Palaeont.* **33,** 157.

Bruton, D. L. (1983). Cambrian origins of the odontopleurid trilobites. *Palaeontology* **26,** 875–85.

Buckeridge, J. S. (1983). Fossil barnacles (Cirripedia: Thoracica) of New Zealand and Australia. *N.Z. Geol. Surv. Pal. Bull.* **50,** 1–151.

Carpenter, F. M. and Burnham, L. (1985). The geological record of insects. *Ann. Rev. Earth Planet. Sci.* **13,** 297–314.

Caster, K. E. and Kjellesvig-Waering, E. N. (1964). Upper Ordovician Eurypterids of Ohio. *Palaeont. Am.* **4,** 300–55.

Chen, Jun-yuan, *et al.* (1985). *Study on Cambrian–Ordovican Boundary Strata and its Biota in Dayangcha, Hunjiang, Jilin, China.* Prospect Publishing House, China.

Chugaeva, M. N. and Apollonov, M. (1982). The Cambrian–Ordovician boundary in the Batyrbaisae section, Malyi Karatau Range, Kazakhstan, USSR. In *The Cambrian–Ordovician boundary: Sections, Fossil Distributions and Correlations* (ed. M. G. Bassett and W. T. Dean), pp. 87–94. The National Museum of Wales, Cardiff.

Cocks, L. R. M. (1985). The Ordovician–Silurian boundary. *Episodes* **8,** 91–101.

Copper, P. (1977). Palaeolatitudes in the Devonian of Brazil and the Frasnian–Famennian mass extinction. *Palaeogeogr. Palaeoclimatol. Palaeoecol.* **21,** 165–207.

Emiliani, C., Kraus, E. B., and Shoemaker, E. M. (1981). Sudden death at the end of the Mesozoic. *Earth Planet. Sci. Lett.* **55,** 317–34.

Feist, R. (1970). Devonische Scutelluida aus der ostlichen Montagne Noire (Sudfrankreich). *Palaeontographica* A **147,** 70–114.

Feist, R. (1976). Systematique, phylogenie et biostratigraphie de quelques Tropidocoryphinae (Trilobita) du Devonien francais. *Geobios* **9,** 47–80.

Feist, R. and Klapper, G. (1985). Stratigraphy and conodonts in pelagic sequences across the Middle–Upper Devonian boundary, Montagne Noire, France. *Palaeontographica* A **188,** 1–18.

Feldmann, R. M. (1981). Paleobiogeography of North American lobsters and shrimps (Crustacea, Decapoda). *Geobios* **14,** 449–68.

Fortey, R. A. (1980). Generic longevity in Lower Ordovician trilobites; relation to environment. *Paleobiology* **6,** 24–31.

Fortey, R. A. (1983). Cambrian–Ordovician boundary trilobites from western Newfoundland and their phylogenetic significance. *Spec. Pap. Palaeont.* **30,** 179–211.

Fortey, R. A. (1984). Global earlier Ordovician transgressions and regressions and their palaeobiological significance. In *Aspects of the Ordovician System* (ed. D. L. Bruton), pp. 37–50. Universitetsforlaget, Oslo.

Glaessner, M. F. (1969). Decapoda. In *Treatise on Invertebrate Paleontology, Part R, Arthropoda* 4(2) (ed. R. C. Moore), pp. 400–566. Geological Society of America and University of Kansas Press, Boulder, Colorado, and Lawrence, Kansas.

Gould, S. J. and Calloway, C. B. (1980). Clams and brachiopods — ships that pass in the night. *Paleobiology* **6,** 383–396.

Guinot, D. (1978). Principes d'une classification évolutive des Crustacés decapodes Brachyoures. *Bull. Biol. de la France et de la Belgique* **112,** 211–92.

Harland, W. B., *et al.* (1967). *The Fossil Record.* Geological Society of London.

Harland, W. B., Cox, A. V., Llewellyn, P. G., Pickton, C. A. G., Smith, A. G., and Walters, R. (1982). *A Geologic Time Scale.* Cambridge University Press, Cambridge.

House, M. R. (1975). Faunas and time in the marine Devonian. *Proc. Yorks. geol. Soc.* **40,** 459–88.

House, M. R. (1985). Correlations of mid-Palaeozoic ammonoid evolutionary events with global sedimentary perturbations. *Nature, Lond.* **313,** 17–22.

Jaanusson, V. (1979). Ordovician. In *Treatise on Invertebrate Paleontology, Part A, Introduction* (ed. R. A. Robinson and C. Teichert), pp. 136–66. Geological Society of America and University of Kansas Press, Boulder, Colorado, and Lawrence, Kansas.

Jablonski, D. (1986). Causes and consequences of mass extinctions: a comparative approach. In *Dynamics of Extinction* (ed. D. K. Elliott), pp. 183–227. Wiley, Chichester.

Lespérance, P. J. (1987). Trilobites. In *Global Analysis of the Ordovician–Silurian Boundary, Bull. Br. Mus. nat. Hist.* (*Geol.*) (ed. L. R. M. Cocks and R. B. Rickards), in press.

Lu Yan-hao, *et al.* (1984a). *Stratigraphy and Paleontology of Systemic Boundaries in China; Cambrian–Ordovician boundary.* Anhui Science and Technology Publishing House, Anhui.

Lu Yan-hao, Lin Huan-ling, Han Nai-ren, Li Luo-zhao, and Ju, Tian-yin. (1984b). On the Cambrian–Ordovician boundary of the Jiangshanhangshan area, W. Zhejiang. In *Stratigraphy and Paleontology of Systemic Boundaries in China; Cambrian–Ordovician boundary* (ed. Lu Yan-hao *et al.*), pp. 9–44. Anhui Science and Technology Publishing House, Anhui.

Ludvigsen, R. (1982). Upper Cambrian and Lower Ordovician Trilobite biostratigraphy of the Rabbitkettle Formation, western District of Mackenzie. *Lifesci. Contrib. Roy. Ontario Mus.* **134,** 1–188.

Ludvigsen, R. and Westrop, S. R. (1983). Trilobite biofacies of the Cambrian–Ordovician boundary interval in northern North America. *Alcheringa* **7,** 301–19.

MacArthur, R. H. (1972). *Geographical Ecology: Patterns in the Distribution of Species.* Harper & Row, New York.

McGhee, G. (1982). The Frasnian–Famennian extinction event. A preliminary analysis of Appalachian marine ecosystems. *Geol. Soc. Am. Spec. Pap.* **190,** 491–500.

McLaren, D. J. (1970). Time, life and boundaries. *J. Paleont.* **44,** 801–15.

McLaren, D. J. (1982). Frasnian–Famennian extinctions. *Geol. Soc. Am. Spec. Pap.* **190,** 477–84.

Miller, J. B. (1986). Conodonts. In *Minutes of Calgary Plenary Session of Cambrian–Ordovican Boundary Working Group,* Fig. 1. Springfield, Missouri.

Newman, W. A. (1987). Evolution of cirripedes and their major groups. In *Barnacle Biology* (ed. A. Southward), *Crustacean Issues* **5,** Balkema, Rotterdam. (In press).

Newman, W. A. (1988). Cirripedia. In *Grassé's Traité de Zoologie.* Masson, Paris. (In press).

Newman, W. A., Zullo, V. A., and Withers, T. H. (1969). Cirripedia. In *Treatise on Invertebrate Paleontology, Part R, Arthropoda* 4(1) (ed. R. C. Moore) pp. 206–95. Geological Society of America and University of Kansas Press, Boulder, Colorado, and Lawrence, Kansas.

Owen, A. W. (1986). The uppermost Ordovician (Hirnantian) trilobites of Girvan, southeast Scotland with a review of coeval trilobite faunas. *Trans. Roy. Soc. Edin. Earth Sci.* **77,** 231–9.

Palmer, A. R. (1984). The biomere problem: evolution of an idea. *J. Paleont.* **58,** 599–611.

Patterson, C. and Smith, A. B. (1987). Periodicity of extinctions: a taxonomic artefact? *Nature, Lond.* (In press.)

Petryk, A. A. (1981). Stratigraphy, sedimentology and palaeogeography of the Upper Ordovician–Lower Silurian of Anticosti Island, Quebec. In *Subcommission on Silurian Stratigraphy, Ordovician–Silurian Boundary Working Group. Field Meeting, Anticosti-Gaspe, Quebec. 1981, Vol. 2: Stratigraphy and Paleontology* (ed. P. J. Lesperance), pp. 11–40.

Pitrat, C. W. (1973). Vertebrates and the Permo-Triassic extinction. *Palaeogeogr. Palaeoclimatol. Palaeoecol.* **14,** 249–64.

Plotnick, R. E. (1983). Patterns in the evolution of the eurypterids. PhD dissertation, University of Chicago (unpubl.).

Plotnick, R. E. (1987). Habitat of Llandoverian–Gedinnian Eurypterids. In *Final Report of Project Ecostratigraphy* (ed. A. Boucot). In press.

Plotnick, R. E. (1988). Eurypterida. In *Richardson's Guide to the Fossil Fauna of Mazon Creek* (ed. C. Shabica, S. Sroka, S. Kruty, and G. Baird) (In press).

Plotnick, R. E., Baumiller, T. and Wetmore, K. L. (1987). Fossilization potential of the mud crab, *Panopeus* (Brachyura; Xanthidae) and temporal variability in crustacean taphonomy. *Palaeogeogr. Palaeoclimatol. Palaeoecol.* (In press.)

Raup, D. M. (1986). Biological extinction in earth history. *Science (Wash.)* **231,** 1528–33.

Raup, D. M. and Sepkoski, J. (1982). Mass extinctions in the marine fossil record. *Science (Wash.)* **215,** 1501–3.

Raup, D. M. and Sepkoski, J. (1984). Periodicity of extinctions in the geologic past. *Proc. nat. Acad. Sci.* **81,** 801–5.

Rolfe, W. D. I. (1980). Early invertebrate terrestrial faunas. *Spec. Publ. Syst. Ass.* **15,** 117–57.

Rong Jia-yu (1984). Distribution of the Hirnantia fauna and its meaning. In *Aspects of the Ordovician System* (ed. D. L. Bruton), pp. 101–12. Universitetsforlaget, Oslo.

Schram, F. R. (1986). *Crustacea*. Oxford University Press, Oxford.

Schram, F. R., Feldmann, R. M., and Copeland, M. J. (1978). The Late Devonian Palaeopalaemonidae and the earliest decapod crustaceans. *J. Paleont.* **52,** 1375–87.

Sdzuy, K. (1955). Der fauna der Leimitz–Schiefer (Tremadoc). *Abh. senckenb. Naturforsch. Ges.* **492,** 1–74.

Selden, P. A. (1985). Eurypterid respiration. *Phil. Trans. Roy. Soc. Lond.* **B309,** 219–26.

Sepkoski, J. (1979). A kinetic model of Phanerozoic taxonomic diversity II. Early Phanerozoic families and multiple equilibria. *Paleobiology* **5,** 222–51.

Sepkoski, J. (1981). A factor analytic description of the Phanerozoic marine fossil record. *Paleobiology* **7,** 36–53.

Sepkoski, J. (1982*a*). Mass extinctions in the Phanerozoic oceans: a review. *Geol. Soc. Am. Spec. Pap.* **190,** 283–89.

Sepkoski, J. (1982*b*). A compendium of fossil marine families. *Milwaukee Public Mus. Contrib. Biol. Geol.* **51,** 1–125.

Sheehan, P. (1975). Brachiopod synecology in a time of crisis (Late Ordovician–Early Silurian). *Paleobiology* **1,** 205–12.

Simpson, G. G. (1953). *The Major Features of Evolution*. Columbia University Press, New York.

Spicer, R. A., Burnham, R. J., Grant, P., and Glicken, H. (1985). *Pityrogramma calomelanos*, the primary post eruption colonizer of Volcan Chichoual, Chiapas, Mexico. *Am. Fern J.* **53,** 1–5.

Stitt, J. H. (1975). Adaptive radiation, trilobite palaeoecology and extinction, Ptychaspid biomere, Late Cambrian of Oklahoma. *Fossils and Strata* **4,** 381–90.

Stitt, J. H. (1977). Late Cambrian and earliest Ordovician trilobites, Wichita Mountains Area, Oklahoma. *Bull. Oklahoma Geol. Surv.* **124,** 1–79.

Struve, W. (1976). Beitrage zur Kenntnis der Phacopina (Trilobita) 9. *Phacops* (*Omegops*) n. sg. (Trilobita, Ober-Devon). *Senck. leth.* **56,** 429–51.

Thomas, A. T., Owens, R. M., and Rushton, A. W. A. (1984). Trilobites in British Stratigraphy. *Spec. Rep. geol. Soc. Lond.* **16,** 1–78.

Thomson, K. S. (1977). The pattern of diversification among fishes. In *Patterns of Evolution as Illustrated by the Fossil Record* (ed. A. Hallam), pp. 377–404. Elsevier, Amsterdam.

Tomlinson, J. T. (1969). The burrowing barnacles (Cirripedia: Order Acrothoracica). *Bull. U.S. Nat. Mus.* **296,** 1–162.

Wolfe, J. A. and Upchurch, G. R., Jr. (1986). Vegetation, climatic and floral change at the Cretaceous–Tertiary boundary. *Nature, Lond.* **324,** 148–52.

Wright, C. W. and Collins, J. S. H. (1972). British Cretaceous crabs. *Palaeont. Soc. Monogr.* **126,** 1–114.

10. Anachronistic, heraldic, and echoic evolution: new patterns revealed by extinct planktonic hemichordates

R. B. RICKARDS

Department of Earth Sciences, University of Cambridge, Cambridge, UK

Abstract

Anagenetic grades and cladogenetic diversifications have been widely identified in graptolite studies. Moreover they have been tied to an accurate bio and chronostratigraphy, especially in strata of Silurian age (438–408 Ma). Anachronistic evolution is identified by distinct lineages demonstrably out of phase with this framework, but which have achieved similar phenotypic expression. Two kinds of anachronistic evolution have been recognized: *heraldic* evolution which precedes some major evolutionary grade or clade, yet heralds it in the sense that the (later) advanced phenotype is fully developed; and *echoic* evolution which post-dates a major evolutionary grade or clade, but achieves only partial phenotypic expression of the earlier successful features. Both heraldic and echoic evolution depend for their recognition upon a well defined stratigraphy and a good fossil record. Heraldic evolution is necessarily an unsuccessful line of evolution for, if successful, would have pre-empted the known, later record of the group. Echoic evolution is also relatively unsuccessful when compared with the main lineages, achieving no new anagenetic grades, but in contrast to heraldic lineages may involve extensive speciation over a period of a few million years, representing impressive cladogenetic change. The time-scale between heraldic and echoic events, and those of the major anagenetic grades which they mimic, provides a rough measure of both the environmental pressures causing the changes and of preadaptive genetic

Extinction and Survival in the Fossil Record (ed. G. P. Larwood), Systematics Association Special Volume No. 34, pp. 211–30. Clarendon Press, Oxford, 1988.

responses. The environmental trigger components are not necessarily immediately obvious, but the establishable duration of them must aid subsequent identification. Establishment of a timetable for preadaptation or predisposition suggests that in future a measure of mutation rates may be possible which is independent of those currently in use. It follows from a study of heraldic evolution, in particular, that 'advanced' characters cannot in themselves be used as either stratigraphic or evolutionary grade indicators.

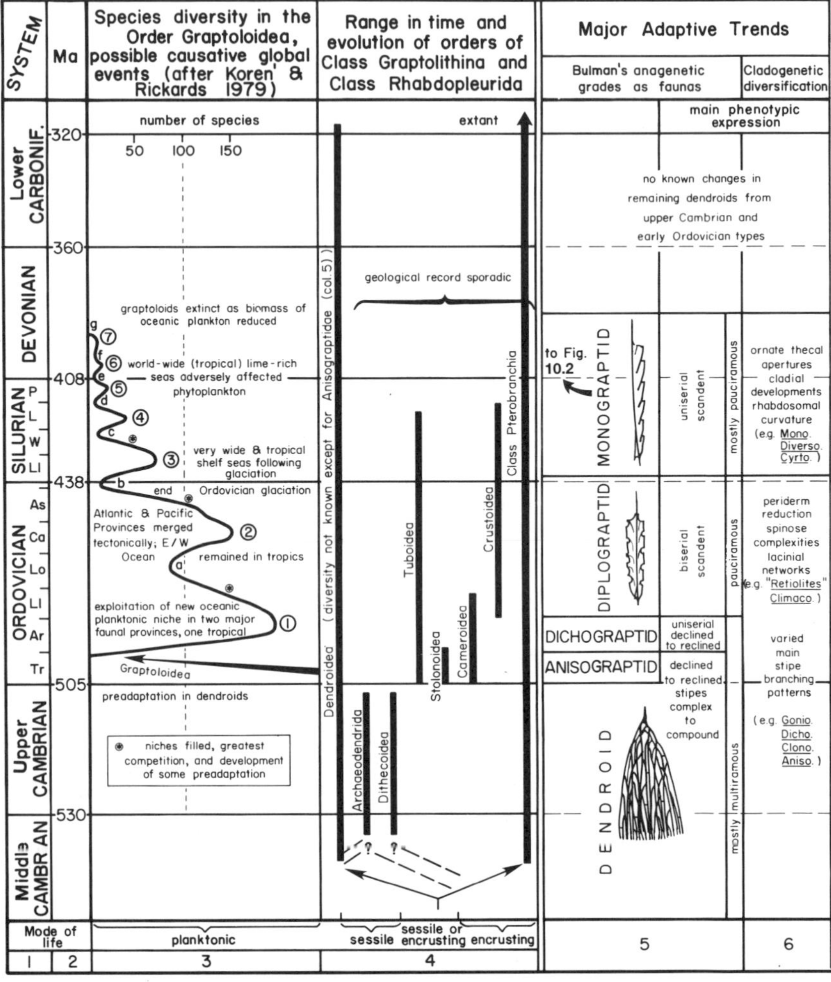

Fig. 10.1. Full explanation in text. Columns 3–10 illustrate the major patterns of graptolite evolution against the stratigraphic and chronostratigraphic scales (columns 1, 2, and 14). Columns 11–13 depict anachronistic evolution as defined in the text, also plotted against the time scales of columns 1, 2, and 4.

Introduction

The broad evolutionary history of graptolites has been documented by many workers since the middle of the last century, but especially by Kozlowski, Bulman, and Urbanek from the 1930s onwards. The crucial papers in this regard, together with much new information, are summarized and appraised by Rickards (1974, 1975, 1976, 1977, 1978, 1979*b*), Rickards *et al.* (1977) and Koren and Rickards (1980). Figure 10.1 in the

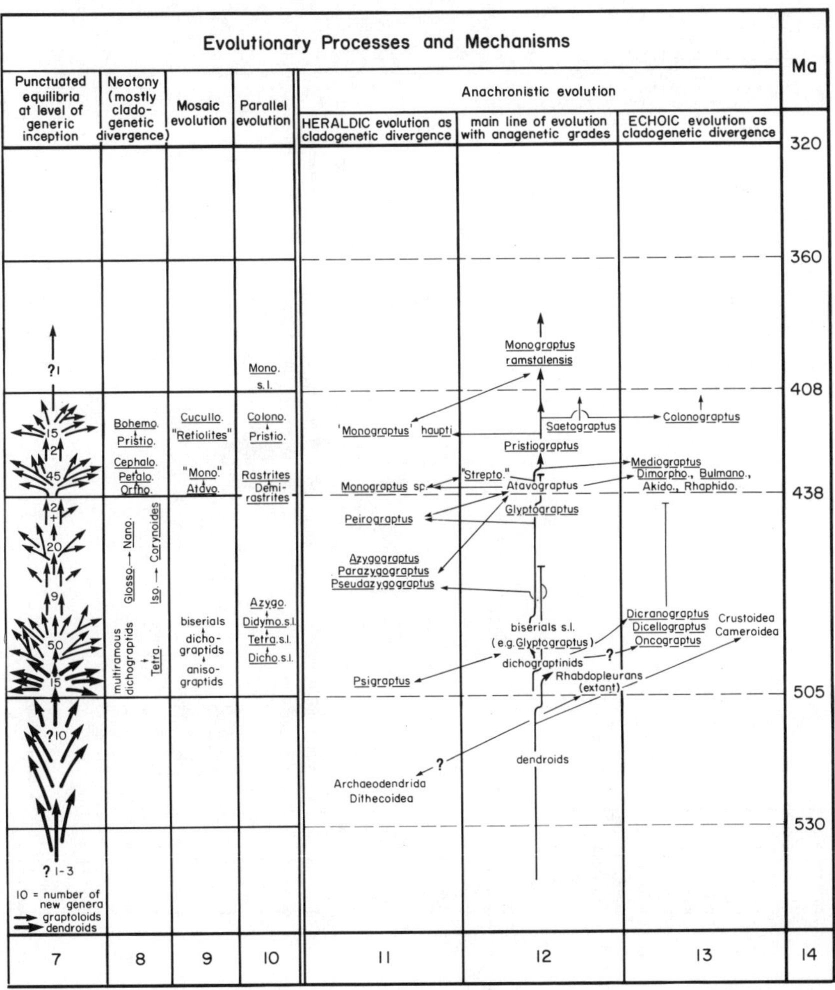

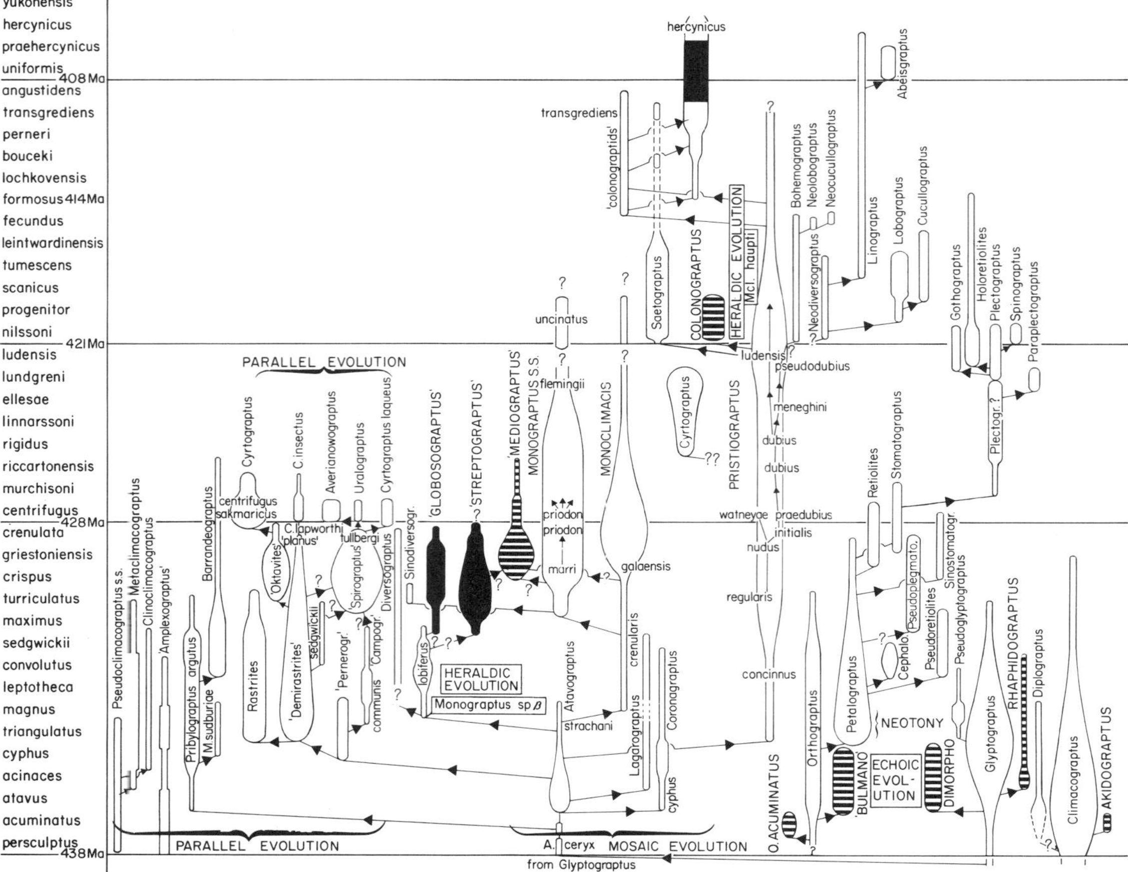

Fig. 10.2. Balloon diagram depicting basic evolution at generic level, modified after Rickards *et al.* (1977).

basic evolution (column 4) against time and likely evolutionary processes and mechanisms. Figure 10.2 is a diversity diagram showing the stratigraphic and absolute time ranges; species diversity; deduced main evolutionary links of the Siluro-Devonian planktonic genera; and evolutionary position of some Silurian heraldic and echoic events. Detail of evolutionary lineages *within* each balloon is given in Rickards *et al.* (1977) and those of Ludlow age in several papers by Urbanek (1958, 1963, 1966, 1970, 1973). Species diversity for benthonic dendroids and other orders is sketchily known, would almost certainly show monographic and preservational peaks and is therefore omitted from each diagram. In biostratigraphic terms, benthonic graptolites are little used and their evolution later than the Tremadoc poorly understood.

It is my opinion that the basic pattern of graptolite evolution at least as far as phenotypic expression is concerned is one of punctuated equilibria (but see sections under Heraldic evolution below). This is shown diagrmatically in Fig. 10.1 (column 7) where the sucesssive bursts of evolution are plotted at generic level against the biostratigraphic and chronological scale (column 1). Clearly, they correlate well with the species diversity (and cladogeny of Fig. 10.1, column 6) though not necessarily with anagenetic grades (column 5). The interpretation portrayed in the left side of Fig. 10.1 may be argued in terms of detail and of modern parlance, but it is not original in fact; many authors have described evolutionary bursts in graptolite evolution and Elles (1922) deduce a sequence of faunas and trends which later became Bulman's sequence of faunas, and anagenetic and cladogenetic stages (Bulman 1933, 1958). What has happened in modern times is that biostratigraphic studies have given greater accuracy to the pattern and some of the bursts of evolution have been related to global events (as in Rickards 1975; and Fig. 10.1, column 3 herein).

Anagensis

The earliest known graptolites are benthonic Dendroidea from the middle Cambrian (550 Ma). The origin of graptolites is far from certain but what evidence there is has been summarized by Rickards (1979*a*) where it was suggested that achievement of coloniality by infaunal phoronid worms may have resulted in the Hemichordata, including the rhabdopleurans, and the obscure orders Archaeodendrida and Dithecoidea. Despite these uncertain beginnings the graptolite record is thereafter well documented, the Cambrian dendroid genera persisting with apparently little change until the upper Carboniferous or Permian, a duration in excess of 200 million years.

In the Tremadoc (Tr, column 1 Fig. 10.1) the benthonic forms gave rise, via the intermediate family Anisograptidae, to planktonic genera. This first major change in graptolite evolution was accompanied by extremely rapid cosmopolitan distribution at specific level and morphological changes which can reasonably be related to the planktonic mode of life (Rickards 1975), such as modifications to assist buoyancy and others to retard sinking often in combination. Such changes are readily encompassed by Huxley's (1958) concept of anagenesis as involving grades of biological improvement. In this event a major ecological niche — the world's tropical oceans — was quite unoccupied by macrozooplankton, and the graptolites moved in as anisograptids and then as dichograptids, thereafter showing rapid species diversification (Fig. 10.1, column 3). Specific diversity presumably represents cladogenetic change, and adaptation to smaller niches (such as depth and palaeolatitude) within the major oceanic niche.

Subsequently two comparable anagenetic grades were achieved: a) the multibranched dichograptid changed in the Arenig to the small, compact biserial colony, and b) the biserial became, in the latest Ordovician, a long, thin, uniserial colony. In contrast to the planktonic event the biserial and uniserial events to not have obvious environmental triggers, although the morphological changes are as striking as that from benthos to plankton. Hence although one is tempted to dub them anagenetic grades, the nature of the 'biological improvement' is obscure, and is implied only by their strikingly successful record thereafter.

Cladogenesis

Only in the Order Graptoloidea is species diversity documented (Fig. 10.1, column 3). The planktonic anagenetic event referred to in section II was followed in the Arenig by considerable speciation recognized, as in all palaeospecies, by morphological characters. These are primarily of colony branching patterns, but also include a measure of thecal variation, especially of proportions, as well as skeletal periderm changes in some genera. The changes are best regarded as cladogenetic in that they are relatively small scale upon the basic planktonic mode which evolved in the Tremadoc; that is, more or less horizontally disposed stipes, uniserial and simpler in composition, lighter and thinner than their dendroid ancestors.

Following the Biserial Event there is only one major species diversity peak (Fig. 10.1, column 3, peak 2) typified by small-scale variation upon the biserial theme. Although the (anagenetic) reason for the appearance of biserial colonies is difficult to understand it is at least clear that they adopted a vertical disposition in the water column (Rickards 1975), and that the development of spinose and lacinial networks, 'float' structures, and meshwork skeletal periderms helped combat the natural tendency of the rather compact colonies to sink more quickly than their ramose ancestors. Those characters, especially in combination, are rather rare in both the Dichograptid and Monograptid faunas.

The Uniserial Event was followed by a later Silurian speciation (Fig. 10.1, column 3, peak 3) probably coinciding with an ameliorated post-glacial climate, the development of widespread, transgressive, shallow, tropical, ocean margins, and the general geographical enlargement of the tropical planktonic niche. The most spectacular feature of the cladogenetic divergence was the development of great variation in thecal type and to a lesser extent the evolution of varied thecal cladia. This particular period in the evolution of the graptolites is probably the best documented in terms of a precise biostratigraphy and is referred to again for that reason in the section Nature of the Geological Record.

Neoteny, mosaic and parallel evolution

Columns 5–10 of Fig. 1 portray what the writer regards as the basic evolutionary patterns and processes of graptolite evolution. Even at generic level a model of punctuated equilibria can be suggested, which is only reinforced when species lineages are considered (see Rickards *et al.* 1977). That those punctuations can be broadly correlated with either the major anagenetic events or with cladogenetic divergence, or both, is apparent from Fig. 10.1. Columns 8–10 (Fig. 10.1) give examples of the three evolutionary processes most widely accepted. It would seem that each is concerned with cladogeny is identified by specific diversity and is traceable through particular lineages. Neoteny in particular seems to give rise to relatively little speciation, which is some contrast to mosaic and parallel evolution. One of the most striking examples of mosaic evolution is that given by Rickards *et al.* (1977) with respect to the evolution of and from the genus *Atavograptus* (see also the lower central part of Fig. 10.2); whilst parallel evolution is strongly exhibited by the polyphyletic origin of *Rastrites* from '*Demirastrites*' (Rickards *et al.* 1977).

Nature of the geological record

It is certainly correct that parts of the geological record are poor and this is naturally true of vertebrate palaeontology where the number of bones are often counted in tens and hundreds, and the number of complete skeletons in proportionately smaller numbers. However, the invertebrate record is strikingly different, especially where it consists of fossils of planktonic or nectonic origin. Graptolites have one of the most impressive and 'complete' records, not of benthos, but of planktonic species which have an almost unquestionably cosmopolitan distribution and commonly occur in large numbers. Being planktonic they may occur in most sedimentary rock types and, hence, are invaluable in international correlation of Ordovician, Silurian, and to a lesser extent Devonian strata: the Silurian System can be subdivided into upwards of 50 internationally recognized and useful assemblage zones, giving an average duration for each unit of well under 1 Ma. A precision in stratigraphy has evolved, during the last 20 years, which is almost as good as that using cephalopods in the Mesozoic rocks.

I estimate that there are half a million graptolite specimens held by major museum depositories throughout the world. In other words, very large, examinable numbers of the major element of macro-zooplankton from the Palaeozoic seas. It is common for PhD students in the United Kingdom to establish accurately localized collections in excess of 20 000 specimens. Sizes of colonies vary from a few millimetres to 1 m 20 cm, whilst the stipes (or branches) have a diameter averaging 1.5 mm: in short, they are macro- not microfossils and they can readily be examined with standard binocular microscopic techniques. In terms of *preserved* numbers of specimens estimates have been made by Rickards (1979*b*) in terms of specimens per unit area of a single bedding plane in the black shale environment and in terms of specimens per unit volume. A figure of 5 billion per km^3 was suggested as an underestimate, for various reasons. For all practical considerations there is an almost unlimited supply of specimens either collectable or already available.

It is these factors which have resulted in the high degree of confidence in internal correlation of, particularly, the Silurian strata and their faunal contents. However, more important from my immediate standpoint is that the sequence of species, the detailed patterns of occurrence, are closely similar in different continents. Thus, not only can one easily correlate zonal assemblages between China and Western Europe, for example, but the order of appearance of particular species enables one to exclude certain (morphological) evolutionary options and to propose, with confidence, alternatives. A close examination of the distribution in time and space enabled Rickards *et al.* (1977) to postulate a framework of evolution, together with actual lineages, for almost the whole of the Silurian graptoloids (briefly summarized at generic level in Fig. 10.2 herein). It was during this work that I detected interesting phenotypic events which were demonstrably out of phase with a firmly established record. I then extended the study to the graptoloid record as a whole, again noticing out of phase evolutionary lineages of basically two kinds. These are examined in the following sections.

Anachronistic evolution

This I define as a line of evolution which is demonstrably out of phase, by at least several million years, with the main line of evolution with which it

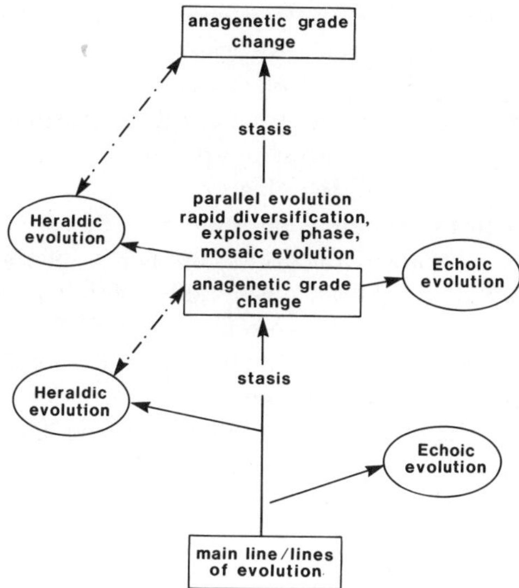

Fig. 10.3. Suggested relationship of heraldic and echoic evolution to main line evolution. Dotted arrows indicate the grade changes mimicked by heraldic evolution; other arrows the principle evolutionary relationships.

connects. Thus, it may be a lineage occurring usually early (Heraldic evolution, section below) or it may occur later than the main anagenetic grade yet achieve only partial phenotypic expression (section on Echoic evolution). Figure 10.3 summarizes the basic relationships envisaged.

Heraldic evolution

Heraldic evolution is defined as a lineage or lineages in which little speciation takes place, and in which the species are uncommon and short-lived, but where full phenotypic expression is seen of morphological features later to become typical of diverse species and genera, themselves evolving by similar morphological stages from the main line of evolution. Thus, an heraldic line has advanced morphological characters too early and in that sense, as well as in its short duration and lack of species diversity, is relatively unsuccessful. Such lineages clearly herald later successful morphological types, but the latter are not themselves necessarily on the main line of evolution for the group as a whole and may represent cases of strong cladogenetic divergence.

Figure 10.1 (columns 11 and 12) shows some examples of heraldic evolution in relation to time, and to the major steps of dendroid and graptoloid evolution. Other cases occur, but these examples are chosen for their distribution in time and their varied taxonomic levels. The earliest example, relating the two orders Archaeodendrida and Dithecoidea to the Dendroidea, is the only speculative case listed. It depends upon the accuracy of the geological record of the pterobranchs. If rhabdopleurans evolved in the Tremadoc then the two short-lived orders (Fig. 10.1, column 4) herald them at least in their gross morphology. If, as has been suggested elsewhere (Rickards 1979*a*), the dendroids and rhabdopleurans had a common ancestor in the middle Cambrian, then the two short-lived orders are best regarded as 'failed' dendroids or 'failed' pterobranchs (that is, cases of echoic evolution). Their considerable speciation suggests, perhaps, that this second interpretation is more likely.

At lower taxonomic levels than graptoloid orders the case for heraldic evolution is strong. Thus, the genus *Psigraptus* (of the intermediate dendroid/graptoloid family Anisograptidae) contains few species, and has an unusually advanced colony structure closely resembling reclined, sparsely branched dichograptids and tetragraptids. These appeared with enormous success in the Arenig as a major element in the Planktonic Event (Fig. 10.1, column 7). Some 7 Ma separate *Psigraptus* from the main line change to the suborders Didymograptina.

Later in the Ordovician the genera *Pseudazygograptus*, *Parazygograptus*, and *Azygraptus* herald the uniserial scandent stipe of *Atavograptus*. In fact, the three didymograptid genera fail to achieve full scandency, but are reduced to a single, uniserial stipe, as in *Atavograptus*; and in the genus *Pseudazygogratpus* that single stipe is strongly reclined. *Peiragraptus* achieved full scandency of a single, uniserial stipe based on theca 1^1 (having suppressed the dicalycal budding of theca 2^1), but was unable to rid itself of theca 1^2. Interestingly, this is the opposite situation to that of *Parazygograptus* where the uniserial stipe is based upon theca 1^2. *Peirograptus* possibly occurred about 5 Ma before *Atavograptus*.

The remarkable similarity of the Silurian forms *Monograptus* sp. B and lobed monograptids has been commented upon elsewhere (Rickards 1977). Figure 10.4 shows the highly advanced form *Monograptus* sp. B (see Hutt 1975; Rickards *et al.* 1977), and its relationship to the *lobiferus* line of evolution and the succeeding groups which have been given generic status (*Globosograptus*, *Stretograptus*) by some workers. *Monograptus* sp. B occurs approximately 5 Ma before the explosion of other similarly lobed forms and, as in the above cited cases, there can be no question of bridging the gap with other graptolites; the record is too complete, and the origin of the later forms reasonably clear (Rickards *et al.* 1977).

Another example is of the manner in which *Monoclimacis haupti* heralds the late Silurian and Devonian monograptid evolutionary bursts (see Rickards and Palmer 1976): *Monograptus ramstalensis* from the Devonian is strikingly similar to the low Ludlow species yet has a different evolutionary origin and occurs in a plexus of related spcies which is some 7–9 Ma later than *Monoclimacis haupti*. Similarly, *M. rickardsi* occurs some 5 Ma before the

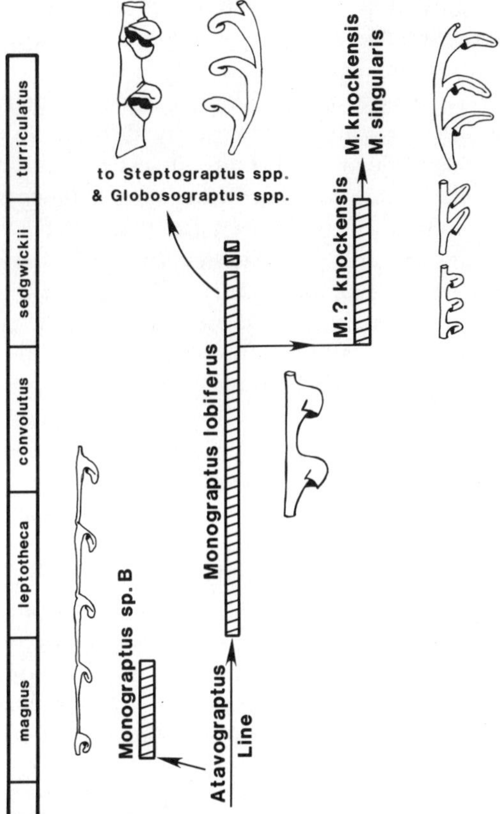

Fig. 10.4. A case of heraldic evolution in which *Monograptus* sp. B. (after Rickards *et al.* 1977) presages the later development of various lobed monograptid groups and genera. Full explanation in text.

successful *M. flemingii* plexus, but is virtually indistinguishable from some of them.

Heraldic evolution need not be confused with parallel evolution or with evolutionary convergence and homeomorphy. Parallel evolution is itself a form of convergence, where similar changes are taking place in different lineages at the same time, presumably as a result of the similar environmental pressures. Convergence and homeomorphy in their general senses need have no time connotation, nor close taxonomic relationship, but merely indicate a similar but recognizably distinct morphology, adapted to similar modes of life. Heraldic evolution by contrast necessitates a demonstrable time difference (seemingly of the order of 5–10 Ma in graptolites) and a close relationship to the major evolutionary lineage. Without the proven time difference the phenotypic expression of the genotypes is indistinguishable, but a difference in time of occurrence can be proved. Furthermore, heraldic evolution represents an attempt to bridge a particular anagenetic grade later achieved.

Iterative evolution seems to mean different things to different workers, but is in essence the repeated appearance through time of similar morphological types as offshoots from more or less related lineages. There is no indication of relationship to a particular anagenetic grade, and many cases of iterative evolution are themselves fairly successful yet in no way herald major (genetic) changes (e.g. Stanley 1979, pp. 156–7). The only similarity between iterative evolution in general, and heraldic evolution, is that labile regulatory systems must be present and that for periods of time these must remain phenotypically unexpressed. However, this similarity is shared by other evolutionary mechanisms, such as punctuated equilibria for example.

The concept of bridging an anagenetic grade has been discussed very thoroughly by Jaanusson (1981) in his account of functional thresholds. He argues that morphological changes are for functional reasons, even though such functions may not be fully appreciated, especially in the graptolites (but see section on Anagenesis above). He considers that some form of adjustment is made prior to the threshold leap, reflected phenotypically in gradual morphological changes termed by him *proximations*. The genetic build-up to such thresholds is what some term preadaptation and others term predisposition. In any event heraldic evolution shows that the necessary genetic changes may occur at least 5–10 Ma before the threshold event.

Echoic evolution

Echoic evolution is defined as a lineage or lineages, sometimes with considerable species diversity, but in which only partial phenotypic expression is given to morphological characters dramatically and successfully achieved at an earlier (anagenetic) grade; each evolves by similar morphological steps either on or from the major evolutionary lineages.

Figure 10.1 (column 13) lists some examples of echoic evolution, chosen for their stratigraphic and taxonomic range. The earliest case and highest taxonomic level exemplified, is that the cameroids and crustoids (Fig. 10.1, column 4) can be regarded as failed rhabdopleurans, having evolved from dendroid graptolites by the development of strong rhabdopleuran tendencies. However, this stratigraphically earliest example is debatable, not only in the implied stratigraphy, but in the evolutionary interpretation: an equal possibility is that the crustoids and cameroids are similar to rhabdopleurans only because they are unusual graptolites in that they adopted an encrusting mode of life. On this interpretation they would be homeomorphic.

At higher stratigraphic levels, and with more closely related taxonomic units, the case for echoic evolution is clear. The genera *Dicranograptus, Dicellograptus,* and possibly *Oncograptus* are 'failed' biserials. The biserial colony appeared in the Arenig with the genus *Glyptograptus* from a presumed but as yet unknown dichograptid *s.l.* ancestor. Glyptograptids formed one of the major lines of evolution through the remainder of the Ordovician giving rise to the next major anagenetic grade (*Atavograptus*) in the very highest Ordovician strata. *Dicranograptus* and *Dicellograptus* range, respectively, into the Caradoc and Ashgill Series, but do not reach the Silurian. *Neodicellograptus,* yet another example of echoic evolution, occurs in the middle Silurian. They also show considerable species diversity and not a little endemism. The biserial protion of *Dicranograptus* may reach several centimetres, but is commonly only a few millimetres; whilst a biserial portion is unusual in described dicellograptids, although the sicula is quite frequently largely embedded in the dorsal wall of one stipe. *Oncograptus* has more obvious dichograptid affinities and at least superficially resembles *Isograptus* stipes placed back to back. The distal uniserial parts are typically dichograptid in form. *Glyptograptus* and *Oncograptus* are not widely separated in time, but *Dicellograptus* and *Dicranograptus* appear approximately 10 Ma later, on current evidence.

Perhaps the most striking case of echoic evolution is the later derivation of uni-biserial stipes represented by the genera *Dimorphograptus, Bulmanograptus, Akidograptus, Parakidograptus,* and *Rhaphidograptus* (Fig. 10.1, column 13, and Figs. 10.2 and 10.5). These appear some 1–3 Ma after the dramatic change from the biserial *Glyptograptus* ex gr. *persculptus* to the uniserial *Atavograptus ceryx,* perhaps involving dithyrial populations in the *persculptus* assemblage zone (see Rickards and Hutt 1970). The uniserial portion varies from one theca to numerous thecae, depending upon the species. These genera and species are truly a pale echo of *Atavograptus* and *Glyptograptus* (Fig. 10.5). The changes cannot be regarded as examples of parallel evolution because the uniserial scandent condition is not achieved in these genera — they fail quite spectacularly in this respect. However, the general aquisition of a diminutive, slim, and thorn-like proximal end *is* an example of parallel evolution which affected a great many lineages at this time, whether uniserial, biserial, or uni-biserial (Rickards *et al.* 1977). Nor are the unibiserial forms involved in the striking mosaic evolutionary pattern displayed by the evolving monograptids, for they represent, by way of contrast, short-lived speciation events without descendants (Fig. 10.2).

At a slightly higher stratigraphic level the genus *Mediograptus* resembles a 'failed' *Monograptus s.s.* or, possibly, a 'failed' lobed monograptid. In either case it follows some 2 Ma after these genera and ranges from the late Llandovery Series into the low Wenlock Series (Fig. 10.2).

The final example concerns the origin of *Saetograptus* and *Colonograptus*

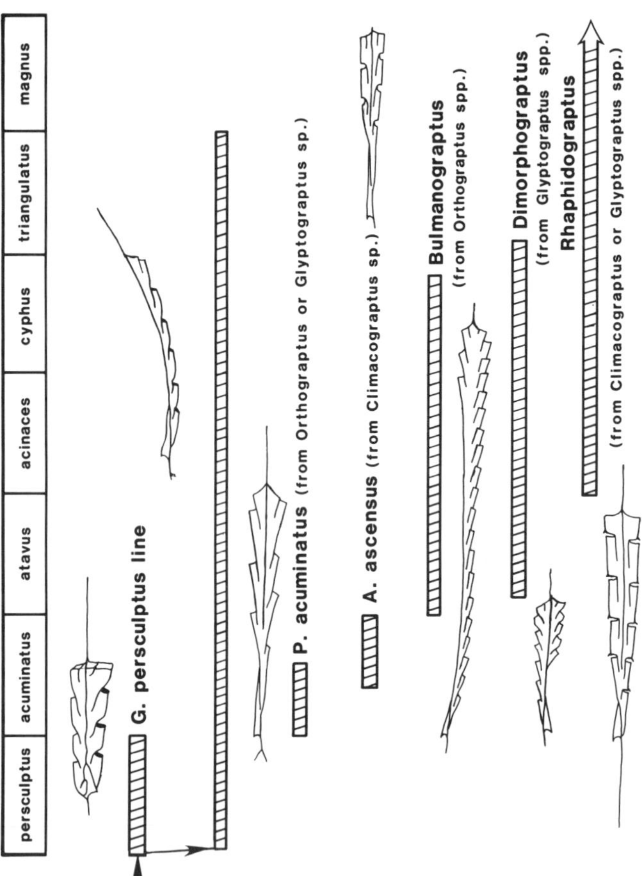

Fig. 10.5. A case of echoic evolution in which various unibiserial genera partially echo the uniserial forms derived from the *Glyptograptus persculptus* group. Full explanation in text.

from *Pristiograptus* in the low Ludlow. In the late Wenlock there is a probable ancestor to the earliest saetograptid (*S. varians*) in *P*? *ludensis*. In *S. varians* the first 1–3 thecal apertures are drawn out into long, enrolled spines (Fig. 10.6). *P*? *ludensis* has paired thecal lappets on the first theca but neither spines nor enrolling (Fig. 10.6). *Colonograptus* follows close upon *Saetograptus*, possibly as much as 1 or 2 Ma later, but never develops past the paried lappet stage. Early colonograptids usually have 2–4 such thecae, but later forms commonly have fewer. Species diversity in both *Saetograptus* and *Colonograptus* is considerable, but the latter is essentially a 'failed' saetograptid.

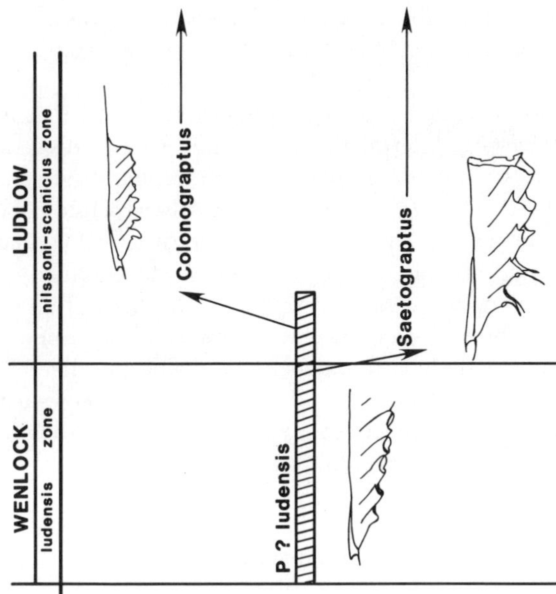

Fig. 10.6. A Ludlow case of echoic evolution in which the paired lappets of *Colonograptus* form a later, partial development of the enrolled, spinose thecal lappets of *Saetograptus*. Full explanation in text.

Possible mechanisms

As I have indicated in the section on Anagenesis cause and effect in graptoloid evolution are not readily understood and at present are best stated in general terms. It is true that some obvious global events, such as the end-Ordovician glaciation, coincided with dramatic changes in the graptolite skeleton, species distribution, and species diversity, and it is tempting to equate them as I have suggested in Fig. 10.1 (column 3). Although such events as the biserial/uniserial change are spectacular, they are only seen as sudden changes in specific lineages, if one ignores heraldic and echoic evolution. It would seem that the event was being built towards, or predisposed in Huxley's (1942) term for up to 5–10 Ma or so, and the origin of related echoic lineages span a further 1–10 Ma approximately. At an heraldic stage genetic preadaptation and environmental niche were, to put it quite crudely, not ready for each other. Either the genotype was incompletely preadapted (Bock 1959) for an available niche, or the niche itself was developing. In any case, it would be difficult to estimate the degree of preadaptation until the niche and its development was identified and defined. However, as a rough guide genetic preadaptation *could* have been established, waiting for a trigger mechanism as it were, or the chance coincidence of preadaptation and suitable niche, for at least 10 Ma in some cases.

Echoic evolution is rather more easily demonstrated in that it is typified by striking *morphological* intermediates and high species diversity. Yet the echoic event occurs after its 'morphological descendant' (see Fig. 10.5, for example) by a rather shorter period of time than heraldic evolutionary events precede the latter, namely 1–10 Ma. It is possible that once the environmental niche became occupied by the product of major anagenetic change (denroid-plankton, dichograptid-biserial, biserial-uniserial) then the niche itself perforce changed rapidly. Because it has a new, successful, biological component the environment is, by definition, different; and any similar preadaptation in other biserial lineages would necessarily result in partial success at best, hence the partial phenotypic expression and short-lived occurrence in the record.

Conclusions

Genetic predaptation before a major anagenetic change may be established for a long period of time, resulting rarely in heraldic evolution of surprisingly 'advanced' types. This process usually takes place essentially in the period of morphological stasis of the main line following a previous period of explosive evolution (Fig. 10.3). Echoic evolution on the other hand usually follows upon major anagenetic changes, and their associated explosive evolution, and is hence part of them. There is an initial period of divergent mosaic evolution during explosive phases (Fig. 10.2, bottom centre, around *Atavograptus*) after which parallel evolution, and relatively minor neotonous lines, are more obvious. Heraldic and echoic evolution appear to be more or less distinct from those processes.

Morphologically advanced types should not be taken as evidence of age in stratigraphic schemes presaging evolutionary studies, without independent evidence of the stratigraphic sequence, and independent dating of the advanced type itself. As I have shown, relatively uncommon advanced forms do occur. Similarly, morphological intermediates should be treated with suspicion if they cannot be accurately dated: they often succeed a major and relatively sudden change of grade. None the less, careful examination of the grade changes themselves may reveal morphological intermediates over a very short time-scale ($<\frac{1}{2}$ Ma) especially in graptolite lineages.

I consider that both heraldic and echoic evolution can be distinguished readily from convergent evolution. The last named is a general case lacking precision or location in closely related lineages. Parallel evolution, an extreme form of convergence applying to very similar lineages during the same period of time, *may* be a component in echoic evolution (as in the case of *Dimorphograptus,* etc.; Figs. 10.2 and 10.5), but it is more typical of periods of stasis with no recognizable echoic component. This is not to claim that there isn't a complex relationship between mosaic, parallel, heraldic, and echoic evolution: such is apparent even on the broad scenario of Fig. 10.2.

The time scale of a three-part relationship (heraldic-anagenetic grade-echoic) may be as much as 20 Ma. Longer periods are possible if the concepts truly can be applied at higher taxonomic levels, though at present I have my doubts about this (sections on Heraldic evolution and Echoic evolution above). Whatever the possible time span in excess of 20 M it seems already too long a period to admit the likelihood of catastrophic events affecting the environment, such as would be provided by meteoritic impact or close passage of a meteor shower. Heraldic and echoic evolutionary patterns are intimately linked both with the punctuated equilibrium model in general, and with 'mass extinctions' in particular, so that even the rough measure they provide of genetic and environmental change is significant. The extinction of the planktonic graptolids (Fig. 10.1, column 3) is not one of mass extinction in any strict sense, but rather follows a series of diversive peaks each one smaller than the last. The peaks and troughs are spaced as shown in Table 10.1. What seems to be indicated is a gradual decline from the initial period of occupation of the new, vast, oceanic niche where they became the Earth's first macrozooplankton. The decline took something of the order of 100 Ma, but was punctuated by a series of evolutionary explosions followed not only by periods of evolutionary stasis, but possibly by adverse environmental factors such as the end-Ordovician glaciation. In the late Silurian and Devonian, as species diversity (and geographical distribution) declined, the adverse environmental processes need not necessarily have been as spectacular in themselves as earlier events. The graptoloid macrozooplankton was then more vulnerable and the impact upon it of less than global factors is reflected in a series of closely spaced troughs. The periodicity of heraldic and echoic evolution seems only roughly related to the periodicity shown in Table 10.1, although as indicated above both processes are related to the explosive evolution model.

The concept of evolutionary stasis punctuated by shorter periods of explosive evolution [the punctuated equilibria model; or Jaanusson's (1981) proximations and functional thresholds] perhaps need further comment, for it could be argued that long periods of genetic preadaptation, proved by heraldic evolution, argues against the punctuated equilibrium model. There can be no doubt about the nature of the short explosive evolutionary periods, nor that they were preceded by a *relatively* quiescent period of longer duration with little phenotypic change, in which many lineages gradually became extinct. However, to claim that nothing happens

Table 10.1 Timing of graptoloid speciation events.

Time (Ma) between troughs	Approximate age (Ma)	Sequence of peaks (nos) and troughs (letters) as on Fig. 10.1	Approximate age (Ma)	Time (Ma) between peaks	Series or system
	387	g			Earliest Middle Devonian
9		7	391		Late Lower Devonian
	396	f		13	Middle Lower Devonian
13		6	404		Low Devonian
	409	e		7	Late Pridoli
6		5	411		Pridoli
	415	d		7	Late Ludlow
7		4	418		Ludlow
	422	c		16	Wenlock
18		3	434		Llandovery
	440	b		20	Ashgill
23		2	454		Caradoc
	463	a		30	Llandeilo
		1	483		Arenig

during the period of 'stasis' is rather like saying that nothing happens in a volcano until it explodes. In fact, genetic preadaptation or predisposition (resulting in proximations before the threshold event) was building up for long periods — heraldic evolution, unusual though it is, not only proves this, but roughly measures it. Although phenotypic expression was unspectacular during 'stasis', evolution at the molecular and genetic levels probably occurred at considerable rates: indeed, if the neutral theory of evolution is correct molecular evolution continued at high and constant rates. Nevertheless, it is phenotypic expression with which the palaeontologist usually deals and in this sense the punctuated equilibrium model is a useful approximation to what is seen in the rocks.

Acknowledgements

I am grateful for help from a number of workers but especially from Professors Charles Holland, John Thoday, and Harry Whittington for their very constructive suggestions.

References

Bock, W. J. (1959). Preadaptation and multiple evolutionary pathways. *Evolution* **13,** 194–211.

Bulman, O. M. B. (1933). Programme evolution in the graptolites. *Biol. Rev.* **8,** 311–34.

Bulman, O. M. B. (1958). The sequence of graptolite faunas. *Palaeontology* **1,** 159–73.

Elles, G. L. (1922). The graptolitic faunas of the British Isles. *Proc. geol. Ass.* **33,** 168–200.

Hutt, J. E. (1975). The Llandovery graptolites of the English Lake District. Pt. 2. *Palaeontogr. Soc.* (*Monogr.*), **129,** 57–137.

Huxley, J. S. (1942). *Evolution: The Modern Synthesis.* George Allen & Unwin, London.

Huxley, J. S. (1958). Evolutionary processes and taxonomy. *Upps. Univ. Arsskr.* **6,** 21–39.

Jaanusson, V. (1981). Functional thresholds in evolutionary progress. *Lethaia* **14,** 251–60.

Koren, T. N. and Rickards, R. B. (1980). Extinction of the graptolites. In *The Caledonides of the British Isles — reviewed* (eds. A. L. Harris *et al.*), pp. 457–66. [*Spec. Publ. Geol. Soc. Lond.* No. 8]. Scottish Academic Press. Geological Society, London.

Rickards, R. B. (1974). A new monograptid genus and the origin of the main monograptid genera. In *Graptolite Studies in Honour of O. M. B. Bulman. Special Paper in Palaeontology* No. 13, (ed. R. B. Rickards *et al.*), pp. 141–47.

Rickards, R. B. (1975). Palaeoecology of the graptolithina, an extinct class of the plylum Hermichordata. *Biol. Rev.* **50,** 397–436.

Rickards, R. B. (1976). The sequence of Silurian graptolite zones. *Geol. J.* **11,** 153–88.

Rickards, R. B. (1977). Patterns of evolution in the graptolites. In Hallam, A. (Ed.). *Patterns of Evolution as Illustrated by the Fossil Record* (ed. A. Hallam), pp. 333–58. Elsevier, Amsterdam.

Rickards, R. B. (1978). Major aspects of evolution of the graptolites. *Acta palaentol. polon.* **23,** 585–94.

Rickards, R. B. (1979*a*). Early evolution of graptolites and related groups. In *The Origin of Major Invertebrate Groups* (ed. M. R. House), Systematic Association Special Volume No. 12, pp. 435–41. Academic Press, London.

Rickards, R. B. (1979*b*). Graptolithina. In *Encyclopedia of Paleontology* (ed. R. W. Fairbirdge and D. Jablonksi), pp. 351–9. Hutchinson & Ross, Dowden.

Rickards, R. B. and Hutt, J. E. (1970). The earliest monograptid. *Proc. geol. Soc. Lond.* **1665,** 115–19.

Rickards, R. B. and Palmer, D. C. (1976). Early Ludlow monograptids with Devonian morphological affinities. *Lethaia* **10,** 59–70.

Rickards, R. B., Hutt, J. E., and Berry, W. B. N. (1977). The evolution of the Silurian and Devonian graptoloids. *Bull. Br. Mus. nat. Hist.* (*Geol.*), **28,** 1–120.

Stanley, S. M. (1979). *Macroevolution: pattern and process.* W. H. Freeman, San Francisco, 1–332.

Urbanek, A. (1958). Monograptidae from erratic boulders of Poland. *Palaentol. polon.* **9,** 1–105.

Urbanek, A. (1963). On generation and regeneration of cladia in some Upper Silurian monograptids. *Acta palaentol. polon.* **8,** 135–254.

Urbanek, A. (1966). On the morphology and evolution of the Cucullograptinae (Monograptidae, Graptolithina). *Acta palaentol. polon.* **11,** 291–544.

Urbanek, A. (1970). Neocucullograptinae n. subfam. (Graptolithina) their evolutionary and stratigraphic bdaring. *Acta palaentol. polon.* **15,** 163–388.

Urbanek, A. (1973). Organization and evolution of graptolite colonies. In *Animal Colonies* (eds. R. S. Boardman *et al.*), pp. 441–514. Hutchinson & Ross, Dowden.

11. Extinction and survival in the Conodonta

R. J. ALDRIDGE

Department of Geology, University of Nottingham, Nottingham, UK

Abstract

The euconodonts, or true conodonts, have a stratigraphic range from the Upper Cambrian to the end of the Triassic. The standing diversity of genera was considerably higher in the post-Tremadoc Ordovician than at any subsequent time in their history, although species diversity occasionally reached comparable levels in the Devonian. The variety of structural styles displayed by the mineralized feeding apparatus of conodonts was at its greatest in the Arenig, and relatively few types survived beyond the Ordovician. Intervals of high generic origination rates occurred in the Arenig, the Llandeilo, the Ashgill, the late Llandovery, the Famennian, and the Scythian. Extinction maxima were at the end of the Arenig, the mid-Caradoc, the end of the Ashgill, the early Wenlock, the end of the Famennian, the mid-Carboniferous, and the early Permian. Some of these events concur with those documented for other groups in the fossil record, but there is not a consistent correspondence. There is no evidence of an overall periodicity in conodont extinctions, but cyclicity is apparent at different scales in different parts of the stratigraphic column. The final extinction in the Norian appears to have been the culmination of a gradual decline, rather than a particularly abrupt event. There is geochemical evidence that evolutionary patterns of conodonts were in part controlled by fluctuations in oceanic water temperature, and of a correlation between extinctions and anoxic events at the end of the Devonian and Permian periods.

Extinction and Survival in the Fossil Record (ed. G. P. Larwood), Systematics Association Special Volume No. 34, pp. 231–56. Clarendon Press, Oxford, 1988.

The data base

Conodont animals were entirely soft-bodied except for the mineralized elements of their feeding apparatuses, by which they are well represented in the fossil record. Although these microscopic phosphatic elements normally became scattered in the sediment following death and decay of the animals, their preservation potential is very high and they may be recovered from a wide variety of lithologies by disaggregation or dissolution techniques. Evolutionary studies of conodonts have, of necessity, been based on morphological changes shown by these elements and on the changing structures of the multi-element apparatuses of which they were components. It may well be that evolutionary developments are faithfully represented by the elements, though the record of specimens with soft tissues preserved is never likely to be good enough to test this hypothesis fully. However, the few known specimens of animals displaying both soft and hard tissues do suggest that the body plan was relatively conservative, with similar animals containing mineralized apparatuses that indicate assignment to separate genera (Aldridge *et al.* 1986).

The affinities of the conodonts have long been enigmatic, but the recent discoveries of fossils of complete animals have led to the suggestion that they belong to a separate group of jawless craniates, with closest relationship to the Myxinoidea (Aldridge *et al.* 1986). The feeding apparatus of each individual comprised several elements arranged in a bilaterally symmetrical array in which the two sides presumably operated in lateral opposition. The simplest apparatuses consisted of coniform elements only, and probably served mainly as grasping structures (Smith *et al.* 1987), although some were provided with sharp edges and costae which could have been used to cut and slice captured food (Aldridge and Jeppsson 1984). More complex apparatuses display a differentiation of the elements into an anterior ramiform group, probably used to grasp prey, and a posterior set of pectiniform pairs that appear designed to process the food by shearing and/or grinding (Aldridge and Briggs 1986; Aldridge *et al.* 1987; Briggs *et al.* 1983). The elements of the true conodonts, or euconodonts, appear in the fossil record in the Upper Cambrian and the last specimens are found at the top of the Triassic.

Despite the excellent fossil record of conodont elements there are considerable difficulties in constructing a data base for the investigation of evolutionary patterns in the group. In common with other palaeontological data, problems arise from differences in taxonomic procedure from country to country, from stratigraphic level to stratigraphic level, and from taxonomist to taxonomist. There are also uncertainties in international stratigraphic correlation, especially in systems such as the Ordovician, where provincial differences have led to the development of endemic faunas. Imprecisions are also introduced where stratigraphic ranges have been only coarsely recorded, perhaps to stage level, although in much of the recent conodont literature distributions have been reported with significantly finer accuracy.

There is an additional difficulty in assembling information on ranges that particularly affects groups like the conodonts whose record largely comprises scattered skeletal parts. To establish species or genus distributions with confidence, the multi-element composition of the complete apparatus must be known; otherwise, component elements may be referred to different taxa and recorded separately. The complete apparatuses of a few, but only a few, species are known from natural assemblages on bedding planes (see e.g. Schmidt 1934; Rhodes 1952; Aldridge *et al.* 1987) and some partial apparatuses are found as fused clusters of elements (see e.g. Pollock 1969; Nicoll 1982, 1985). However, the vast majority of apparatuses have to be reconstructed from morphological and distributional evidence provided by the scattered elements themselves and, although considerable progress in this direction has been made in recent years, the full composition of many taxa is still uncertain. The problem may be circumvented to a degree in apparatuses with conservative ramiform elements and distinctive, characteristic, and rapidly-evolving pectiniform elements, such as are common in the Upper Palaeozoic. In the earlier part of the conodont record, though, a wide variety of apparatus styles was developed and poorly known or rare elements currently of uncertain status may well truly represent separate taxa.

A thorough analysis of conodont evolution and extinction patterns would require compilation of a data base of stratigraphic ranges at the specific level. Although specialists have produced adequate information to do this for some parts of the stratigraphic column, it is currently wiser in an overview to deal with the distributions shown by genera. A tabulation of generic ranges was given by Keim (in Clark *et al.* 1981), but this included only genera reported before 1980 and recorded their stratigraphic ranges only to the series level. For this paper, I have endeavoured to construct a more refined data base of generic distributions (Appendix), adding genera defined since 1979 and determining their stratigraphic ranges to at least stage level and mostly to zonal level. I have largely been conservative in my inclusion of genera, omitting those described only from single ramiform elements and those reported in open nomenclature; this conservatism will have decreased the apparent generic diversity of conodonts through their record, probably with the greatest effects in the Lower–Middle Ordovician. There are undoubtedly other omissions, errors, and inaccuracies in this data base, but it should serve as a basis for future refinement. One common problem encountered in its compilation was that first appearances of genera are often easier to abstract from the literature than final occurrences; this is particularly true for long-ranging genera that have minimal biostratigraphic importance.

Another factor exerting considerable influence on the analysis of evolutionary/extinction events is the choice of absolute time-scale to which the distributional data is related. There are marked differences in the various time-scales determined for the Palaeozoic and Triassic (see e.g. Harland *et al.* 1982; Forster and Warrington 1985; Gale 1985; McKerrow *et al.* 1985; Odin 1982, 1985*a*,*b*), so the interim time-scale proposed by Snelling (1985) from a consideration of all the available data is employed here.

The record

1. *General considerations*

The standing diversity of euconodont genera through the stratigraphic range of the group is plotted in Fig. 11.1a, and rates of origination and extinction are shown in Fig. 11.1b,c. Clark (1983) noted that rates of origination were highest in the Ordovician and in the Devonian to early Carboniferous, a pattern which is also evident from these plots. Clark's data, however, were compiled on a period-by-period basis, and the more refined picture given here shows that the Ordovician maxima were separated by episodes of low origination rates and that the later peak was concentrated on rapid origination events in the Famennian and earliest Dinantian. Indeed, there are two intervals within the Ordovician when extinction rates significantly exceeded origination rates (Fig. 11.1d), prior to the major extinction at the end of the period. It is, however, clear that generic diversity in the post-Tremadoc Ordovician was constantly higher than at any subsequent time in conodont history. The number of genera present at any one time in the Silurian to Triassic interval fluctuates around 10, with significant peaks above this in the late Llandovery to earliest Wenlock and in the Famennian. Significant low points in the generic diversity curve are apparent in the early Permian, across the Permo-Triassic boundary and at the final demise of the Conodonta at the end of the Triassic.

It is instructive to examine how the broad patterns shown by the generic diversity curve compare with data that are available for species diversity (Fig. 11.2). The three generic diversity peaks in the Ordovician are matched by specific diversity peaks in the data plotted by Sweet (1985), but it is evident that at a specific level the Arenig peak is considerably enhanced relative to the other two. Sweet (1985, p. 488) considered that this level of diversity was not to be rivalled in subsequent conodont history, yet Ziegler and Lane (1987) recorded a very similar number of species in the Famennian. This may in part reflect differences in taxonomic approach, but is also the result of wide radiation in particular genera, such as

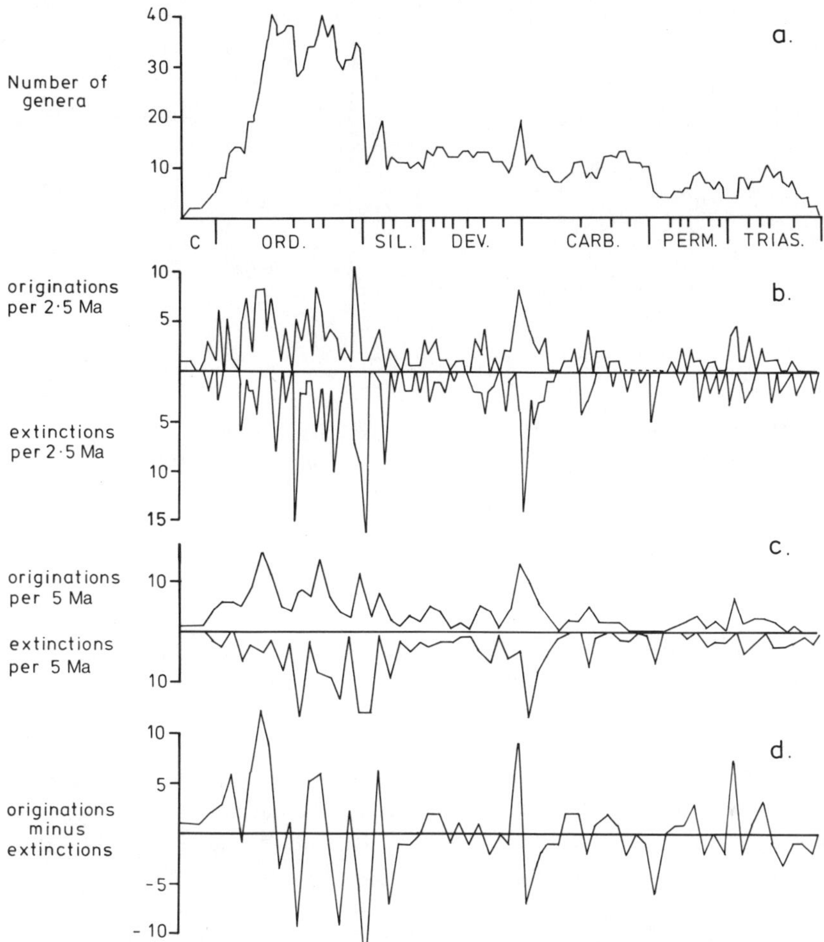

Fig. 11.1. Variations in conodont diversity, origination rates and extinction rates through time. (a) Standing diversity of genera, sampled at 40 points per 100 Ma (b) Origination and extinction rates plotted at 2.5-Ma intervals. (c) Origination and extinction rates plotted at 5-Ma intervals. (d) Originations minus extinctions plotted at 5-Ma intervals.

Palmatolepis, in the late Devonian. Ziegler and Lane's (1987) data also show high peaks in the early Eifelian and Givetian, which are present, but less obvious, in the generic curves. Indeed, the clear pattern of cyclicity apparent from the specific data for the Devonian is much more muted in its expression at a generic level. It is more difficult to compare the curves given by Clark (1987) with the generic curve for the late Permian and Triassic, as he plotted the total number of species in each stage as a point

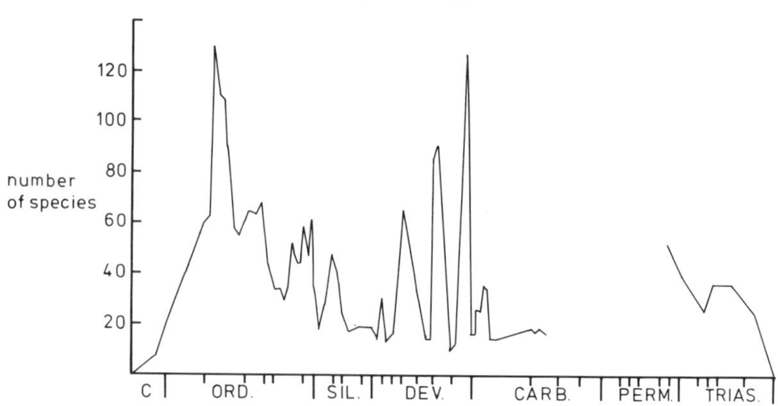

Fig. 11.2. The standing diversity of conodont species through time. Cambrian and Ordovician data from Sweet (1985); Silurian data compiled by the author; Devonian and early Carboniferous data from Ziegler and Lane (1987), slightly modified to include some allowance for apparatuses lacking pectiniform elements; late Permian and Triassic data from Clark (1987). Data compiled in this form are not available for the late Carboniferous or early to middle Permian.

in the middle of the stage, rather than sampling total standing diversity at successive points in time. However, the late Permian diversity drop, the mid-Triassic recovery and the steady late Triassic decline are evident from both curves.

At present it is difficult to find a basis for analysing these diversity patterns in terms of the groups of conodonts affected. A suprageneric classification is given in the *Treatise on Invertebrate Paleontology*, but it is incomplete and exploratory, and is regarded by its authors as being provisional (Clark 1981). Many of the families included are not clades (see Fåhræus 1984) and, thus, disappearance of a family from the record does not represent the extinction of a clade. Sweet (1985, fig. 8) recognized that much of the story of conodont evolution was recorded in changes in apparatus styles and divided his species diversity graph for the Cambrian to early Devonian into four areas, labelled 1, 2, 3–4 and 5–7. The figures relate to the numbers of different types of element recognized in the apparatus; in the simplest, unimembrate apparatuses there may have been several elements, but these were morphologically identical or nearly so, while in septimembrate apparatuses seven clearly distinctive element types are recognized. One problem with this approach is that the categories are demonstrably polyphyletic: a bimembrate apparatus with two types of coniform element is not closely related to one comprising two types of platform-like pectiniform element. Other difficulties may occur where variation between elements is subtle, as these may be recognized as separate entities in some taxa (or by some taxonomists), but combined into a single type in others. The incomplete knowledge of the apparatuses of many species introduces further complications. However, Sweet's approach does serve to emphasize his point that complex apparatus differentiation was a major evolutionary feature of the conodonts. His figure illustrates the fact that primitive unimembrate and bimembrate apparatuses had largely disappeared by the beginning of the Silurian and that quinque- to septi-membrate apparatuses or their derivatives dominated post-Ordovician conodont history.

Assessment of apparatus styles is clearly important in interpreting the patterns shown in conodont evolution. In the outline that follows, an attempt is made to relate patterns of evolution and extinction to the record of apparatus types, but it should be appreciated that there remains a considerable amount of basic taxonomic work to be completed before such an analysis could be considered authoritative.

2. *The late Cambrian and Ordovician*

The *Treatise on Invertebrate Paleontology* (Robison 1981) includes within the scope of the Conodonta elements of three histological types, distinguished as protoconodonts, paraconodonts, and euconodonts by Bengtson (1976). Szaniawski (1982, 1983) has demonstrated striking morphological and structural similarities between protoconodont elements and the grasping spines of chaetognaths, and it is probable that these elements represent the fossilized remains of Chaetognatha. The paraconodonts may be more closely related to the euconodonts, and there is a considerable body of evidence that the euconodonts were derived from the paraconodonts by the acquisition of a crown (Bengtson 1983). An alternative origin is from the phosphatic cones of *Fomitchella,* which Dzik (1986) regarded as plausible fore-runners of the euconodonts. Whatever the ancestry, euconodonts appeared in the late Cambrian and, after displaying relatively slow evolution and diversification in that epoch, underwent an impressive radiation in the early Ordovician.

Miller (1980, 1984) considered that early euconodonts were polyphyletically derived from paraconodont ancestors and that early stocks included the *Proconodontus* and *Teridontus* lineages, originating in the late Cambrian, and the *Fryxellodontus* and *Chosonodina* lineages, which appeared in the early Ordovician. *Proconodontus* and *Teridontus* apparatuses comprised curved coniform elements that were probably unimembrate or had rather subtle variation between elements. *Cordylodus,* the earliest genus with denticulated processes on the elements, is an early Ordovician descendant of *Proconodontus,* but it is the *Teridontus* lineage that apparently gave rise to the majority of post-Cambrian conodonts (Sweet 1985). The radiation that began in the Tremadoc is accompanied by the differentiation of elements within apparatuses to give multimembrate types. This differentiation is displayed principally in the distribution of costae or of denticulated processes, forming suites of elements in which symmetry-transition series can frequently be recognized. Barnes *et al.* (1979) reviewed the structure and evolution of Ordovician apparatuses, identifying five main apparatus types (I–V), each with sub-types. Although several details of their summary need to be modified in the light of subsequent work, many of their general conclusions remain valid. All five of their major apparatus types were represented by the Arenig, at which time the variety of apparatus styles was probably at its peak. There was a major period of extinction at the end of the Arenig, at a rate of six genera per Ma, affecting all groups, but particularly the type I apparatuses of Barnes *et al,* which display simple transition series of coniform elements.

The Middle Ordovician was a time of diversification in the more complicated apparatus types, particularly the type IV apparatuses, which are commonly sexi- or septimembrate with well-developed ramiform elements and sometimes show development of platform-like pectiniform elements. The differentiation of functions that the separation of pectiniform and ramiform sets in the apparatus implies was clearly successful for conodonts, and introduced a pattern that was to dominate later faunas. The Llandeilo also saw expansion of type III coniform apparatuses, in which curvature variation accompanies the transition in cross-sectional symmetry of elements.

In the late Ordovician there was a period of stabilization of conodont faunas (Barnes *et al.* 1979), with a continued flowering of type III and IV apparatuses, and a decline in the simpler type I and II styles. Type V, in which the apparatus comprises only one or two platform elements, also became rare, although this structure had only ever been represented in a few genera. The Caradoc was punctuated by an interval in which extinction rates exceeded origination rates, reaching four genera per Ma in the middle of the epoch. Following a brief late Caradoc–early Ashgill recovery, the latest Ordovician saw severe faunal changes, with extinctions reaching a rate of seven genera per Ma, the highest in conodont history. At this time, the last type I, II and V apparatuses disappeared from the record.

3. *The Silurian*

The Silurian survivors of the Ashgill extinction mostly fall into two groups. The first includes sexi- or septimembrate apparatuses comprising well-developed ramiform elements (e.g. *Oulodus*) or ramiform and pectiniform elements (e.g. *Ozarkodina*). The second comprises essentially bimembrate or trimembrate apparatuses of coniform elements in which some of the members may show subtle symmetry transition (e.g. *Panderodus*). There was relatively little innovation in the earliest Silurian, but a steady accumulation of new genera took place in the late Llandovery. These originations are of cryptic ancestry and introduce several new apparatus styles, including quinquemembrate structures with a reduced symmetry transition series (e.g. *Pterospathodus*) and other multimembrate types with two or more platform-like pectiniform elements (e.g. *Apsidognathus*).

This diversification was curtailed by an extinction event in the earliest Wenlock. At the scale plotted on Fig. 11.1b the rate is three genera per Ma, but it may in fact have been much more severe than this, as many genera disappear within a very small interval at the base of the Wenlock (Aldridge 1976; Jeppsson 1987). It is possible that the importance of this event has been underestimated, as it eliminated most, if not all, of the innovative late Llandovery stocks that might otherwise have gone on to enrich later faunas. As it is, the dominant faunas of the middle and late Silurian record an enhancement of the polarization of apparatus styles that began at the Ordovician–Silurian boundary, with sexi- or septimembrate structures of the *Ozarkodina* plan and coniform *Panderodus* being particularly characteristic. Generic diversity remained relatively constant for the rest of the period, but species diversity was a little higher in the Wenlock and late Ludlow than in the early Ludlow and Pridoli (Fig. 11.2).

4. *The Devonian and early Carboniferous*

Patterns of conodont species diversity through this period have been documented by Ziegler and Lane (1987), who recognized seven evolutionary cycles spanning the late Silurian to late Carboniferous. These cycles, particularly in the Devonian, are clearly shown on the species diversity curve (Fig. 11.2) and are also apparent at generic level when originations minus extinctions are plotted against time (Fig. 11.1d). Each cycle is documented as beginning with a low-diversity episode, which proceeds through innovative and radiative phases to a high-diversity episode in which gradualistic evolution predominates. The high-diversity episode is curtailed by an extinction event, followed by the low-diversity episode and innovative phase of the succeeding cycle. Ziegler and Lane (1987) concentrated their analysis on pectiniform elements, which are the characteristic components of the majority of middle and late Palaeozoic apparatuses. They noted that most low-diversity episodes were characterized by large-cavity pectiniform elements whereas high-diversity episodes were marked by radiation of pit-bearing forms. Pectiniform elements with restricted cavities, or pits, disappeared in the early Viséan, after which time comparable high-diversity episodes were not developed.

The nadir of conodont diversity in the Devonian was at the Frasnian–Famennian boundary (Figs 11.1a and 11.2), following a sharp extinction event in the latest Frasnian that removed almost all previous species of palmatolepids, ancyrodellids, ancyrognathids, and polygnathids (Ziegler and Lane 1987). The mid-Famennian was a time of rapid diversification, both at generic level and within several generic stocks, but towards the end of the age extinctions at the rate of five genera per Ma caused a significant depletion, returning generic diversity levels to those that had been maintained for much of the pre-Famennian Devonian. Although extinction was the dominant feature of the latest Devonian, originations of several genera (e.g. *Siphonodella*) that were to become important in the early Carboniferous occurred at this time. A small peak in originations occurred in the early Dinantian (Fig. 11.1b), reflecting a radiation in genera referred to the Bactrognathodontidae, but otherwise this was a time of slow generic diversity decline. A gradual recovery was staged during the late Viséan to early Namurian, with origination rates exceeding extinction rates through this interval (Fig. 11.1d), but was curtailed by extinctions at the level of the proposed mid-Carboniferous boundary (see Lane and Manger 1985). Although the extinction rate did not reach two genera per Ma, it is more severe when considered at a per taxon level, as diversity was already relatively low. Ziegler and Lane (1987) considered that this was an extinction event of equivalent importance to those of the late Ordovician, late Frasnian, and late Famennian.

Although not considered in Ziegler and Lane's (1987) survey, an important feature of the Devonian record is the decline and demise of the true coniform apparatuses that had formed abundant constituents of most early Palaeozoic faunas. Several genera (e.g. *Panderodus, Belodella*) had survived from the Silurian, but coniform species were never abundant after the early Devonian and all had disappeared by the end of the period.

5. *The late Carboniferous and Permian*

A modest recovery followed the mid-Carboniferous extinction and diversity levels remained more-or-less constant through the late Carboniferous and into the earliest Permian. As reported by Clark (1972), the early Permian was a time of crisis for the conodonts, with extinctions of several characteristic late Carboniferous genera reducing diversity to its lowest post-Cambrian level. A slight diversification in the later Permian was eclipsed at the end of the period, when the number of genera again dropped to four.

Apparatus styles in the late Carboniferous were conservative, most genera conforming to a sexi- or septimembrate *Ozarkodina* plan. It is likely that most Permian genera had similar structures, but knowledge of several apparatuses is incomplete.

6. *The Triassic*

Despite being reduced to a diversity of five or six species in the latest Permian, conodonts survived the end Permian extinction event and at least five species continued into the earliest Triassic (Clark 1987). A burst of originations during the early Triassic, particularly in the Smithian, was consolidated by steady middle Triassic origination rates (Fig. 11.1c) to give a Ladinian generic diversity of similar level to the late Carboniferous. From the early Carnian onwards, though, extinction rates outpaced origination rates and after some 25 Ma the conodonts became extinct. The generic diversity curve (Fig. 11.1a) shows this to be a steady decline rather than an abrupt or catastrophic event. At a species level, Clark (1987) has contrasted early and late Triassic origination rates: 63 originations in the combined Scythian, Anisian and Ladinian (3.2 per Ma) against 27 in the Carnian and Norian (1.1 per Ma).

There is still considerable uncertainty about the construction of many Triassic apparatuses. Several have apparent seximembrate structures of ramiform elements, developed from *Ellisonia* or *Xaniognathus* stocks. Others (e.g. *Gladigondolella, Pseudofurnishius*) have been recorded as unimembrate, with only platform-like pectiniform elements. It is possible that relationships exist between these groups, and Kozur and Mostler (1971), for example, reconstructed the apparatus of the type species of *Gladigondolella* to include a suite of ramiform elements of *Ellisonia* type. Sweet (in Robison 1981), however, concluded that *Gladigondolella* and similar genera were indeed unimembrate, and later suggested (Sweet 1985) that such apparatuses may have developed from standard multimembrate types by non-mineralization of the ramiform elements.

Discussion

Considerable attention has been focused recently on the hypothesis that major extinction events occurred periodically, driven by a single causal agent (Fischer and Arthur 1977; Raup and Sepkoski 1984, 1986; Sepkoski and Raup 1986). The statistical analyses of the fossil record adduced to support this idea have largely concentrated on the mid-Permian to Pleistocene interval, but Sepkoski and Raup (1986) tentatively extended their analysis back into the Palaeozoic. While admitting that the stratigraphical resolution of their data in this part of the column was too coarse to determine the existence of any periodicity, they did identify seven extinction maxima as significantly higher than neighbouring background extinction rates. Two of these were within the Cambrian, the others occurred at the end of the Ordovician, in the late Silurian, in the Famennian Stage of the Devonian, and in the Namurian and Stephanian series of the Carboniferous. A hint of possible significance was also detected in the Llanvirn. These events, together with the mass extinction long recognized at the end of the Permian, all occurred within the range of the conodonts and, although they would not all necessarily be reflected in the evolutionary patterns shown by any particular group of organisms, it is of interest to examine how they relate to episodes of extinction and radiation recognized in the Conodonta.

The Llanvirn, end Ordovician, and Namurian events match well with extinction maxima in conodonts, but the Famennian event appears out of phase as that was a time of conodont diversification. However, the diversity high is bracketed by late Frasnian and end Famennian extinctions, and the apparent discrepancy may simply be a result of the coarse stratigraphic scale on which the Sepkoski and Raup (1986) data were compiled. For the late Silurian and Stephanian, there is no evidence of above average extinction rates on the conodont generic plots, although there is a minor dip in diversity in late Silurian species (Fig. 11.2). In contrast, the important conodont extinctions in the Caradoc, early Wenlock, and early Permian are not matched in Sepkoski and Raup's data. There is a diversity low across the Permian–Triassic boundary, but conodonts survived this event better than many other groups (Sweet 1973; Clark 1987), even though there is some evidence of a drop in overall abundance (Clark *et al.* 1986).

In general the correspondence between the conodont record and the data presented by Sepkoski and Raup (1986) is not impressive. Nor does the pattern of conodont extinctions, with a concentration of major events in the early Palaeozoic (Fig. 11.1d), suggest any overall periodicity. Indeed, the recognition of extinction/radiation cycles at different scales in the stratigraphical record suggests that several factors may have been of influence in driving conodont evolution, rather than a single, regularly repeated, catastrophic cause. Even the final demise of the group at the end of the Triassic does not appear to have been particularly abrupt, but rather the culmination of a long period of decline. There is a possibility, though, that some event delivered the final blow, as Clark (1983) has shown that the extinction occurred sooner than would be predicted from a theoretical examination of survivorship curves.

Some cyclicity is apparent in conodont evolution throughout their history, but is evident on different scales and with varying intensity at different times. In the Ordovician to mid-Silurian, four or five extinction/radiation cycles occurred in 70 Ma (Fig. 11.1c,d), an average separation of 14 or 17.5 Ma. The Devonian cycles documented by Ziegler and Lane (1987) have a mean duration of 12.5 Ma, whereas Clark (1987) recognized cyclicity at intervals of less than 2 Ma in the Triassic. These cycles have variously been related to fluctuations in global tectonic activity, in sea-level, and in oceanic temperature; they could also have been affected by extraterrestrial causes. Ethington *et al.* (1987) related an abrupt change in Tremadoc (Fauna C/D) faunas in the Midcontinent Province to environmental changes during shoaling, and noted that the conodont extinctions occurred higher in the stratigraphic succession than trilobite extinctions, presumably caused by the same changes. Patterns of regression and transgression through the Ordovician have been summarized by Fortey (1984), who attributed most of the variation in sea level to fluctuations in the extent of south polar ice-sheets. Eustatic falls in sea level at the end of the Arenig and in the late Ashgill correlate with conodont extinction maxima, but the mid-Caradoc was a time of major transgression, and there is no compelling comparison between the sea level curve and that of conodont diversity. The same is true of the Devonian, where the pattern of eustatic sea level changes documented by Johnson *et al.* (1985) does not correspond closely with the conodont cycles of Ziegler and Lane (1987). Sweet (1985) suggested that oceanic temperatures played a part in controlling conodont distribution and development throughout their history and pointed out that this hypothesis may be tested by oxygen isotope analysis of the apatite of the conodont elements themselves. Preliminary results reported by Geitgey (1985) indicate that the early Ordovician and late Devonian origination peaks do correspond to the two 'warmest' peaks on the palaeotemperature curve, and that the late Ordovician and much of the Carboniferous were intervals of 'cooler' oceanic water. He further concluded that conodonts were more susceptible to changes in marine temperature than other contemporary taxa.

The field of conodont geochemistry in general holds immense promise for generating and testing hypotheses of the controls on conodont distribution and evolution. Wright *et al.* (1984) showed that variation in cerium anomalies in conodont elements reflected changing oxidation-reduction levels in ancient seas and allowed recognition of anoxic events. It has been suggested by Jeppsson (1987), among others, that such events could profoundly influence conodonts, and Wright *et al.* (1984) correlated the end Devonian and end Permian extinction events with anoxic signals retained in conodont elements. Fossil apatite also carries signatures of variation in a wide variety of other geochemical parameters, including rare earth elements, strontium and neodymium isotopes and metallic trace elements (Kovach 1980, 1981; Wright *et al.* 1984, 1985). These variations all reflect environmental changes that would have had potential effects on the whole marine biota. Hence, results of geochemical work on conodonts have wide implication for interpretation of the relationships between oceanic geochemistry and evolutionary patterns not just in the conodonts themselves, but in all coeval taxa.

Acknowledgements

This work on conodont palaeobiology is financed by N.E.R.C. Research Grant GR3/5105A. My knowledge of conodont distribution and evolution has been gained from papers by and discussions with numerous conodont specialists, none of whom is responsible for my lapses. I particularly thank

A. Swift and S. J. Tull for assistance and advice at various stages in the preparation of this paper. The text-figures and data base were drafted by Mrs J. Wilkinson.

References

Aldridge, R. J. (1976). Comparison of macrofossil communities and conodont distribution in the British Silurian. In *Conodont paleoecology* (ed. C. R. Barnes), Geological Association of Canada Special Paper 15, pp. 91–104.

Aldridge, R. J. and Briggs, D. E. G. (1986). Conodonts. In *Problematic Fossil Taxa* (ed. A. Hoffman and M. H. Nitecki), pp. 227–39. Oxford University Press, New York.

Aldridge, R. J. and Jeppsson, L. (1984). Ecological specialists among Silurian conodonts. *Spec. Pap. Palaeont.* **32,** 141–9.

Aldridge, R. J., Briggs, D. E. G., Clarkson, E. N. K., and Smith, M. P. (1986). The affinities of conodonts — new evidence from the Carboniferous of Edinburgh, Scotland. *Lethaia* **19,** 279–91.

Aldridge, R. J., Smith, M. P., Norby, R. D., and Briggs, D. E. G. (1987). The architecture and function of Carboniferous polygnathacean conodont apparatuses. In *Palaeobiology of Conodonts* (ed. R. J. Aldridge), pp. 63–75. Ellis Horwood, Chichester.

Barnes, C. R., Kennedy, D. J., McCracken, A. D., Nowlan, G. S., and Tarrant, G. A. (1979). The structure and evolution of Ordovician conodont apparatuses. *Lethaia* **12,** 125–51.

Bengtson, S. (1976). The structure of some Middle Cambrian conodonts, and the early evolution of conodont structure and function. *Lethaia* **9,** 185–206.

Bengtson, S. (1983). The early history of the Conodonta. *Fossils and Strata* **15,** 5–19.

Briggs, D. E. G., Clarkson, E. N. K., and Aldridge, R. J. (1983). The conodont animal. *Lethaia* **16,** 1–14.

Clark, D. L. (1972). Early Permian crisis and its bearing on Permo-Triassic conodont taxonomy. *Geol. Palaeontol.* **SB1,** 147–58.

Clark, D. L. (1981). Classification. In *Treatise on Invertebrate Paleontology, Part W, Supplement* 2, *Conodonta* (ed. R. A. Robison), pp. W102–3. Geological Society of America and University of Kansas Press, Lawrence, Kansas.

Clark, D. L. (1983). Extinction of conodonts. *J. Paleontol.* **57,** 652–61.

Clark, D. L. (1987). Conodonts: the final fifty million years. In *Palaeobiology of Conodonts* (ed. R. J. Aldridge), pp. 165–74. Ellis Horwood, Chichester.

Clark, D. L. *et al.* (1981). Conodonta. In *Treatise on Invertebrate Paleontology, Part W, Supplement* 2 (ed. R. A. Robison), pp. W1–202. Geological Society of America and University of Kansas Press, Lawrence, Kansas.

Clark, D. L., Wang, C.-Y., Orth, C. J., and Gilmore, J. S. (1986). Conodont survival and low iridium abundances across the Permian–Triassic boundary in South China. *Science* **233,** 984–6.

Dzik, J. (1986). Chordate affinities of the conodonts. In *Problematic Fossil Taxa* (ed. A. Hoffman and M. H. Nitecki), pp. 240–54. Oxford University Press, New York.

Ethington, R. L., Engel, K. M., and Elliott, K. L. (1987). An abrupt change in conodont faunas in the Lower Ordovician of the Midcontinent Province. In *Palaeobiology of Conodonts* (ed. R. J. Aldridge), pp. 111–27. Ellis Horwood, Chichester.

Fåhræus, L. E. (1984). A critical look at the Treatise family-group classification of Conodonta: an exercise in eclecticism. *Lethaia* **17,** 293–305.

Fischer, A. G. and Arthur, M. A. (1977). Secular variations in the pelagic realm. In *Deep-water carbonate environments* (ed. H. E. Cook and P. Enos), Society of Economic Paleontologists and Mineralogists. Special Publication No. 25, pp. 19–50.

Forster, S. C. and Warrington, G. (1985). Geochronology of the Carboniferous, Permian and Triassic. In *The Chronology of the Geological Record* (ed. N. J. Snelling), Geological Society of London Memoir 10, pp. 99–113. Blackwell, Oxford.

Fortey, R. A. (1984). Global earlier Ordovician transgressions and regressions and their biological implications. In *Aspects of the Ordovician System* (ed. D. L. Bruton), pp. 37–50. Universitetsforlaget, Oslo.

Gale, N. H. (1985). Numerical calibration of the Palaeozoic time-scale; Ordovician, Silurian and Devonian periods. In *The Chronology of the Geological Record* (ed. N. J. Snelling), Geological Society of London Memoir 10, pp. 81–8. Blackwell, Oxford.

Geitgey, J. E. (1985). Temperature as a factor affecting conodont diversity and distribution. In *Fourth European Conodont Symposium* (*ECOS IV*), *Nottingham* 1985, *Abstracts* (ed. R. J. Aldridge, R. L. Austin, and M. P. Smith), p. 12. University of Southampton.

Harland, W. B., Cox, A. V., Llewellyn, P. G., Pickton, C. A. G., Smith, A. G., and Walters, R. (1982). *A Geologic Time Scale.* Cambridge University Press, Cambridge.

Jeppsson, L. (1987). Lithological and conodont distributional evidence for episodes of anomalous oceanic conditions during the Silurian. In *Palaeobiology of Conodonts* (ed. R. J. Aldridge), pp. 129–45. Ellis Horwood, Chichester.

Johnson, J. G., Klapper, G., and Sandberg, C. A. (1985). Devonian eustatic fluctuations in Euramerica. *Geol. Soc. Am. Bull.* **96,** 567–87.

Kovach, J. (1980). Variations in the strontium isotopic composition of seawater during Paleozoic time determined by the analysis of conodonts. *Geol. Soc. Am. Abstr. Programs* **12,** 465.

Kovach, J. (1981). The strontium content of conodonts and possible use of the strontium concentrations and strontium isotopic composition of conodonts for correlation purposes. *Geol. Soc. Am. Abstr. Programs* **13,** 285.

Kozur, H. and Mostler, H. (1971). Holothurien-Sklerite und Conodonten aus der Mittel- und Obertrias von Köveskal (Balatonhochland, Ungarn). *Geol. Palaeontol. Mitt. Innsbruck* **1** (10), 1–36.

Lane, H. R. and Manger, W. L. (1985). Toward a boundary in the middle of the Carboniferous (1975–1985): ten years of progress. *Cour. Forsch. Inst. Senckenberg.* **74,** 15–34.

McKerrow, W. S., Lambert, R. St. J., and Cocks, L. R. M. (1985). The Ordovician, Silurian and Devonian periods. In *The Chronology of the Geological Record* (ed. N. J. Snelling), Geological Society of London Memoir 10, pp. 73–80. Blackwell, Oxford.

Miller, J. F. (1980). Taxonomic revisions of some Upper Cambrian and Lower Ordovician conodonts with comments on their evolution. *Univ. Kansas Paleontol. Contr. Pap.* **99,** 1–44.

Miller, J. F. (1984). Cambrian and earliest Ordovician conodont evolution, biofacies, and provincialism. In *Conodont Biofacies and Provincialism* (ed. D. L. Clark), Geological Society of America Special Paper 196, pp. 43–68.

Nicoll, R. S. (1982). Multielement composition of the conodont *Icriodus expansus* Branson & Mehl from the Upper Devonian of the Canning Basin, Western Australia. Bureau of Mineral Resources *J. Austr. Geol. Geophys.* **7,** 197–213.

Nicoll, R. S. (1985). Multielement composition of the conodont species *Polygnathus xylus xylus* Stauffer, 1940 and *Ozarkodina brevis* (Bischoff and Ziegler, 1957) from the Upper Devonian of the Canning Basin, Western Australia. Bureau of Mineral Resources *J. Austr. Geol. Geophys.* **9,** 133–47.

Odin, G. S. (ed.) (1982). *Numerical Dating in Stratigraphy.* John Wiley, Chichester.

Odin, G. S. (1985*a*). Remarks on the numerical scale of Ordovician to Devonian times. In *The Chronology of the Geological Record* (ed. N. J. Snelling), Geological Society of London Memoir 10, pp. 93–8. Blackwell, Oxford.

Odin, G. S. (1985*b*). Comments on the geochronology of the Carboniferous to Triassic times. In *The Chronology of the Geological Record* (ed. N. J. Snelling), Geological Society of London Memoir 10, pp. 114–7. Blackwell, Oxford.

Pollock, C. A. (1969). Fused Silurian conodont clusters from Indiana. *J. Paleontol.* **43,** 929–35.

Raup, D. M. and Sepkoski, J. J., Jr (1984). Periodicity of extinctions in the geologic past. *Proc. Nat. Acad. Sci. USA* **81,** 801–5.

Raup, D. M. and Sepkoski, J. J., Jr (1986). Periodic extinction of families and genera. *Science* **231,** 833–6.

Rhodes, F. H. T. (1952). A classification of Pennsylvanian conodont assemblages. *J. Paleontol.* **26,** 886–901.

Robison, R. A. (ed.) (1981). *Treatise on Invertebrate Palaeontology, Part W, Supplement* 2, *Conodonta.* Geological Society of America and University of Kansas Press, Lawrence, Kansas.

Schmidt, H. (1934). Conodonten-Funde in ursprünglichen Zusammenhang. *Paläont. Z.* **16,** 76–85.

Sepkoski, J. J., Jr and Raup, D. M. (1986). Periodicity in marine extinction events. In *Dynamics of Extinction* (ed. D. K. Elliott), pp. 3–30. Wiley, New York.

Smith, M. P., Briggs, D. E. G., and Aldridge, R. J. (1987). A conodont animal from the lower Silurian of Wisconsin, U.S.A., and the apparatus architecture of panderodontid conodonts. In *Palaeobiology of Conodonts* (ed. R. J. Aldridge), pp. 91–104. Ellis Horwood, Chichester.

Snelling, N. J. (1985). An interim time-scale. In *The Chronology of the Geological Record* (ed. N. J. Snelling), Geological Society of London Memoir 10, pp. 261–5. Blackwell, Oxford.

Sweet, W. C. (1973). Late Permian and Early Triassic conodont faunas. In *The Permian and Triassic Systems and their mutual boundary* (ed. A. Logan and A. V. Hills), Canadian Society for Petroleum Geology Memoir 2, pp. 630–46.

Sweet, W. C. (1985). Conodonts: those fascinating little whatzits. *J. Paleontol.* **59,** 485–94.

Szaniawski, H. (1982). Chaetognath grasping spines recognised among Cambrian protoconodonts. *J. Paleontol.* **56,** 806–10.

Szaniawski, H. (1983). Structure of protoconodont elements. *Fossils and Strata* **15,** 21–7.

Wright, J., Miller, J. F., and Holser, W. T. (1985). Chemostratigraphy of conodonts across the Cambrian-Ordovician boundary. In *Fourth European Conodont Symposium* (*ECOS IV*), *Nottingham* 1985, *Abstracts* (ed. R. J. Aldridge, R. L. Austin, and M. P. Smith), p. 31. University of Southampton.

Wright, J., Seymour, R. S., and Shaw, H. F. (1984). REE and Nd isotopes in conodont apatite: variations with geological age and depositional environment. In *Conodont Biofacies and Provincialism* (ed. D. L. Clark), Geological Society of America Special Paper 196, pp. 325–40.

Ziegler, W. and Lane, H. R. (1987). Cycles in conodont evolution from Devonian to mid-Carboniferous. In *Palaeobiology of Conodonts* (ed. R. J. Aldridge), pp. 147–63. Ellis Horwood, Chichester.

Appendix

Data base of the stratigraphic ranges of conodont genera, used in the compilation of the graphs in Fig. 11.1. The time-scale is that of Snelling (1985).

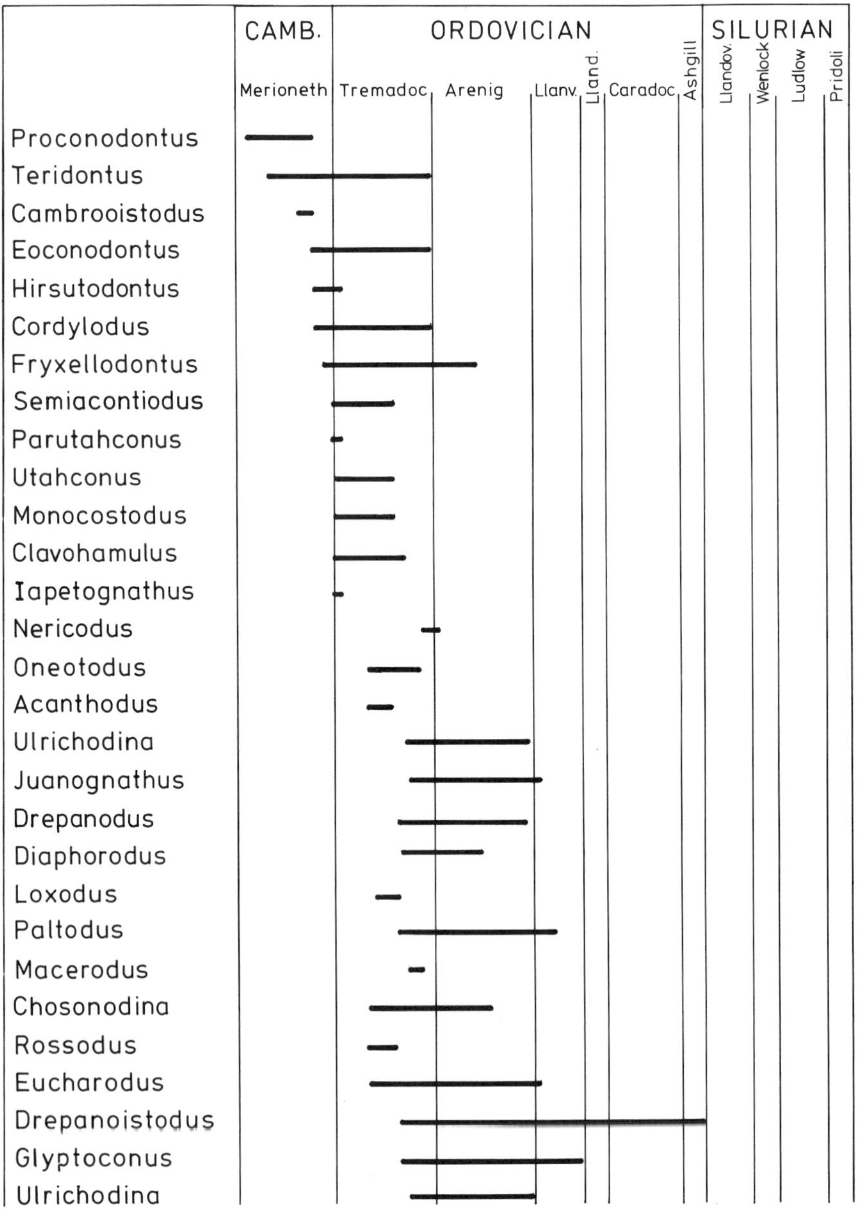

CAMB. ORDOVICIAN SILURIAN

Merioneth Tremadoc Arenig Llanv. Lland. Caradoc Ashgill Llandov. Wenlock Ludlow Pridoli

Proconodontus
Teridontus
Cambrooistodus
Eoconodontus
Hirsutodontus
Cordylodus
Fryxellodontus
Semiacontiodus
Parutahconus
Utahconus
Monocostodus
Clavohamulus
Iapetognathus
Nericodus
Oneotodus
Acanthodus
Ulrichodina
Juanognathus
Drepanodus
Diaphorodus
Loxodus
Paltodus
Macerodus
Chosonodina
Rossodus
Eucharodus
Drepanoistodus
Glyptoconus
Ulrichodina

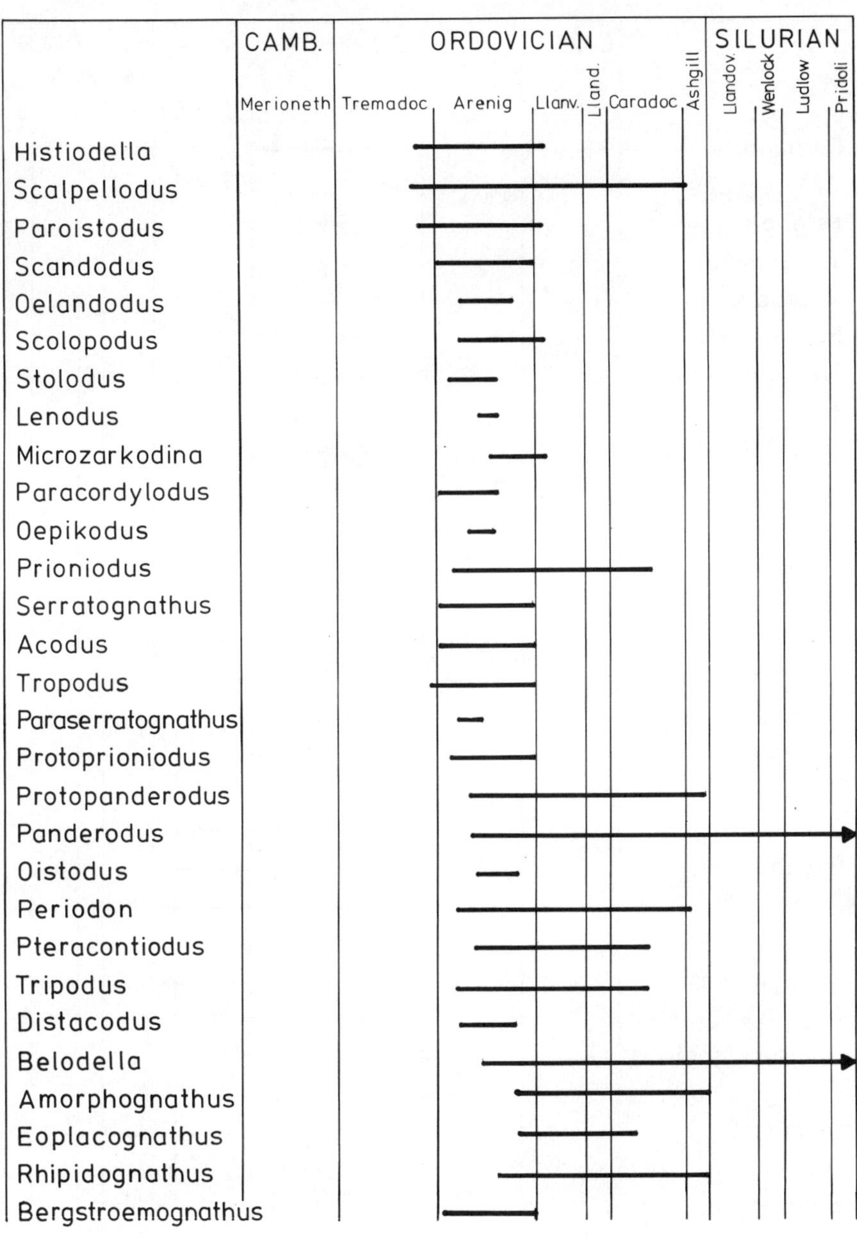
CAMB.
ORDOVICIAN
SILURIAN
Merioneth
Tremadoc
Arenig
Llanv.
Lland.
Caradoc
Ashgill
Llandov.
Wenlock
Ludlow
Pridoli
Histiodella
Scalpellodus
Paroistodus
Scandodus
Oelandodus
Scolopodus
Stolodus
Lenodus
Microzarkodina
Paracordylodus
Oepikodus
Prioniodus
Serratognathus
Acodus
Tropodus
Paraserratognathus
Protoprioniodus
Protopanderodus
Panderodus
Oistodus
Periodon
Pteracontiodus
Tripodus
Distacodus
Belodella
Amorphognathus
Eoplacognathus
Rhipidognathus
Bergstroemognathus

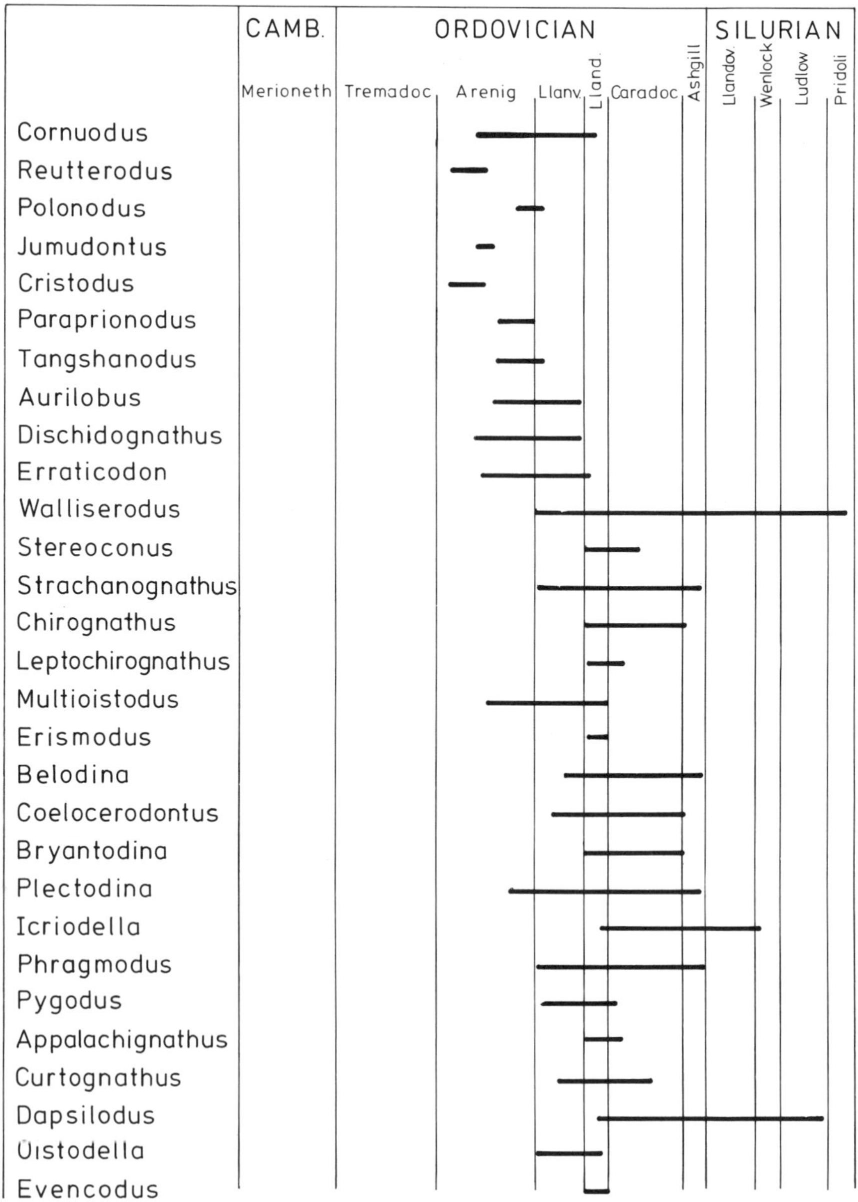

CAMB. ORDOVICIAN SILURIAN

Merioneth Tremadoc Arenig Llanv. Lland. Caradoc Ashgill Llandov. Wenlock Ludlow Pridoli

Cornuodus
Reutterodus
Polonodus
Jumudontus
Cristodus
Paraprionodus
Tangshanodus
Aurilobus
Dischidognathus
Erraticodon
Walliserodus
Stereoconus
Strachanognathus
Chirognathus
Leptochirognathus
Multioistodus
Erismodus
Belodina
Coelocerodontus
Bryantodina
Plectodina
Icriodella
Phragmodus
Pygodus
Appalachignathus
Curtognathus
Dapsilodus
Oistodella
Evencodus

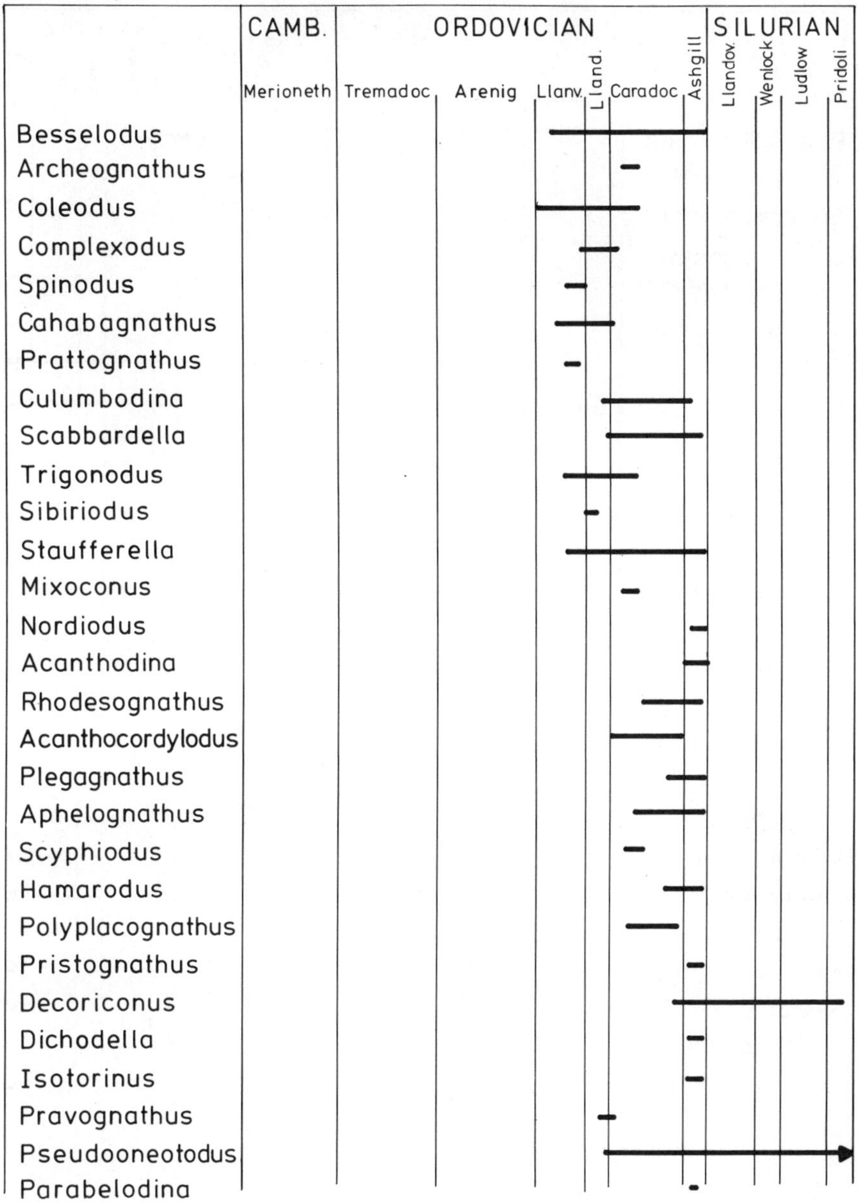

CAMB. | ORDOVICIAN | SILURIAN

Merioneth | Tremadoc | Arenig | Llanv. | Lland. | Caradoc | Ashgill | Llandov. | Wenlock | Ludlow | Pridoli

Besselodus
Archeognathus
Coleodus
Complexodus
Spinodus
Cahabagnathus
Prattognathus
Culumbodina
Scabbardella
Trigonodus
Sibiriodus
Staufferella
Mixoconus
Nordiodus
Acanthodina
Rhodesognathus
Acanthocordylodus
Plegagnathus
Aphelognathus
Scyphiodus
Hamarodus
Polyplacognathus
Pristognathus
Decoriconus
Dichodella
Isotorinus
Pravognathus
Pseudooneotodus
Parabelodina

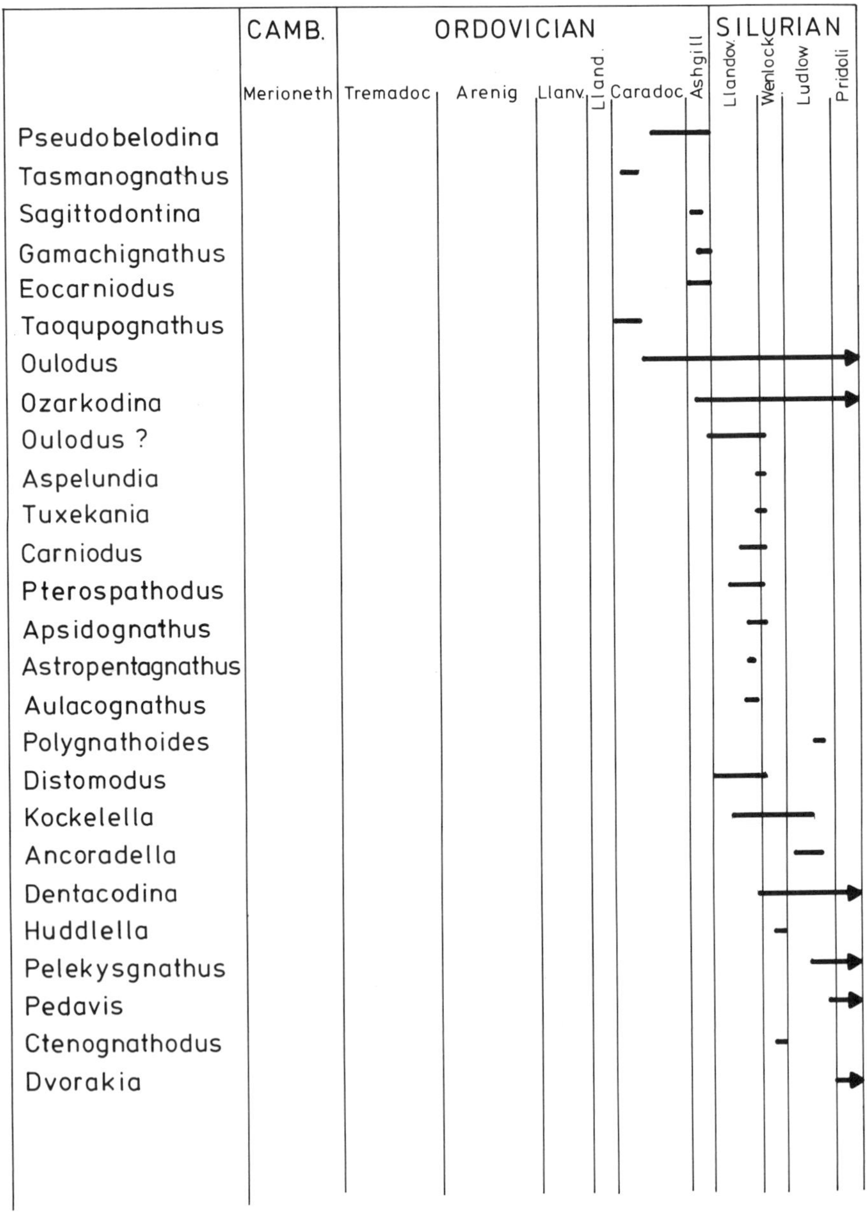

CAMB. | ORDOVICIAN | SILURIAN

Merioneth | Tremadoc | Arenig | Llanv. | Lland. | Caradoc | Ashgill | Llandov. | Wenlock | Ludlow | Pridoli

Pseudobelodina
Tasmanognathus
Sagittodontina
Gamachignathus
Eocarniodus
Taoqupognathus
Oulodus
Ozarkodina
Oulodus ?
Aspelundia
Tuxekania
Carniodus
Pterospathodus
Apsidognathus
Astropentagnathus
Aulacognathus
Polygnathoides
Distomodus
Kockelella
Ancoradella
Dentacodina
Huddlella
Pelekysgnathus
Pedavis
Ctenognathodus
Dvorakia

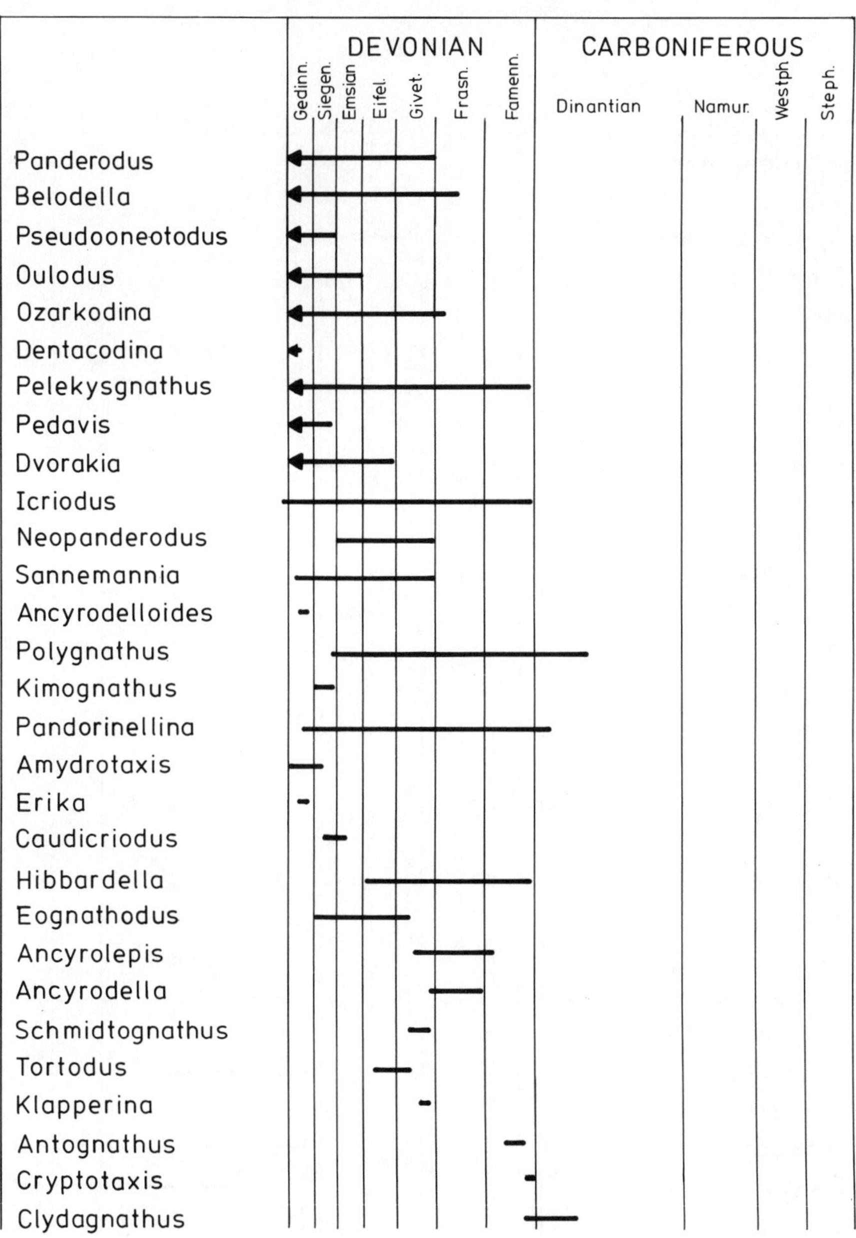
DEVONIAN
CARBONIFEROUS
Gedinn.
Siegen.
Emsian
Eifel.
Givet.
Frasn.
Famenn.
Dinantian
Namur.
Westph.
Steph.
Panderodus
Belodella
Pseudooneotodus
Oulodus
Ozarkodina
Dentacodina
Pelekysgnathus
Pedavis
Dvorakia
Icriodus
Neopanderodus
Sannemannia
Ancyrodelloides
Polygnathus
Kimognathus
Pandorinellina
Amydrotaxis
Erika
Caudicriodus
Hibbardella
Eognathodus
Ancyrolepis
Ancyrodella
Schmidtognathus
Tortodus
Klapperina
Antognathus
Cryptotaxis
Clydagnathus

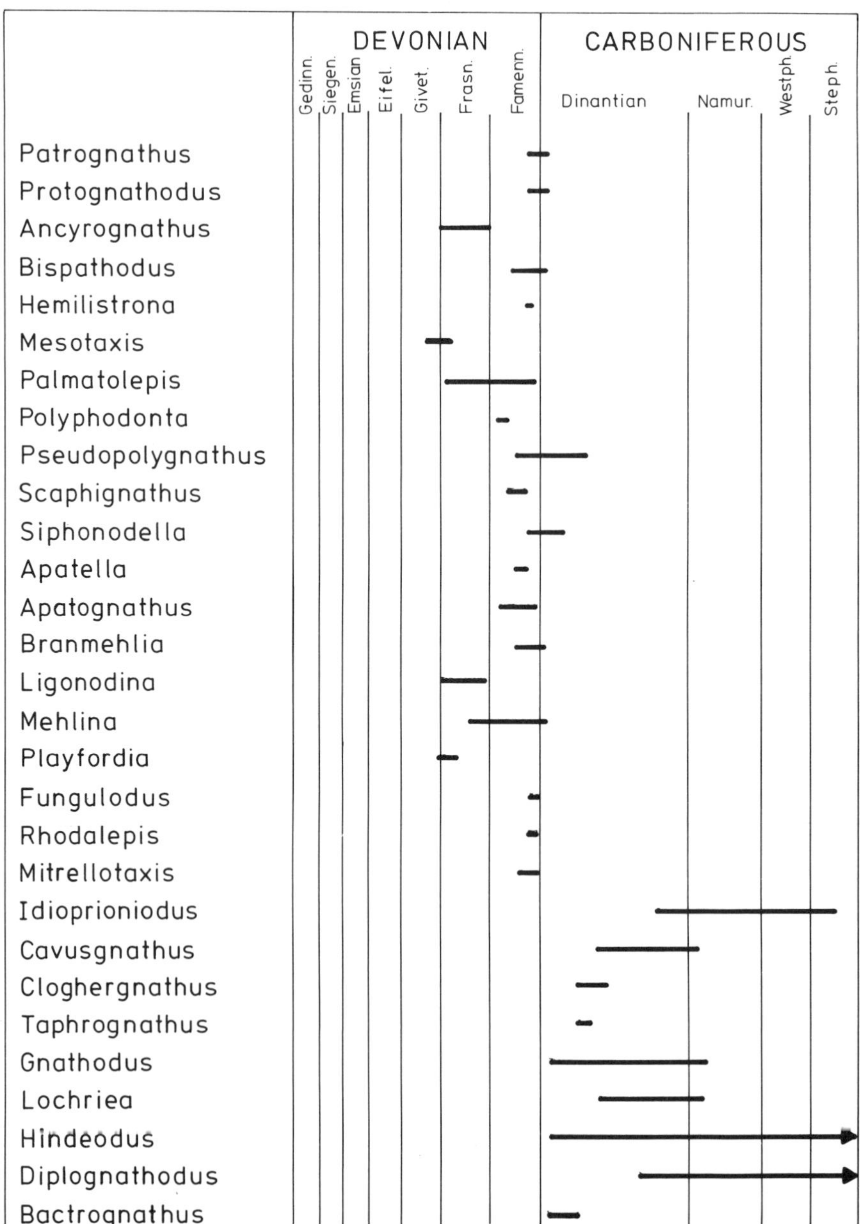
DEVONIAN
CARBONIFEROUS
Gedinn.
Siegen.
Emsian
Eifel.
Givet.
Frasn.
Famenn.
Dinantian
Namur.
Westph.
Steph.
Patrognathus
Protognathodus
Ancyrognathus
Bispathodus
Hemilistrona
Mesotaxis
Palmatolepis
Polyphodonta
Pseudopolygnathus
Scaphignathus
Siphonodella
Apatella
Apatognathus
Branmehlia
Ligonodina
Mehlina
Playfordia
Fungulodus
Rhodalepis
Mitrellotaxis
Idioprioniodus
Cavusgnathus
Cloghergnathus
Taphrognathus
Gnathodus
Lochriea
Hindeodus
Diplognathodus
Bactrognathus

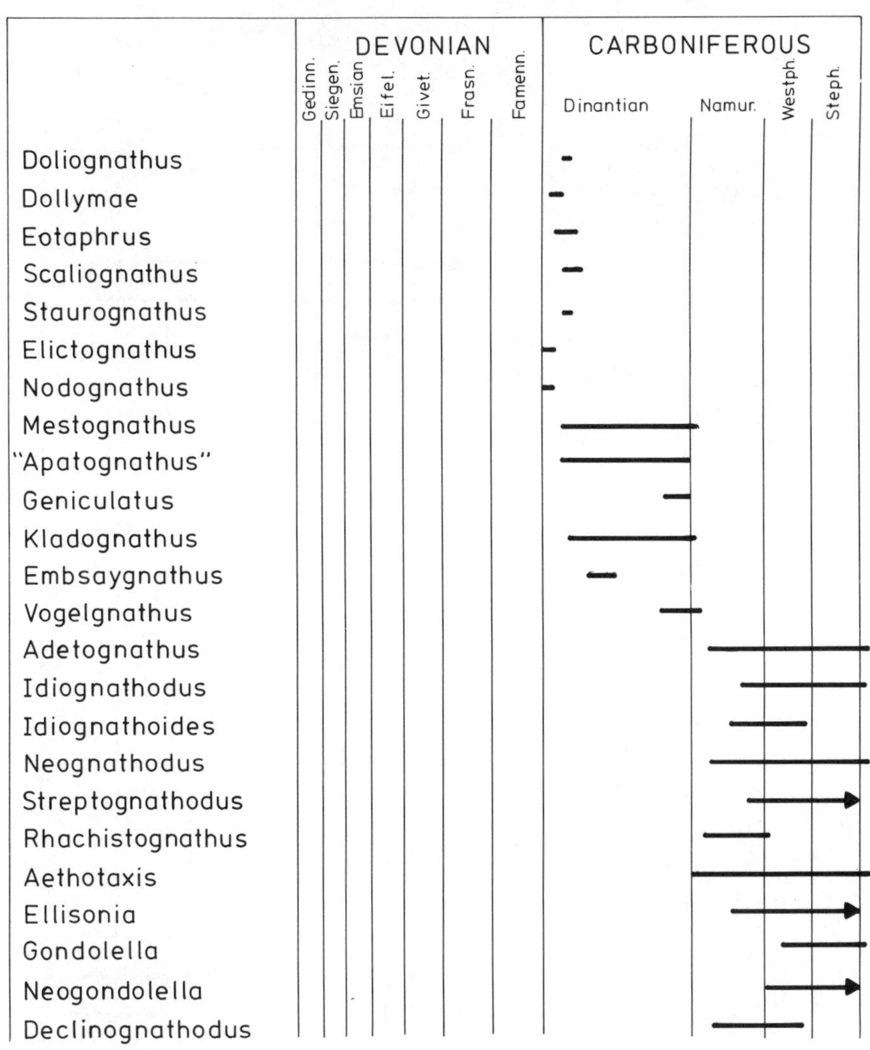
DEVONIAN
CARBONIFEROUS
Gedinn.
Siegen.
Emsian
Eifel.
Givet.
Frasn.
Famenn.
Dinantian
Namur.
Westph.
Steph.
Doliognathus
Dollymae
Eotaphrus
Scaliognathus
Staurognathus
Elictognathus
Nodognathus
Mestognathus
"Apatognathus"
Geniculatus
Kladognathus
Embsaygnathus
Vogelgnathus
Adetognathus
Idiognathodus
Idiognathoides
Neognathodus
Streptognathodus
Rhachistognathus
Aethotaxis
Ellisonia
Gondolella
Neogondolella
Declinognathodus

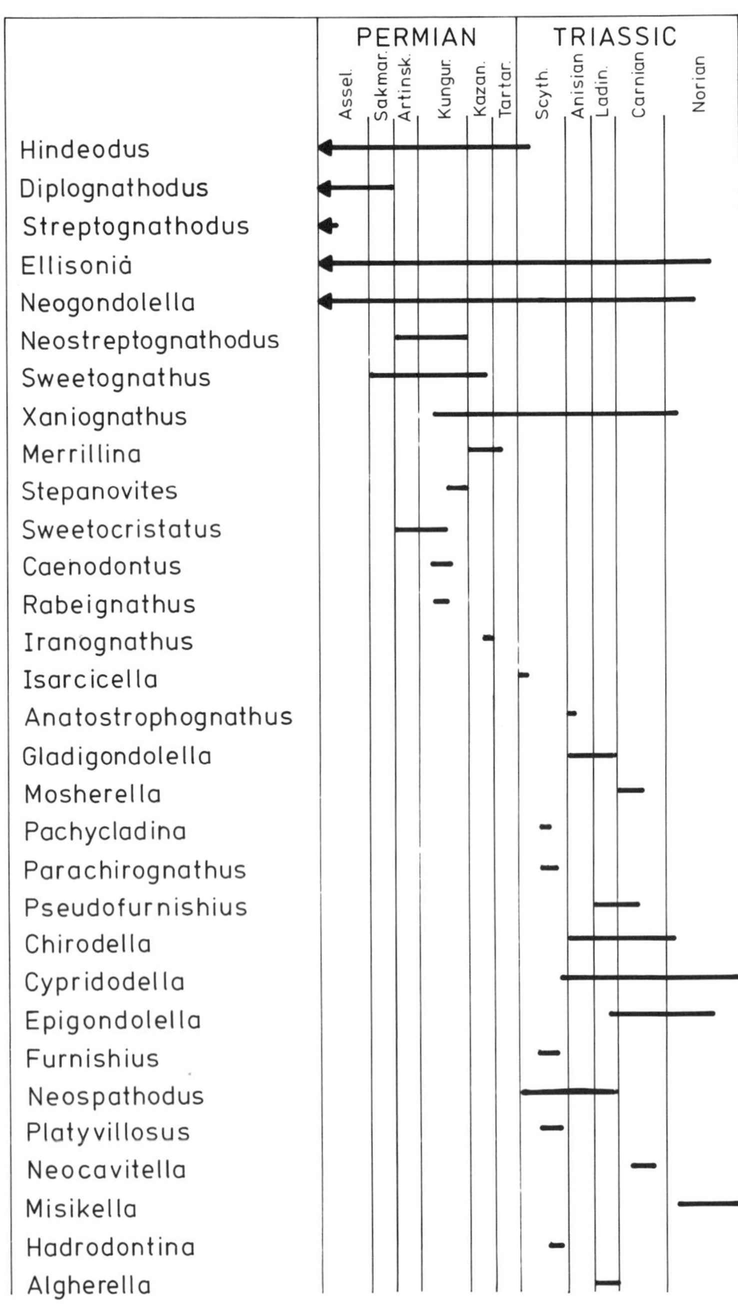
PERMIAN
TRIASSIC
Assel.
Sakmar.
Artinsk.
Kungur.
Kazan.
Tartar.
Scyth.
Anisian
Ladin.
Carnian
Norian
Hindeodus
Diplognathodus
Streptognathodus
Ellisonia
Neogondolella
Neostreptognathodus
Sweetognathus
Xaniognathus
Merrillina
Stepanovites
Sweetocristatus
Caenodontus
Rabeignathus
Iranognathus
Isarcicella
Anatostrophognathus
Gladigondolella
Mosherella
Pachycladina
Parachirognathus
Pseudofurnishius
Chirodella
Cypridodella
Epigondolella
Furnishius
Neospathodus
Platyvillosus
Neocavitella
Misikella
Hadrodontina
Algherella

12. Extinction and survival of the jawless vertebrates, the Agnatha

L. BEVERLY HALSTEAD

Departments of Geology and Zoology, University of Reading, Reading, UK

Abstract

The jawless vertebrates, the agnathan fishes, are first known in the Ordovician of Australia, North America, South America (Gagnier *et al.* 1986), and Siberia; thereafter, they become an important element of the fauna in marine Silurian and, subsequently, in freshwater Devonian rocks. There was a major extinction event at the end of the Lower Devonian (Emsian) and a further more terminal extinction at the end of the Frasnian. During Carboniferous times the naked lampreys made their appearance and these survive to the present day as ectoparasites. The living hagfish, which are also naked and closely related to the lampreys, are important marine scavengers. The agnathan survivors are either parasites or scavengers.

Nature of data base

The primitive jawless vertebrates of Agnatha represent a primitive grade of organization, they naturally have a number of features in common, the most striking of which is the absence of jaws and teeth. The consequence of this is that they are restricted in their mode of life. A life of active predation is denied them and, in general, they must have been microphagous (i.e. mud grubbers).

The agnathans fall into two contrasted groups, which do not seem to be directly related to one another. One the Monorhina, to which the living cyclostomes clearly belong, is characterized by the possession of a single

Extinction and Survival in the Fossil Record (ed. G. P. Larwood), Systematics Association Special Volume No. 34, pp. 257–67. Clarendon Press, Oxford, 1988.

median nasal organ and pouch-like gills; three exclusively fossil groups, the cephalaspids, anaspids, and galeaspids belong here. The other group, the Diplorhina, has paired nasal organs and fish-like gills; the thelodonts and heterostracans seem to be related to the basic stock of all the higher vertebrates.

The data base for the agnathan fishes comes from the *Handbook of Palaeoichthyology* Volume 1 (now in preparation) in which all species are briefly described, the main emphasis being on genera. There are approximately some 750 species known at present and they fall into several major groupings: the cephalaspids, some 150 species being dealt with by J. C. Collins, 15 anaspids, 37 galeaspids and four lampreys by L. B. Halstead, and 100 thelodonts by Susan Turner and 450 heterostracans by L. B. Halstead. This last group is divided into several major parts, the more important of which are the cyathaspids, amphiaspids, pteraspids, and psammosteids.

Particular advances have been made recently as a consequence of the publication of a number of monographic studies by Larissa Novitskaya on the pteraspids (1983, 1986), by Blieck (1984) also on the pteraspids, Janvier (1985) on cephalaspids and Turner 1973 on thelodonts (see below). There has also been a continuing stream of papers from the Chinese on the newly discovered galeaspids (see below). A consequence of recent studies is that many species have disappeared, not by extinction, but in synonymies. By the same token, there has been a proliferation of new genera, although at the time of writing many of these still remain unpublished.

A major factor which introduces a bias in the data base is the distribution of palaeontologists. The workers on fossil fishes have been concentrated in Western Europe and North America, where the faunas of the Old Red Continent have been intensively studied. The translation of Dineley to Canada, led to major contributions from the Canadian Arctic. The work of Karatajute-Talimaa and Novitskaya in the Soviet Union brought to light major discoveries in Siberia and the settling of Australia by Susan Turner has resulted in a revival of Australian agnathans. However, one of the most fascinating aspects of current fish palaeontology is the role of the peripatetic French, who seem to be involved in all the major discoveries all over the world, in such places as the United States, Iran, Thailand, and Bolivia, as well as in their more traditional areas of research such as Algeria. The significance of the French discoveries is that they indicate that agnathan distribution and evolution is likely to be more complex than the research in Euroamerica had previously suggested.

The data base is comparatively excellent for North America and Europe as far east as the Urals. However, Siberia, China, and Australia are the three areas where there is now renewed interest, and major discoveries are being continually made. The isolated scales from South America simply indicate the possibility of more important discoveries in the future the most recent being the record of a new heterostracan from the Ordovician of Bolivia (Gagnier *et al.* 1986).

Patterns of extinction

The jawless vertebrates are such a small group that they tend to be incorporated into larger taxonomic groupings such as 'fish'. When any range chart of the agnathans is examined it is evident that the last cephalaspid, anaspid, galeaspid, thelodont, and heterostracan becomes extinct during the Upper Devonian at the end of the Frasnian and before the advent of the Famennian. It is generally assumed that the microphagous agnathans, generally considered to be detritus feeders or mud grubbers, were replaced by benthonic detritivore placoderms, the antiarchs such as *Bothriolepis* which became ubiquitous at the end of the Devonian. This period of major faunal change can be linked to a major transgressive event that must have a profound effect on the physical conditions where the agnathans had once flourished. From a general overview the pattern of extinction appears simple and straightforward, and coincides with the major well recognized Frasnian–Famennian extinction event. However, when the detailed history of the different agnathan groups are examined a different and distinctive pattern begins to emerge.

1. Monorhina—Cephalaspida (Osteostraci)

The cephalaspids were clearly adapted for a benthonic mode of life. With the exception of *Trematapsis* and its allies, the ventral part of the carapace was flat, the dorsal convex; the tail was heterocercal and there were paired pectoral flaps. In the later genera there were lateral extensions of the carapace lateral to the pectoral flaps — the cornua. In the more advanced forms the headshield was filled by bone so that the courses of the nerves and blood vessels can be traced in considerable detail.

The first possible record is a minute fragment of bone seen in a microscope section from the Middle Ordovician Harding Sandstone of Colorado, USA (Ørvig 1965). Cephalaspids first appear in numbers in the Wenlock of Scotland, and Oesel, Estonia. Thereafter, cephalaspids flourish in the late Silurian and early Devonian throughout the Old Red Continent, generally dying out during Pragian—Emsian times. They can only be said to have really flourished in post-Lochkovian times in Spitsbergen. The cephalaspids were effectively extinct at the end of the Lower Devonian and were really no longer a significant part of the vertebrate faunas throughout the world (Halstead and Turner 1973).

However, a single species, *Meteoraspis magnifica,* from the Caithness Flags at Spital, near Thurso, Scotland, marks the single record of an undisputed

Middle Devonian cephalaspid at about the boundary between the Eifelian and Givetian, that is in the very middle of the Middle Devonian. This would seem to represent relict fauna surviving in a isolated locality.

There are two further yet unrelated cephalaspid genera, *Escuminaspis* and *Alaspis,* known from the Upper Devonian, Frasnian, of Scaumenac Bay, Quebec, Canada. These late Devonian genera are the sole representatives from the Upper Devonian and again must be considered 'living fossils', surviving as relicts.

To state that the cephalaspids survived into the Frasnian of the Upper Devonian which is indeed the case, obscures the important fact that the mass of cephalaspids actually died out during the Lower Devonian and only in Spitsbergen did they remain successful to the very end of the Lower Devonian. Thereafter, it seems merely fortuitous that the odd form simply happened to survive in a couple of isolated locations.

2. *Monorhina—Anaspida*

The anaspids ranging in size from 5 to 30 cm in length were the most fish-like in appearance of all the ostracoderms. The head region was covered by small scales, the trunk by deep scales. The tail was reversed heterocercal (i.e. hypocercal) and was associated with elongated ventrolateral fins. On the dorsal surface of the head there was a pineal eye in front of which was situated a single nasohypophysial opening, the gill pouches had separate openings. The cartilaginous branchial basket supporting the gill pouches is preserved in the genus *Jamoytius* and is comparable to that of the lamprey — a living agnathan.

The anaspids are recorded from the Llandovery, Lower Silurian of Cornwallis Island in the Canadian Arctic and thence in the Wenlock of Scotland and Estonia, and the Ludlovian and Pridolian of Norway. This group was restricted geographically to the Canadian–Scotland–Norway–Baltic province. A single fragment of anaspid squammation is known from the Lochkovian of the Welsh Borderland.

There is no further record until the Upper Devonian, Frasnian, of Scaumenac Bay, Quebec, Canada, where the two general *Euphanerops* and *Endeiolepsis* occur. Once again, as with the cephalaspids, this group seems to have vanished from its original range, in this instance at the beginning of the Lower Devonian, and survived only in a single region (Halstead and Turner 1973).

3. *Monorhina—Galeaspida*

The galeaspids were first described from the Lower Devonian of Yunnan, China. These forms known only from the South China province were characterized with a carapace in which the dorsal surface was pierced by two orbits and there was a large median perforation in the midline anterior to the orbits, there were separate gill openings. These animals appear to be related to the cephalaspids, but do not possess the characteristic lateral and dorsal sensory fields (see Halstead *et al.* 1979; Halstead 1985). Wang *et al.* (1980), Pan and Zeng (1985), and Wang (1986) extended the stratigraphical range into the Silurian down to the Llandovery. Pan (1984) recorded incomplete headshields of a polybranchiaspid from red sandstones in Ningxia, West China, associated with the antiarch *Remigolepis,* suggesting a late Upper Devonian, Fammenian age.

In view of the normal early appearance of the Chinese forms, it is likely that *Remigolepis* and the last galeaspid are actually of Frasnian age, but further stratigraphic work is needed before the age can be confirmed. The South China region was inhabited by galeaspids from Lower Silurian through to the Upper Devonian, but the main evolutionary development was confined to the Lower Devonian.

4. *Monorhina—Petromyzones*

The living lampreys with their elongated eel-like shape, round sucker mouth, rasping tongue, and horny teeth, are important ectoparasites at the present time, and appear to be closely related to the primitive *Jamoytius* type of anaspid. The earliest record is *Hardistiella* from the Lower Carboniferous Namurian, Bear Gulch Limestone of Montana described from a single specimen by Janvier and Lund (1983). The first discovered lamprey *Mayomyzon* was recovered from the Upper Carboniferous Westfalian in Illinois by Bardack and Zangeri (1968). Subsequently, Bardack and Richardson (1977) described two curious genera *Pipiscius* and *Gilpichthys* from the same deposits, both of which seem to have been highly specialized lampreys. None of these four fossil genera provides evidence of being an ectoparasite. It may well be that this mode of life evolved subsequently.

The Upper Devonian Frasnian *Scaumenella* was tentatively identified as a larval lamprey by Tarlo (1967), but recently Béland and Arsenault (1985) demonstrated that this genus was in fact an acanthodian that had undergone degradation during fossilization, a process for which they coined the term scaumenellization.

5. *Monorhina — Myxines*

There is no known record of fossil hagfish although Tarlo (1967) tentatively suggested that *Palaeospondylus* from the Middle Devonian, Caithness flags might have been a large hag — a suggestion that now seems unlikely. The extant hags are important scavengers of the in-fauna.

6. *Diplorhina — Thelodonti*

Thelodonts were equipped for swimming, possessing dorsal, anal, and triangular lateral fins, a hypocercal tail to balance the large head. They possessed a squamation of non-imbricating dentine scales, tooth-like in structure, formed all over the body.

Upper Silurian (Ludlovian and Pridolian) and Lower Devonian (Lochkovian) thelodonts are well known from Britian and the Baltic. Furthermore, Lower Silurian (Llandovery) forms have been recorded from the Welsh borderland, as well as in Siberia where they made their first appearance in the Ordovician. In the Old Red Continent, the thelodonts seem to have died out in most places during the Pragian. The major exception was in Spitsbergen where thelodonts occur in the Wood Bay Series which marks the Emsian–Eifelian boundary. It appears, therefore, that in Spitsbergen the thelodonts survived in isolation probably into the Middle Devonian (Eifelian) (Turner, *in* Halstead and Turner 1973).

Thelodonts were described from the later Lower Devonian of Australia by Gross (1971) and Turner *et al.* (1981) from sediments of Emsian age. Middle Devonian thelodonts were described by Blieck and Goujet (1977) from south east Iran and also from beds of presumed similar age in Thailand. Further Middle Devonian thelodonts were described from north east Iran by Turner and Janvier (1979). Thelodonts have now been recorded from the Lower Devonian of Indonesia (Turner and Dring 1981) and China (Wang 1984) as well as from South America (Goujet *et al.* 1984).

It is only recently that, from the isolated discoveries of thelodonts in China and the component parts of Gondwanaland, it has been realized that the thelodonts continued much later in the southern continents than in the traditional northern Old Red Sandstone provinces. The description by Turner and Dring (1981) of Upper Devonian (Frasnian) thelodonts from Australia demonstrated that their survival in Australia went far beyond the Middle Devonian (Eifelian).

The dermal denticles from the Upper Devonian Beacon Sandstone of Granite Harbour, Antarctica, collected on Scott's Antarctic expedition and described by Woodward (1921) as 'primitive ostracoderms or elasmobranchs' were later identified as either thelodont scales or heterostracan psammosteid turbercles, but were identified as the latter primarily because thelodonts were not known after Eifelian times, whereas psammosteids were well known from the Upper Devonian (Halstead Tarlo 1966). It now seems more likely that this material should be attributed to thelodonts and if this attribution is correct, it would confirm Turner and Dring's view that the thelodonts may have survived into the Upper Devonian in other regions of Gondwanaland besides Australia.

The history of the thelodonts indicates that their main extinction was during the Lower Devonian in the northern continents with their survival in isolated provinces such as Spitsbergen into the base of the Middle Devonian. However, it is now apparent that they continued to flourish in the southern continents well into the Upper Devonian, finally succumbing during the Frasnian.

7. Diplorhina — Heterostraci

The heterostracans were by far the most variable of all the agnathans. The head was encased in a bony armour of tesserae or large discrete plates, in some forms a single ossification. Although there were no moveable lateral fins, extensions of the carapace served as stabilizing organs. Wherever preserved the tails were reversed heterocercal (hypocercal). The pattern of plates making up the carapace is the basis of their classification (see Halstead 1973). The early examples range from 10 to 30 cm, but some of the later forms achieved lengths of 2 m. The heterostracans appear to have been by far the most successful of all the ostracoderms and they certainly have been the most intensely studied.

The 'simplest' types are the cyathaspids which first appear in the Llandovery of the Canadian Arctic becoming generally extinct at the end of the Lochkovian in Western Europe. However, in the western United States in Wyoming, Utah, and Ohio they continued well into the Pragian.

The pteraspids which first appeared in the Ludlovian in the Canadian Arctic were an important aspect of the agnathan faunas and are used as important zonal indices in non-marine sediments of the Lower Devonian age. At the end of Lower Devonian times there was a marked reduction in diversity in western Europe where they survived only until the end of the Emsian. Føyn and Heintz (1943) and Dineley (1955) described pteraspid remains from the Wijde Bay Formation in Spitsbergen, which is dated as Middle Devonian (Givetian). This indicated that the pteraspids managed to survive in an isolated province long after the group had become extinct elsewhere.

That the situation is more complex than this is evidenced from Blieck's (1984) recent monograph, where he records Middle Devonian (Eifelian) pteraspids from Severnaya Zemlya, Taimyr, and Norilsk of Siberia, as well as from the Ahretouche Formation of the Western Dra of Morocco. There is also a possible record from near Lake Baikal in Soviet Kazakhstan of reputed Givetian age.

Siberia, in this case, seems to have been the province where the pteraspids were able to continue in a sense comparable to the situation in Australia with regard to the thelodonts. The other late pteraspids come from the Jefferson Formation in Idaho, USA, which are also of Givetian age, but which were at first identified as psammosteids on the grounds of their tubercular ornamentation and late age (see Halstead Tarlo 1965, 1966).

The heterostracan group that was endemic to Siberia during the Lower Devonian was the amphiaspids which flourished during the Lower Devonian (Pragian) in Taimyr, Kureyka-Norilsk, and the new Siberian Islands. Two genera, *Amphiaspis* and *Pelurgaspis,* however, managed to continue in the Kureyka region into the overlying late Emsian stage (see Novitskaya 1976*a,b*; and Halstead 1985).

The heterostracans that flourished during the Middle (Givetian) and Upper (Frasnian) Devonian were the psammosteids. This group is best documented from the Baltic Province, where it appears to have undergone its major evolution (Obruchev and Mark-Kurik 1965). The psammosteids continued until the Frasnian in the Baltic, Timan, Poland and Scotland (Halstead Tarlo 1965, 1966). However, in the Baltic a few examples of the most advanced psammosteid *Psammosteus tenuis* (syn. *P. waltergrossi, P. grossi*) survived to the very end of the Frasnian. This species was the last heterostracan so far recorded.

Conclusions

A main event in the history of the fossil agnathans was the colonization of non-marine habitats during the late Silurian and early Devonian. This is well documented in Euramerica and is related to the closure of the Iapetus Ocean to form the Old Red Continent. Further invasions of non-marine conditions took place in Siberia and in China but the details and exact timing are less well documented (see Halstead 1985).

The main evolutionary development of the agnathans was during the Lower Devonian, toward the end of which there was a gradual decline culminating in a general extinction of the agnathans (Fig. 12.1). The cyathaspid heterostracans and the anaspids seem to have died out during the early part of the Lower Devonian in the Lockhovian (Gedinnian). The anaspids occur in the Upper Devonian Frasnian of Quebec, but this seems to be a surviving 'living fossil' in an isolated refuge. The cephalaspids in the main died out at the end of the Lower Devonian, but a single genus is known from the Middle Devonian Caithness Flags of the Orcadian Basin of Scotland and in the Upper Devonian of Scaumenac Bay, Quebec, Canada, again a 'living fossil' remnant. The thelodonts appear to have died out before the end of the Lower Devonian, but in Spitsbergen the thelodonts continued into the early Middle Devonian. Subsequently, the thelodonts are recorded in the Middle Devonian from Iran, Thailand, and Australia. In Australia they seem to have continued until the end of the Frasnian, having died out elsewhere with the possible exception of Antarctica.

The pteraspids again seem to have died out towards the end of the Lower Devonian, but in Spitsbergen they survived into the latter part of the Middle Devonian Givetian. Apart from the Middle and Upper Devonian thelodonts of Australia the only group of agnathans to flourish during the latter part of the Devonian were the psammosteid heterostracans, but these too finally became extinct at the end of the Frasnian.

The history of the agnathans is marked by two important periods of

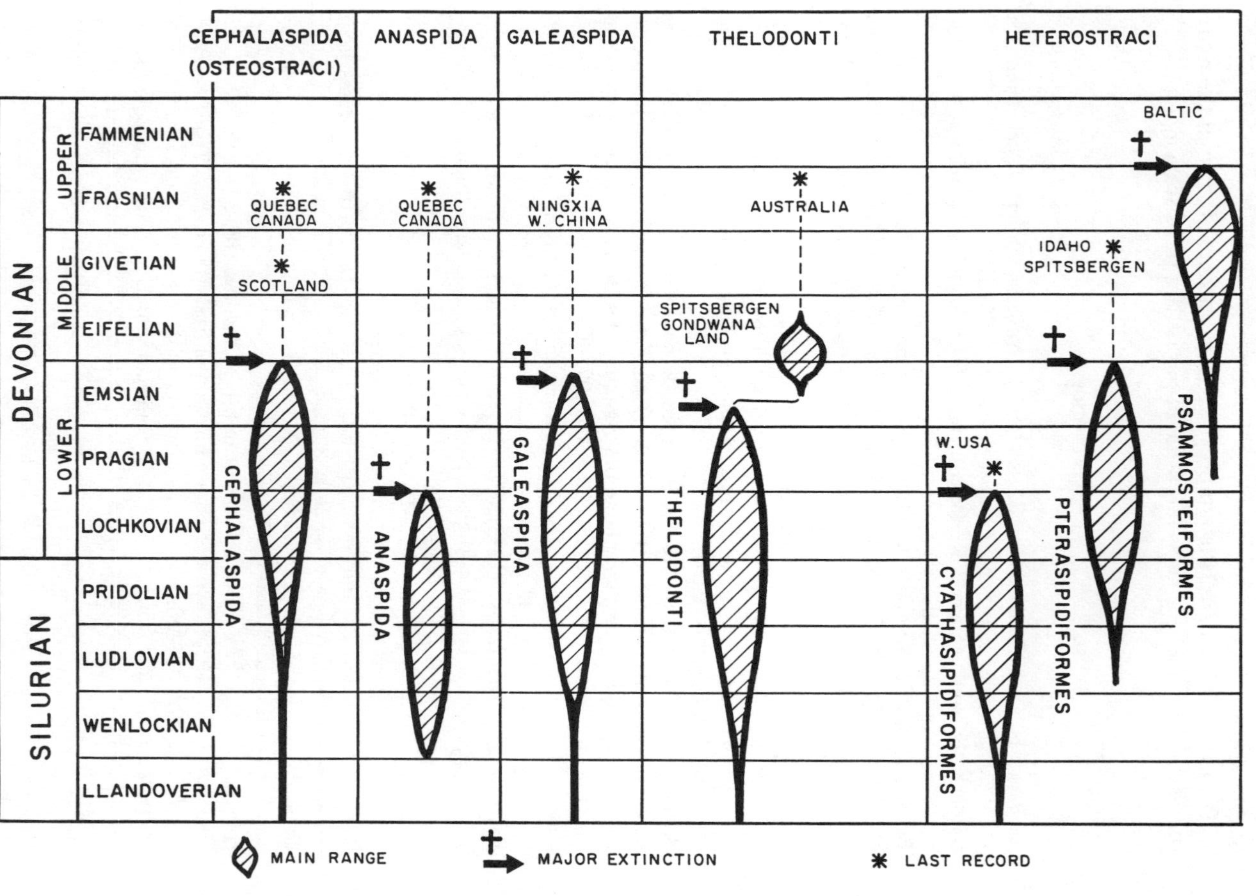

Fig. 12.1. Stratigraphic range of fossil agnathans.

extinction: the earlier end-Emsian and the later end-Frasnian events (Fig. 12.1). Both seem to coincide with major global marine transgressions. marine transgressions.

The extant agnathans, the lampreys and hags, exemplify two of the more successful strategies for survival, a life of parasitism or of scavenging.

References

Bardack, D. and Richardson, E. S. (1977). New agnathous fishes from the Pennsylvanian of Illinois. *Fieldiana* (*Geol.*) **33,** 489–510.

Bardack, D. and Zangerl, R. (1968). First fossil lamprey: a record from the Pennsylvanian of Illinois. *Science* **1962,** 1265–7.

Béland, P. and Arsenault, M. (1985). Scaumenellisation de l'Acanthodii Trizeugacanthus affinis (Whiteaves) de la formation d'Escuminac (Devonien Superieur de Miguasha, Quebec): revision du *Scaumenella mesacanthi* Graham-Smith. *Can. J. Earth. Sci.* **22,** 514–24.

Bleick, A. (1984). Les Héterostracés Pteraspidiformes, Agnathes du Silurien–Dévonien du Continent Nord — Atlantique et des Blocs Avoisiments: Révision systèmatique, phylogenie, biostratigraphie, biogeographic. *Cahiers paléont.* (*sect. vert.*). Éd. C.N.R.S. Paris.

Blieck, A. and Goujet, D. (1977). A propos de nouveau materiel de Thelodontes (Vertébrés Agnathes) d'Iran et de Thailande: aperçu sur la repartitition geographique et stratigraphque des Agnathes des 'regions gondwaniennes' au Paleozoique moyen. *Ann. Soc. Géol. Nord.* **97,** 363–72.

Dineley, D. L. (1955). Some Devonian fish remains from North Central Vestspitsbergen. *Geol. Mag.* **92,** 255–60.

Føyn, S. and Heintz, A. (1943). The Downtonian and Devonian vertebrates of Spitsbergen. VIII. The English–Norwegian–Swedish Expedition 1939. Geological Results. *Skr. Svalb. Ishavet* **85,** 1–51.

Gagnier, P.-Y., Blieck, A. R. M., and Rodrigos, G. (1986). First Ordovician vertebrate from South America. *Geobios* **19,** 629–34.

Goujet, D., Janvier, P., and Suarez-Riglos, M. (1984). Devonian vertebrates from south America. *Nature, Lond.* **312,** 311.

Gross, W. (1971). Unterdevonische Thelodontier- und Acanthodier-Schuppen aus Westaustralien. *Paläont. Z.* **45,** 97–106.

Halstead, L. B. (1973). The heterostracan fishes. *Biol. Rev.* **48,** 279–332.

Halstead, L. B. (1985). The vertebrate invasion of fresh water. *Phil. Trans. Roy. Soc. Lond.* **B309,** 243–58.

Halstead, L. B. and Turner, S. (1973). Silurian and Devonian ostracoderms. In *Atlas of Paleobiogeography* (ed. A. Hallam), pp. 67–79. Elsevier, Amsterdam.

Halstead, L. B., Liu, Y-H., and P'an, K. (1979). Agnathans from the Devonian of China. *Nature, Lond.* **282,** 831–3.

Halstead Tarlo, L. B. (1965). Psammosteiformes (Agnatha), a review with descriptions of new material from the Lower Devonian of Poland. I General Part. *Palaeontologia Polonica* **13,** 1–135.

Halstead Tarlo, L. B. (1966). Psammostiformes (Agnatha), a review with descriptions of new material from the Lower Devonian of Poland. II. Systematic Part. *Palaeontologia Polonica* **15,** 1–168.

Harland, W. B., *et al.* (1967). *The Fossil Record.* Geological Society of London.

Janvier, P. (1985). Les Cephalaspides du Spitsberg. *Cahiers paléout.* (*Sect. vert.*). Ed. C.N.R.S., Paris, 1–244.

Janvier, P. and Lund, R. (1983). *Hardistiella montanensis.* n. gen. et. sp. (Petromyzontida) from the Lower Carboniferous of Montana, with remarks on the affinities of the lampreys. *J. Vert. Paleont.* **2,** 407–13.

Novitskaya, L. I. (1976*a*). The question of the separation of the Kureykan Suite on the amphiaspid (Agnatha) complex. *Isvest. Akad. Nauk SSR. Ser. Geol.* **1,** 47–55.

Novitskaya, L. I. (1976*b*). The palaeoecological precursors for the formation of faunas of amphiaspidians in Siberia. In *Palaeontology*: *Marine Geology* (ed. B. S. Sokolov pp. 95–103. Ed. 'Nauka', Moscow.

Novitskaya, L. I. (1983). The morphology of ancient agnathans. *Trud. Paleont. Inst. Moscow* **196,** 1–182.

Novitskaya, L. I. (1986). Ancient agnathans of the U.S.S.R. *Trud. Paleont. Inst. Moscow* **219,** 1–160.

Obruchev, D. V. and Mark-Kurik, E. (1965). *Psammosteids* (*Agnatha, Psammosteidae*) *from the Devonian of the U.S.S.R.* Tallin, Estonia.

Ørvig, T. (1965). Palaeohistological notes. 2. Certain comments on the phyletic significance of acellular bone tissue in early lower vertebrates. *Ark. f. Zool.* **16,** 551–6.

Pan, J. (1984). The phylogenetic position of the Eugaleaspida in China. *Proc. Linn. Soc. N.S.W.* **107,** 309–19.

Pan, J. and Zeng, X. (1985). Dayongaspidae, a new family of Polybranchiaspiformes (Agnatha) from early Silurian of Hunan, China. *Vertebrata Pal. Asiatica* **23,** 207–13.

Tarlo, L. B. H. (1967). Agnatha In *The Fossil Record* (ed. W. B. Harland *et al.*), pp. 629–36. Geol. Soc. London.

Turner, S. (1973). Siluro-Devonian thelodonts from the Welsh Borderland. *Q. J. geol. Soc. Lond.* **127,** 557–84.

Turner, S. and Dring, R. S. (1981). Late Devonian thelodonts (Agnatha) from the Gneudna Formation, Carnarvon Basin, Western Australia. *Alcheringa* **5,** 39–48.

Turner, S. and Janvier, P. (1979). Middle Devonian Thelodonti (Agnatha) from the Khush-Yeilagh Formation, north-east Iran. *Geobios.* **12,** 889–92.

Turner, S., Jones, P. J., and Draper, J. J. (1981). Early Devonian thelodonts (Agnatha) from the Toko Syncline, Western Queensland, and a review of other Australian discoveries. Bureau of Mineral Resources *J. Austr. Geol. Geophys.* **6,** 51–69.

Wang, N. (1984). Thelodont, acanthodian and chondrichthyan fossils from the Lower Devonian of southwest China. *Proc. Linn. Soc. N.S.W.* **107,** 419–41.

Wang, N. Z. (1986). Notes on two genera of Middle Silurian Agnatha (*Hanyangaspis* and *Latirostraspis*) of China. *Palaeont. Soc. China, 13–14 Conv.* 49–57.

Wang, S., Xia, S., Chen, L., and Du, S. (1980). On the discovery of Silurian agnathans and Pisces from Chaoxian County, Anhui Province and its stratigraphical significance. *Bull. Chinese Acad. geol. Sci.* **1,** 101–12.

Woodward, A. S. (1921). Fish remains from the Upper Old Red Sandstone of Granite Harbour, Antarctica *Br. Antarct.* ('*Terra Nova*') *Exped. 1910* **1,** 51–62.

13. Mass extinctions in the fossil record of reptiles: paraphyly, patchiness, and periodicity (?)

M. J. BENTON

Department of Geology, The Queen's University of Belfast, Belfast, UK.

Abstract

The fossil record of reptiles has been cited frequently in discussions of mass extinction events and, in particular, the event at the end of the Cretaceous when the dinosaurs, pterosaurs, and marine reptiles died out. There are a number of problems in analysing the fossil record of reptiles, not least the fact that 'Reptilia' is a paraphyletic group; that is, it is partially a human invention. One solution to that problem is to expand the study to include all Amniota (i.e. reptiles, birds, and mammals). The fossil record of reptiles spans 300 Ma and includes 233 families, of which 200 are non-marine and only 43 are still living. The fossil record of reptiles is no poorer than that of other tetrapod groups, according to a Simple Completeness Metric (SCM), but the completeness varies greatly from stage to stage.

The diversity of families of fossil amniotes remained relatively low until the late Cretaceous, when levels rose rapidly from a total of about 50 families world-wide to the present figure of 329. There is evidence for at least six, and possibly as many as thirteen, mass extinctions in the fossil record of amniotes, and the intensity of these events varies greatly. Taken at face value, the amniote fossil record does not support a model of periodic mass extinctions.

Extinction and Survival in the Fossil Record (ed. G. P. Larwood), Systematics Association Special Volume No. 34, pp. 269–94. Clarendon Press, Oxford, 1988.

Introduction

In considering mass extinctions, the fossil record of reptiles is often cited. The disappearance of the dinosaurs, as well as the flying reptiles, the pterosaurs, and the marine reptiles, the ichthyosaurs, plesiosaurs, and mosasaurs, at the end of the Cretaceous period comes to most people's minds as the best known example of a mass extinction. However, on closer analysis, it turns out that less is known about this event than is commonly assumed. Recent papers offer diametrically opposed views of how the dinosaurs disappeared — suddenly and catastrophically (Russell 1984), or gradually, over many millions of years (Schopf 1982; Carpenter 1984; Sloan *et al.* 1986). The Cretaceous–Tertiary boundary mass extinction event (the 'K–T event') is only one of several mass extinctions that have affected the history of reptiles, but data are just as confused for most of these as well.

In this paper, some aspects of the quality of the fossil record of reptiles will be considered — how incomplete is it and is it uniformly incomplete? Then, the patterns of diversification and mass extinction for the amniotes in general, and the various reptile groups in particular, will be outlined with comments on the extinct forms and the survivors. The nature of each event will be considered in turn, and the suggestion of a large-scale pattern, such as regular cyclicity, of mass extinction events will be discussed. However, first of all, we must consider whether there is any meaning in the phrase 'rates of evolution or extinction of the reptiles'. Is this a biologically meaningful concept?

Rates of evolution of paraphyletic groups

The 'Reptilia', as understood by most biologists and palaeontologists (see e.g. Romer 1966) is a paraphyletic group. That is, all reptiles derive from a single common ancestor (or the group could be defined in that way by juggling some early groups between 'Amphibia' and 'Reptilia' — the relationships of the earliest reptiles, and of the 'reptile-like amphibians', are controversial, and it is still not clear where the boundary should be drawn: Carroll 1969, 1970, 1982; Heaton 1980; Heaton and Reisz 1986). However, the group 'Reptilia' does not include all of the descendants of that ancestor (Aves and Mammalia are excluded).

Rates of evolution, extinction, origination, and so on, should apply to monophyletic groups (*sensu* Hennig 1966; that is, holophyletic groups, *sensu* Ashlock 1971) only, and not to paraphyletic groups. This is because monophyletic groups (clades) have a unique history that exists and is to be discovered, whereas paraphyletic groups may start off with a unique history, but their boundaries are adjusted *a posteriori* and they are in part a human invention (Cracraft 1981).

In the case of the reptiles, the Carboniferous, Permian, or early to middle Triassic systematist of reptiles would be observing a monophyletic group, whether he were a cladist or not. He could measure the rate of evolution of Reptilia with impunity since it is a monophyletic group in all senses. However, by the end of the Triassic period and in the Jurassic a contemporary taxonomist would have had a problem — should he recognize the two or three families of small hairy reptiles with their advanced jaws and inner ears as the new Class Mammalia, or not? Probably not, since their future key role in the history of life could not have been predicted. Likewise, the end-Jurassic or early Cretaceous taxonomist would have had to decide where to assign the first few feathered reptiles — whether to a new Class Aves, or whether to leave them alone.

The decision to extract Mammalia and Aves from 'Reptilia' is clearly an *a posteriori* decision. In a study of the rates of evolution and extinction of reptiles, these two groups should be included in order to make the clade Amniota complete. Their exclusion artificially dents the diversity of the clade, and would clearly lower the values of any calculated measures of rates of origination or diversification. Until the origin of the Mammalia (generally reckoned to have occurred in the latest Triassic), of course, 'Reptila' = Amniota, but after that, 'Reptilia' ceases to be a monophyletic group, and thus it cannot be used as a meaningful entity in macroevolutionary analysis. Nevertheless, several authors (e.g. Simpson 1952; Cutbill and Funnell 1967; Charig 1973; Pitrat 1973; Thomson 1977; Olson 1982) have done so.

No doubt, many other paraphyletic, and even polyphyletic, groups have been included in recent macroevolutionary analyses (see also Cracraft 1981). It is to be hoped that serious attempts will be made to use only clades in future studies — monophyletic groups that include all the descendants of a single common ancestor. The best way to identify clades is, of course, by means of cladistic analysis.

The fossil record of reptiles

1. Scope

The first reptiles, and thus the first amniotes, according to most recent classifications (e.g. Romer 1966; Anderson and Cruickshank 1978; Carroll 1982; Heaton and Reisz 1986) are the Protorothyrididae (= Romeriidae), known first from the Moscovian Stage (*c.* 300 Ma) of the Late Carboniferous. During the remaining 15 Ma or so of the Carboniferous, the early amniotes diversified into a number of additional lineages — the Araeoscelidia (Reisz *et al.* 1984) and the 'Pelycosauria' (Kemp 1982), the most primitive groups of the Diapsida and Synapsida, respectively. These

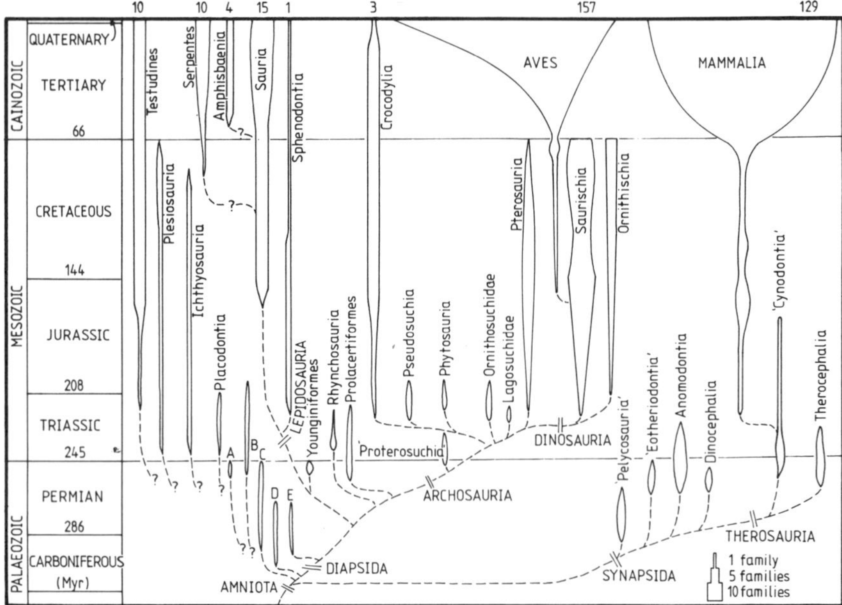

Fig. 13.1. Phylogenetic tree of the Amniota, showing relationships, stratigraphic duration, and diversity of each group. The major groups are indicated as balloons that show the known stratigraphic range by their height, and the relative numbers of families present by their width (see scale in bottom right-hand corner). Abbreviations: A, Pareiasauria; B, Procolophonia; C, Captorhinidae; D, Protorothyrididae; E, Araeoscelidia. Relationships of the groups are indicated by dashed lines on the basis of recent cladistic analyses (e.g. Gaffney 1980; Kemp 1982; Benton 1984, 1985*a*; Gauthier and Padian 1985; Heaton and Reisz 1986).

two amniote clades dominated tetrapod evolution from the late Carboniferous to the present day. The Diapsida radiated during the Permian and Triassic, and gave rise to sphenodontians, the ancestors of lizards and snakes, crocodiles, and dinosaurs in the Triassic, and birds in the Jurassic. The Synapsida radiated in the Permian and Triassic as the mammal-like reptiles, and gave rise to the mammals towards the end of the Triassic.

The diversity and importance of the Diapsida and Synapsida are indicated in the phylogenetic tree in Fig. 13.1. On the left-hand side are a number of lineages whose relationships are uncertain — the Permo-Triassic procolophonians and pareiasaurs, the Testudines (turtles), and the marine plesiosaurs, ichthyosaurs, and placodonts.

In all, there are 233 families of fossil and living reptiles, at a conservative estimate (Benton 1987), of which 200 are non-marine and 33 are exclusively marine. These figures exclude monospecific, monogeneric, and other doubtful or paraphyletic families. Of these, only 43 families still survive. There are 702 families of fossil and living amniotes, of which 644 are non-marine, 58 are exclusively marine (see Appendix), and 329 are still living.

2. *Quality of the fossil record*

The relative incompleteness of the fossil record of tetrapods has been described by many authors (e.g. Pitrat 1973; Bakker 1977; Carroll 1977; Olson 1982; Padian and Clemens 1985; Benton 1985*b*,*c*, 1987). The record of the non-marine tetrapods, which make up the vast bulk of all tetrapods, is particularly poor. Some stratigraphic stages, for example the Aalenian (Middle Jurassic), have yielded no identifiable tetrapod fossils at all anywhere in the world, and other stages [e.g. Gzelian (Carboniferous); Toarcian, Bajocian, Callovian, Oxfordian (Jurassic); Berriasian — Aptian, Cenomanian — Santonian (Cretaceous)] have yielded very few remains.

The incompleteness of the fossil record of terrestrial tetrapods has been characterized in another way by Padian and Clemens (1985, p. 82). Most dinosaur genera are known only from a single stratigraphic stage which would suggest, in a literal reading of the fossil record, that the dinosaurs experienced total generic mass extinction 24 or 25 times during their history. However, at the family level, there is only the one final K–T mass extinction event since dinosaur families generally span more than one stage.

The completeness of a fossil record can be estimated according to a Simple Completeness Metric (SCM), by assessing the relative numbers of taxa that are known to be present compared to the numbers that ought to be present (Paul 1982). In the present study, families are the taxa of interest, and each tetrapod family generally spans several stratigraphic stages. The family may be represented by fossils throughout its entire duration, or there may be gaps spanning one or more stratigraphic stages where fossils are absent. Jablonski (1986) has termed this the Lazarus Effect, where a taxon apparently disappears, and then reappears higher up in the sequence. The more incomplete the fossil record is for a particular stage, the more Lazarus (hidden) taxa there will be. If a stratigraphic stage is entirely devoid of fossils, the SCM (numbers of Lazarus taxa/numbers of taxa represented by fossils) will equal 0 per cent. If every taxon is represented by fossils, the SCM will equal 100 per cent.

Benton (1987) has assessed the completeness of the fossil record of non-marine tetrapods, stage by stage, and by taxonomic classes. According to the SCM, the Aalenian (Jurassic) had a value of 0 per cent, while the Visean (Carboniferous), Ufimian (Permian), and Scythian (Triassic) had values of 100 per cent. These values are only estimates of course, and the 100 per cent values are probably spurious since both the numbers of Lazarus taxa *and* the numbers of taxa represented by fossils are very small, and they are likely to be equally underestimated. For 'Reptilia' alone, the completeness of the fossil record matches that for tetrapods as a whole, with particularly weak areas (SCM < 50 per cent) in the early — middle Jurassic (Toarcian–Bajocian), the late Jurassic (Oxfordian), the late Cretaceous (Turonian – Santonian), and the early Palaeocene (Danian).

Overall, the fossil record of reptiles compares well with that of other tetrapods. For all stratigraphic stages, the SCM is 78.2 per cent, compared to 84.3 per cent for mammals, 56.9 per cent for birds, and 56.3 per cent for amphibians. For the major reptile groups, the values are: Testudines, 69.5 per cent; Diapsida, 59.7 per cent (Lepidosauria, 48.6 per cent, Archosauria, 63.1 per cent); and Synapsida, 94.5 per cent (Benton 1987).

3. *Family diversity data*

Several authors have recently plotted graphs of the diversity of reptile families and orders through time (e.g. Charig 1973; Pitrat 1973; Bakker 1977; Thomson 1977; Olson 1982; Padian and Clemens 1985; Colbert 1986). However, these graphs have been based largely on data from Romer (1966) and Harland *et al.* (1967), the classic source works. More recent studies (Benton 1985*b*,*c*, 1986*a*,*b*) have been based on a new compilation of data on families of non-marine tetrapods (Benton 1987). This is supplemented by a compilation of data on the marine families given here in the Appendix. These new compilations differ significantly from those derived from Romer (1966) and Harland *et al.* (1967) in several ways:

(1) New records up to the end of 1985 are included. This has affected the date of origination or extinction of as many as 50 per cent of families.

(2) The latest cladistic classifications have been incorporated, as far as possible, and attempts have been made to test that all families are clades. This has caused significant rearrangements of families of late Palaeozoic and Mesozoic reptiles in particular, by amalgamations and redistributions of genera into monophyletic taxa. Analyses of the diversity of orders and other higher taxa have not been carried out on the basis of the new compilation, but these would probably produce very different results from those in Olson (1982), Padian and Clemens (1985), and Colbert (1986).

(3) The stratigraphic resolution of family distributions has been improved. As far as possible, the dates of origination and extinction of each family have been determined to the nearest stratigraphic stage, usually by examination of the primary literature. The stage is the smallest practicable division of geological time for this compilation (relevant stage lengths vary from 2–19 Ma in length, with a mean duration of 6 Ma). This allows more detailed analysis than simply relying on the Lower, Middle, and Upper divisions of geological periods in Romer (1966) and elsewhere.

4. *Family diversity analysis*

The new compilations of data on fossil tetrapod family diversities have been used for a variety of graphs and calculations. A small number of

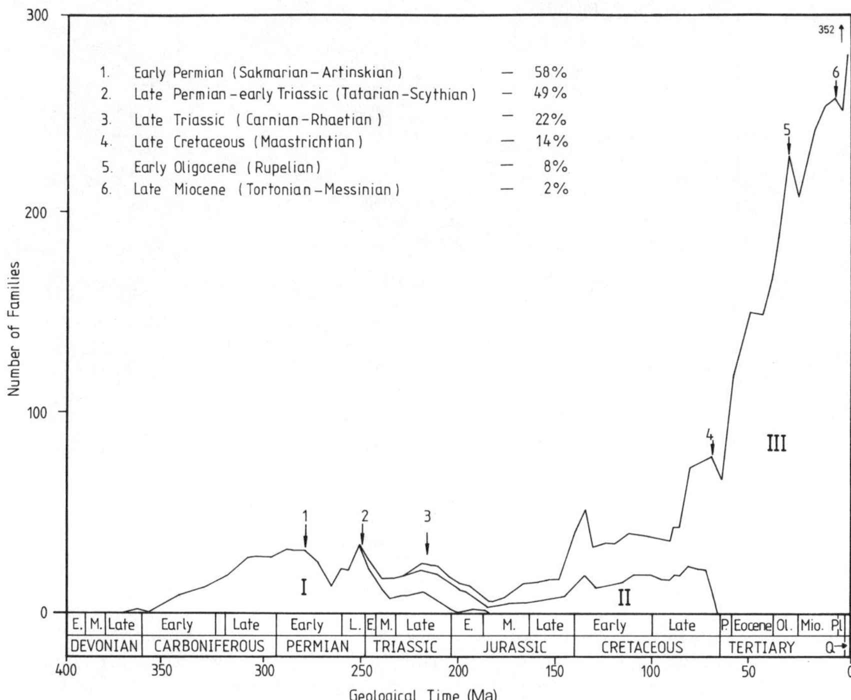

Fig. 13.2. Standing diversity with time for families of non-marine terrestrial tetrapods. The upper curve shows total diversity with time, and six apparent mass extinctions are indicated by drops in diversity, numbered 1–6. The relative magnitude of each drop is given in terms of the percentage of families that disappeared. The time-scale is that of Palmer (1983). Three assemblages of families succeeded each other through geological time: I, II, III (see text for details). Abbreviations: Carb., Carboniferous; Dev., Devonian; Mio., Miocene; Ol., Oligocene; P., Palaeocene; Pl., Pliocene.

families that are based on single species or single genera have been omitted. This adjustment to the data set does not materially affect the results, and it excludes a number of problematic specimens, as well as plesion 'families' that lack autapomorphies (full details in Benton 1985*c*,1987).

The diversity of non-marine tetrapods has increased through time, with a particularly rapid acceleration in the rate of increase from the late Cretaceous (Campanian) onwards (Fig. 13.2; Benton 1985*b*,*c*). Three major diversity assemblages have been identified (Benton 1985*c*), which appeared to dominate for a time, and then gave way to another: I (labyrinthodont amphibians, 'anapsids', mammal-like reptiles) dominated from late Devonian to early Triassic times; II (early diapsids, dinosaurs, pterosaurs) dominated during the Mesozoic; and III (the 'modern' groups — frogs, salamanders, lizards, snakes, turtles, crocodiles, birds, mammals) have dominated from late Cretaceous times to the present day.

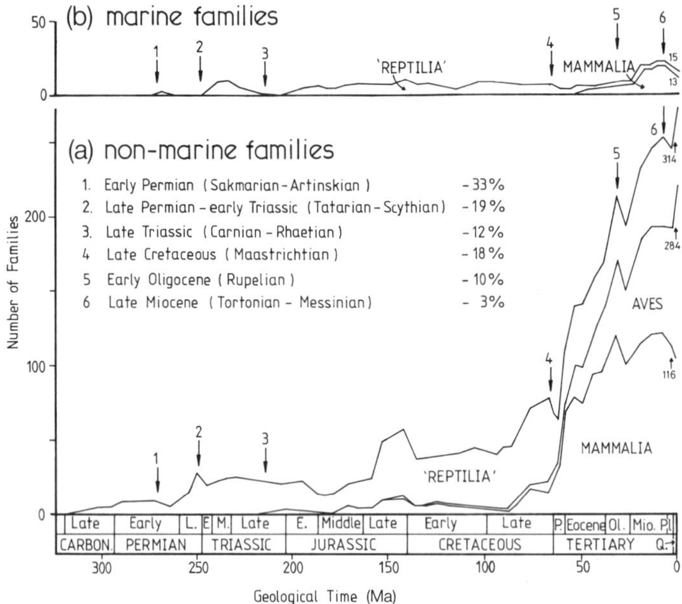

Fig. 13.3. Standing diversity with time for families of (a) non-marine and (b) marine amniotes. The curves for 'Reptilia', Mammalia, and Aves are shown separately. The relative magnitudes of the six drops in diversity that were probably caused by mass extinctions are shown for all amniotes together. Other conventions and abbreviations as in Fig. 13.2.

There appear to be six declines in diversity (Fig. 13.2, nos 1–6) that are attributable to mass extinction events. The other drops (early Jurassic, end-Jurassic, mid-Cretaceous) probably indicate mainly a change in the quality of the fossil record (Benton 1985*b*,*c*), and mass extinctions cannot be assumed here. These three episodes correspond to times when the SCM described above gives particularly low values.

The diversity of amniotes (i.e. tetrapods minus the amphibians) through time is plotted in Fig. 13.3, with data for non-marine and marine families separated. The six mass extinction events are indicated, although their effects are rather different from those for tetrapods as a whole (Fig. 13.2). The early Jurassic, end-Jurassic, and mid-Cretaceous falls in the non-marine record (Fig. 13.3a) are not matched by obvious declines in the marine record (Fig. 13.3b). This confirms, to some extent, the suggestion that these declines represent gaps in the fossil record of terrestrial amniotes since the marine record is largely independent. The problem with pressing this point too far is that the total diversity of families of marine amniotes is so low at all times that fluctuations probably mean very little.

The patterns of family diversity of the major reptilian groups are very different (Fig. 13.4). Most of the major groups shown here are assumed to

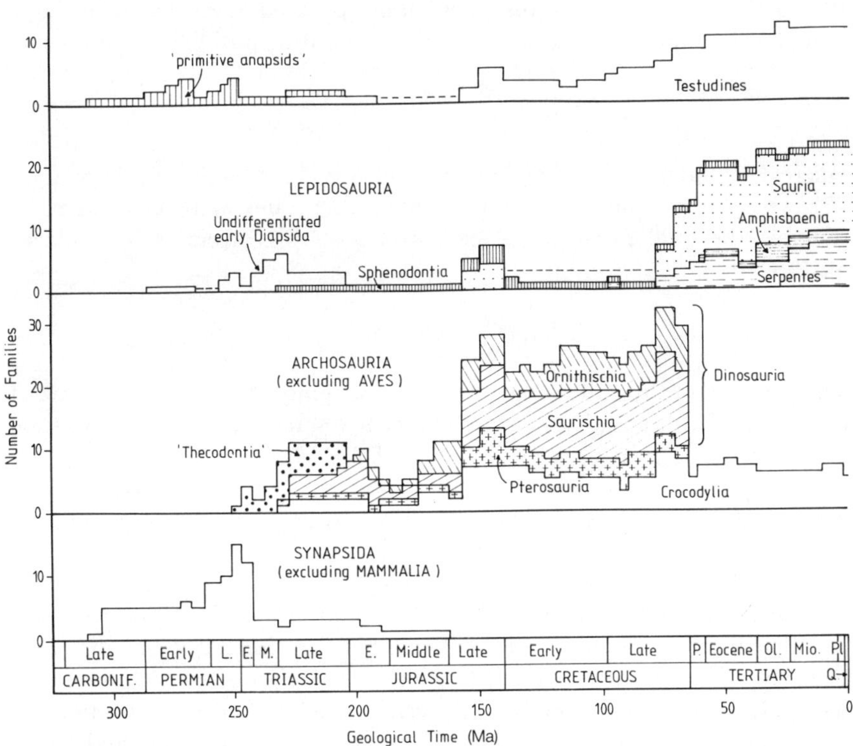

Fig. 13.4. Patterns of family diversity of the major reptilian groups. Diversity is plotted stage by stage, and each group is indicated by a different pattern of shading.

be monophyletic (Testudines, Lepidosauria, Sphenodontia, Sauria (?), Amphisbaenia, Serpentes, Ornithischia, Pterosauria, Crocodylia), but others are paraphyletic ('primitive anapsids', early Diapsida', Archosauria, 'Thecodontia', Saurischia, Synapsida). these graphs confirm the diversification of 'modern' groups from late Cretaceous times onwards, although the crocodiles have retained roughly constant diversity since the late Jurassic. Of the extinct groups, the dinosaurs (Saurischia + Ornithischia) and the pterosaurs maintained roughly level diversities from the late Jurassic until their extinction at the end of the Cretaceous. The Synapsida (mammal-like reptiles) clearly peaked in the late Permian, and their diversity crashed at the end of the Permian, and after the early Triassic, although they continued at low diversity until the end of the middle Jurassic.

Extinction and origination rates were calculated stage by stage for amniote families (marine and non-marine together), based on the new data set. Total extinction (R_e) and total origination (R_s) rates were calculated as the number of families that disappeared or appeared, respectively, during a stratigraphic stage, divided by the estimated duration of that stage (Δt):

$$R_e = \frac{E}{\Delta t} \quad \text{and} \quad R_s = \frac{S}{\Delta t},$$

where E is the number of extinctions and S is the number of originations. Per-taxon extinction (r_e) and origination (r_s) rates were calculated by dividing the total rates by the end-of-stage family diversity D (Sepkoski 1978):

$$r_e = \frac{1}{D} \cdot \frac{E}{\Delta t} \quad \text{and} \quad r_s = \frac{1}{D} \cdot \frac{S}{\Delta t}.$$

The per-taxon rates can be seen as the 'probability of origin' or the 'risk of extinction'. In these calculations, the recent summary geological time scale of Palmer (1983) was used for stage lengths in Ma.

The graphs of total rates for amniote families show great fluctuations in both origination and extinction rates. There is no clear correlation of high extinction rates with all mass extinction events. Of the highest rates, those in the Tatarian, 'Rhaetian', Maastrichtian, Rupelian, and late Miocene correspond to mass extinctions 2, 3, 4, 5, and 6 (Fig. 13.3), respectively. Equally high or higher total extinction rates in the Ufimian (Late Permian), Tithonian (late Jurassic), Coniacian (late Cretaceous), Thanetian (late Palaeocene), Ypresian (early Eocene), Bartonian–Priabonian (middle–late Eocene), Chattian (late Oligocene), and Pleistocene do not match any of the drops in amniote diversity that have been ascribed to mass extinctions. Conversely, the total extinction rate in the Artinskian (no. 1, Fig. 13.3) is not very high.

The total origination rates (Fig. 13.5) generally track the total extinction rates quite closely. Peaks in both rates may be produced, in part, by episodes when the fossil record is better than usual, corresponding to particular Fossil–Lagerstätten, such as the Sakamena Group (late Permian), the Solnhofen Limestone (Tithonian), the Messel deposits (Lutetian), Quercy Phosphorites (Bartonian–Rupelian); or the Oeningen Molasse (middle Miocene). The improvement in the record boosts the apparent number of family originations and extinctions (Hoffman and Ghiold 1985).

origination rates are re-calculated relative to the numbers of taxa available (Fig. 13.6), the rates do not track each other so closely, although 'Lagerstätten peaks' remain in the Ufimian, Tithonian, and Coniacian. There are particularly high per-taxon extinction rates at times of mass extinctions corresponding to the Artinskian, Tatarian, and 'Rhaetian' events (nos 1, 2, 3: Fig. 13.3). Per-taxon extinction rates are barely elevated at the times of the Maastrichtian, Rupelian, or late Miocene mass extinctions (nos 4, 5, 6: Fig. 13.3). These mass extinctions correspond to

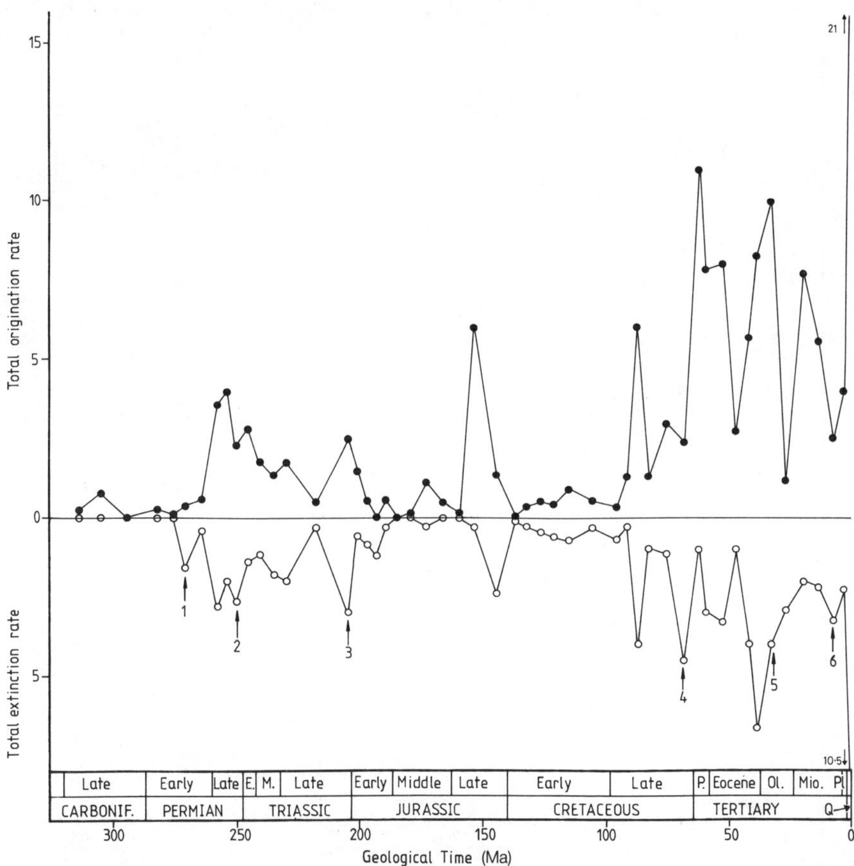

Fig. 13.5. Total rates of origination and extinction for families of amniotes, calculated stage by stage for 51 stages between the late Carboniferous and the Pleistocene. The Miocene was divided into early, middle, and late units only, and the Pliocene was treated as a single time unit.

depressed per-taxon origination rates (Fig. 13.6), a phenomenon noted also for non-marine tetrapods as a whole (Benton 1985*c*).

The graphs of total rates and of per-taxon rates show broad trends. The total rates (Fig. 13.5) apparently *increase* on average towards the present day, while the per-taxon rates (Fig. 13.6) tend to *decrease*. Benton (1985*c*) found the same phenomenon for non-marine tetrapods. Raup and Sepkoski (1982) found, on the other hand, that total extinction rates for marine animals declined markedly through time, while Van Valen (1984), Van Valen and Maiorana (1985), and Kitchell and Pena (1985) found the same for per-taxon rates. In all these cases, the rates of decline were very

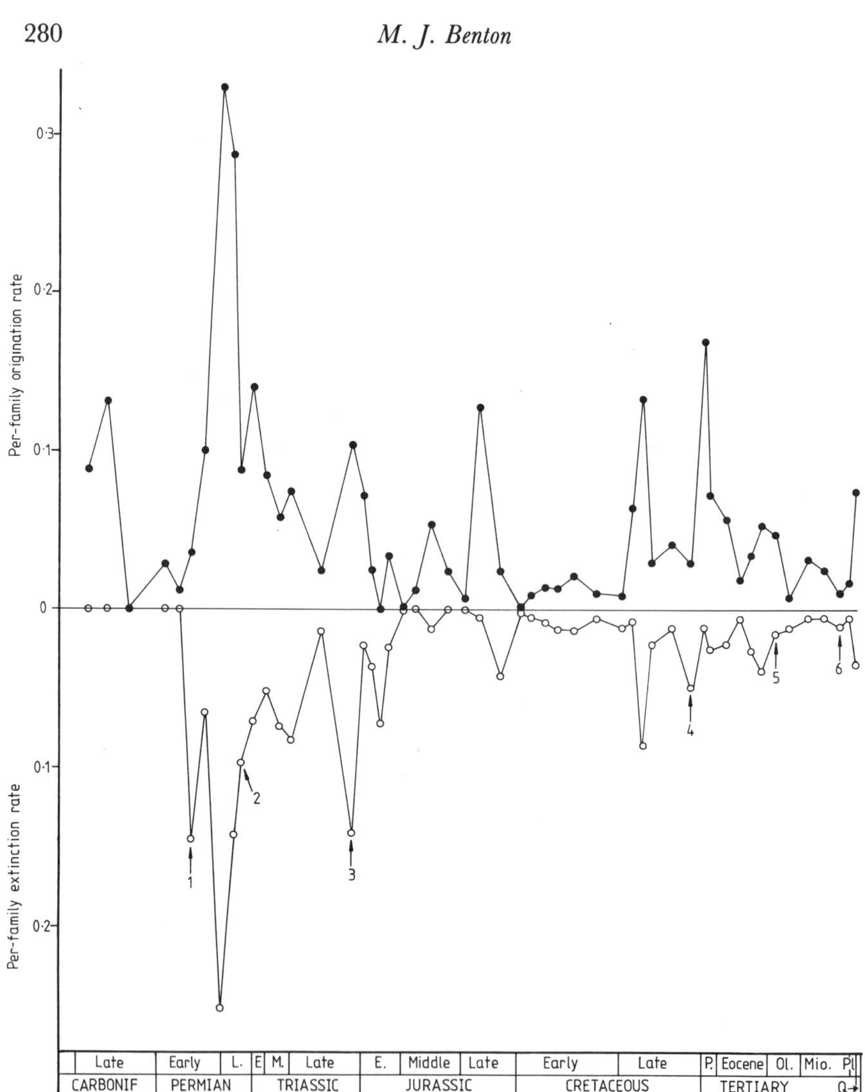

Fig. 13.6. Per-taxon rates of origination and extinction for families of amniotes. Conventions as in Fig. 13.5.

marked, while for non-marine tetrapods and for amniotes the declines in per-taxon rates are small.

The present data do not lend strong support to the idea that family extinction and origination rates indicate optimization of fitness through evolutionary time (Raup and Sepkoski 1982) or to relative decreases in diffuse competition between taxa within communities (Van Valen 1984).

Indeed the declines in extinction and origination rates in the marine record may be largely accounted for by the increase in the mean number of species per family through time (Flessa and Jablonski 1985). The fact that the per-taxon rates for families of amniotes or tetrapods do not decline so rapidly through time as those for marine animals in general may imply that the species: family ratio has not altered so much in these groups.

5. *Mass extinctions*

The history of non-marine tetrapods has apparently been punctuated by at least six mass extinction events (Fig. 13.2; Benton 1985*c*), together with up to seven other possible extinction events. These had widely differing effects, ranging from a 58 per cent drop in family diversity for the early Permian event to a 2 per cent drop for the late Miocene event.

These events also affected the amniotes, but to different extents after the removal of the amphibian data (Fig. 13.3). The late Triassic event hardly shows up at all in the non-marine amniote data (Fig. 13.3a, '3'), and the marine graph (Fig. 13.3b) is unconvincing for all six extinction events, except possibly the K–T ('4'), and the late Miocene ('6'). The relative declines in family numbers are summed for marine and non-marine amniotes together (Fig. 13.3), and the values range from 33 per cent (early Permian) to 3 per cent (late Miocene). The first three extinction events apparently had less marked effects on amniote family diversity as a whole (Fig. 13.3) than on non-marine tetrapods (Fig. 13.2), probably since the excluded amphibian families were heavily affected each time.

The six extinction events mentioned here, as well as other possible events, will be briefly described with notes on the groups that became extinct, and those that survived.

Early Permian (Artinskian)

Eight families died out during this event (no. 1: Figs 13.2 and 13.3):

'Anapsida': Protorothyrididae, Bolosauridae, Mesosauridae;
Diapsida: Araeoscelididae;
Synapsida: Eothyrididae, Edaphosauridae, Ophiacodontidae, Sphenacondontidae.

There were only three families that are known to have survived into the succeeding Kungurian Stage — the Captorhinidae ('Anapsida'), and the Caseidae and Varanopidae (Synapsida). The bulk of this mass extinction event affected the early synapsids, the pelycosaurs (loss of four families), and the two families of pelycosaurs that survived became extinct during the Kungurian Stage. The total extinction rate for amniote families in the Artinskian is not particularly high (Fig. 13.5), although the per-taxon extinction rate (Fig. 13.6) is higher than most others. The total diversity of early Permian amniotes is too low ($n = 6–11$) to read a great deal into these data.

Late Permian (Tatarian)

Twenty-one families of amniotes died out at the end of the Permian (event no. 2: Figs 13.2 and 13.3):

'Anapsida': Captorhinidae, Millerettidae, Pareiasauridae;
Diapsida: Weigeltisauridae, Younginidae, Tangasauridae;
Synapsida: Ictidorhinidae, Gorgonopsidae, Dromasauridae, Endothiodontidae, Cryptodontidae, Aulacocephalodontidae, Dicynodontidae, Pristerodontidae, Cistecephalidae, Diictodontidae, Moschorhinidae, Whaitsiidae, Silphestidae, Procynosuchidae, Dviniidae.

The 'anapsids' and diapsids that died out range in size from 'small lizard' (Millerettidae) to massive herbivore (Pareiasauridae), and they include the first gliding reptiles (Weigeltisauridae). These families occurred world-wide. None of the other 'anapsid' families actually spans the Permian–Triassic boundary, although the Procolophonidae might. Likewise, there may have been a slight hiatus in the evolution of the diapsids — the Proterosuchidae, the earliest archosaurs, is the only diapsid family known to have survived the end-Permian extinction event. However, several other larger diapsid lineages are known to have crossed the Permo-Triassic boundary (Fig. 13.1). The three diapsid families that died out were not restricted geographically, occurring as far apart as England and Madagascar.

The end-Permian extinction event decimated the mammal-like reptiles. Fifteen families out of the 20 that existed at the start of the Tatarian Stage died out (75 per cent decline). The groups going extinct range in size from the small dromasaurids to the hefty aulacocephalodontids. The extinctions span all groups of synapsids (Eotheriodontia, Anomodontia, Therocephalia, 'Cynodontia'), although the majority of families going extinct (eight) belong to the Anomodontia (eight out of ten extinct: 80 per cent decline). There seems to be no geographic control on extinction and survival.

Several families of mammal-like reptiles survived across the Permo-Triassic boundary (Emydopidae, Kingoriidae, Ictidosuchidae, Scaloposauridae, Galesauridae). However, the widespread '*Lystrosaurus* faunas' of South Africa, India, China, Russia, Antarctica, and Australia, which are generally dated as lowermost Triassic in age (Anderson and Cruickshank 1978; Tucker and Benton 1982), are apparently depleted. The best known examples of this fauna, in South Africa and Antarctica, are heavily dominated by one dicynodont genus, *Lystrosaurus* (*c.* 90 per cent of all individual tetrapod specimens collected; Benton 1983). There seems to be little doubt that these are 'post-extinction faunas' in which diversity is low and a great deal of ecospace has been emptied by a major event. Four or five million years later, apparently more balanced faunas are found in which no single genus is so dominant, and in which other lineages have diversified.

The total (Fig. 13.5) and per-taxon (Fig. 13.6) extinction rates for the Tatarian are fairly high, but not as high as those in the Ufimian, the first stage in the late Permian, when there was not a drop in amniote family diversity.

Early Triassic (Scythian)

There was another smaller extinction event about 5 Ma later, at the end of the Scythian Stage, when seven reptile families died out:

Diapsida: Proterosuchidae, Euparkeriidae;
Synapsida: Emydopidae, Kingoriidae, Ictidosuchidae, Scaloposauridae, Galesauridae.

These extinctions did not cause an overall decline in amniote diversity (Fig. 13.3) because of the matching rate of origination of new families in the Scythian. This event had a greater effect on the amphibians (Fig. 13.2). Several authors have suggested that there was a mass extinction event amongst tetrapods at this time (Olson 1982; Benton 1985*b*,*c*, 1986*a*,*b*; 1987, Bray 1985), while others have pointed to a small mass extinction of marine invertebrates (Raup and Sepkoski 1984, 1986; Hoffman 1985; Hoffman and Ghiold 1985; Benton 1986*a*).

Late Triassic (Carnian–'Rhaetian')

The three stages of the late Triassic, the Carnian, Norian, and 'Rhaetian' (or two, if the 'Rhaetian' is included in the Norian) span 18–25 Ma, depending upon which of the current time scales is employed. Raup and Sepkoski (1984, 1986) have argued that the late Triassic extinction consisted of a single event, but Benton (1986*a*,*b*) has identified at least two phases of extinction in the fossil record of tetrapods (no. 3: Figs 13.2 and 13.3), as well as in that of ammonoids and other groups.

The first and larger extinction event occurred at the end of the Carnian Stage. Ten families of amniotes died out:

Diapsida: Thalattosauridae, Trilophosauridae, Rhynchosauridae, Proterochampsidae;
Synapsida: Kannemeyeriidae, Chiniquodontidae;
'Euryapsida': Nothosauridae, Simosauridae, Cymatosauridae, Henodontidae.

These families include a broad range in ecological and taxonomic terms. Two of the families (Rhynchosauridae, Kannemeyeriidae) were important herbivores with nearly global distributions, but other similarly widespread herbivores (Aetosauridae) survived. Most other thecodontian groups, as well as the earliest dinosaurs and a few mammal-like reptiles, survived as well.

The marine reptiles were heavily affected, with four out of six families going extinct. The last of the nothosaurs disappeared (Nothosauridae, Simosauridae), as well as one plesiosaur family (Cymatosauridae), and one placodont family (Henodontidae). The Placochelyidae (placodonts) and the Shastasauridae (ichthyosaurs) survived the event.

The second, smaller, late Triassic extinction event, at the Triassic–Jurassic boundary ('Rhaetian') was marked by the loss of six families:

'Anapsida': Procolophonidae;
Diapsida: Phytosauridae, Aetosauridae, Rauisuchidae, Ornithosuchidae;
'Euryapsida': Placochelyidae.

These extinctions, although few in number, do seem to have some significance. The last of the thecodontians (four families) disappeared on land, as did the last of the placodonts in the sea. Most of the 'modern' groups of amniotes had appeared during the preceding 12–17 Ma of the Norian Stage: the Testudines, the Crocodylia, and the Mammalia, as well as the Pterosauria, and the Dinosauria in the Carnian. At one time it was thought that the end of the Triassic saw the termination of the formerly abundant labyrinthodont amphibians and the mammal-like reptiles, but these two groups continued in greatly reduced diversity until the Bathonian and Callovian (both Middle Jurassic), respectively.

Both the Carnian and the 'Rhaetian' events are associated with peaks in total and per-taxon extinction rates of amniote families (Figs 13.5 and 13.6), but the peaks are higher for the latter event.

Jurassic–Cretaceous events

Raup and Sepkoski (1984, 1986) and Sepkoski and Raup (1986) have identified a number of extinction events that affected marine animals between the 'Rhaetian' and the K–T events. These additional extinction events, with dates of the ends of the stages from Palmer (1983), are:

Jurassic:	Pliensbachian	193 Ma
	Bajocian	176 Ma
	Callovian	163 Ma
	Tithonian	144 Ma
Cretaceous:	Hauterivian	124 Ma
	Aptian	113 Ma
	Cenomanian	91 Ma

Raup and Sepkoski (1986), and Sepkoski and Raup (1986) express some doubt about the Bajocian, Callovian, Hauterivian, and Aptian peaks, but regard the Pliensbachian, Tithonian, and Cenomanian events as 'significant'.

The data on amniote families, and on reptiles in particular, are particularly weak during parts of this time interval (see above). There are indeed declines in family diversity in the non-marine tetrapod (Fig. 13.2) and the amniote (Fig. 13.3) data after the Pliensbachian, Tithonian, and Cenomanian, with the decline after the Tithonian standing out best. There are also slight peaks in total (Fig. 13.5) and per-taxon (Fig. 13.6) extinction rates in the Pliensbachian and Cenomanian, with a more marked peak in the Tithonian. However, the extinction rate peaks in the Coniacian are even larger, but this could be an artefact of the short estimated duration (1 Ma) of that stage (Sepkoski and Raup 1986).

There could be mass extinction events amongst amniotes at the times found by Raup and Sepkoski (1984) in the record of marine animals, but the present data are not good enough to decide either way.

Late Cretaceous (Maastrichtian)

The Cretaceous–Tertiary boundary (K–T) event is surely the best known mass extinction, and not least for its effect on the reptiles (dinosaurs, pterosaurs and plesiosaurs all died out then). However, in relative terms at least, the percentage loss of families of amniotes as a whole (no. 4: Fig. 13.3) was less than for the two Permian events already described, and for non-marine tetrapods (no. 4: Fig. 13.2), the K–T event was apparently less significant than the late Triassic events as well. The total extinction rate for the Maastrichtian (Fig. 13.5) is higher than any before it, but the per-taxon rate for amniotes (Fig. 13.6) is not so impressive, being lower than the 'Rhaetian', Pliensbachian, and Coniacian rates, for example. The decline in amniote family diversity at the K–T boundary is caused partly by a slightly elevated extinction rate, and partly by a low origination rate (Benton 1985*c*).

Thirty-six families of amniotes died out at the K–T boundary:

Diapsida
- Crocodylia: Uruguaysuchidae, Notosuchidae, Goniopholididae;
- Pterosauria: Pteranodontidae, Azhdarchidae;
- Dinosauria: Coeluridae, Ornithomimidae, Dromaeosauridae, Saurornithoididae, Oviraptoridae, Elmisauridae, Megalosauridae, Dryptosauridae, Tyrannosauridae, Camarasauridae, Diplodocidae, Titanosauridae, Hypsilophodontidae, Hadrosauridae, Pachycephalosauridae, Nodosauridae, Ankylosauridae, Protoceratopsidae, Ceratopsidae;
- Sauria: Mosasauridae;

Aves: Baptornithidae, 'Enantiornithes', Lonchodytidae, Torotigidae;

Mammalia
  Marsupialia: Pediomyidae, Stagodontidae;
  Eutheria: Leptictoidea n. fam., Perutheriidae;

'Euryapsida'
  Plesiosauria: Elasmosauridae, Cryptocleididae, Polycotylidae.

The K–T event was clearly taxonomically selective and certain major groups became completely extinct during Maastrichtian times: the Pterosauria (two families), the Dinosauria (19 families) and the Plesiosauria (three families). Other groups were less affected — turtles, crocodiles, lizards, snakes, birds, and mammals (although two out of three marsupial families died out). Indeed, the mammals continued to radiate without any obvious pause right through the K–T boundary.

There has been much debate about whether the K–T event was sudden or not. Many of the reptile groups that had been dominant in the Mesozoic were declining well before the end of the Cretaceous. The ichthyosaurs died out in the Cenomanian, 30 Ma earlier (Baird 1984), and the pterosaurs were reduced from five or six families to only two in the Maastrichtian. Nevertheless, 19 families of dinosaurs existed at the start of Maastrichtian times and died out before the K–T boundary. Russell (1984) argued that they nearly all died out catastrophically right at the K–T boundary, while Schopf (1982), Carpenter (1984), and Sloan *et al.* (1986) suggested that the dinosaurs declined in diversity and abundance during the 8–9 Ma of the Maastrichtian, and that only a few genera remained by the time of the K–T boundary. More palaeontological and stratigraphic work is required on this important aspect of the K–T event.

Late Eocene (Priabonian)

Raup and Sepkoski (1984, 1986), and others, identify a mass extinction amongst marine animals at the Eocene–Oligocene boundary, but this does not appear in the summary tetrapod or amniote data (Figs 13.2 and 13.3). This corresponds to the 'Grande Coupure' of French vertebrate palaeontologists, a marked extinction event amongst amphibians, reptiles, birds, and mammals in Europe. A detailed analysis of amphibians and reptiles at this time (Rage 1984) shows extensive extinctions amongst species and genera. However, the event seems to have had less effect at the family level, and it was apparently largely restricted to Europe. On a global scale, there were slight declines in the diversity of non-marine tetrapod (Fig. 13.2) and amniote (Fig. 13.3) families, and extinction rates were high (Figs 13.5 and 13.6). However, there were no extinctions of families of amphibians or reptiles.

Early Oligocene (Rupelian)

Tetrapod, and amniote, diversity declined markedly in the mid-Oligocene (no. 5: Figs 13.2 and 13.3), after the extinction of 28 families during the early Oligocene. The main losses occurred amongst the Mammalia (26 families of non-marine and marine mammals), with only one bird family and one reptile family (the lizards Necrosauridae) becoming extinct at that time. Extinction rates of amniote families were not particularly high (Figs 13.5 and 13.6), but they were matched by low origination rates. This event has been recognized also by Prothero (1985) for North American land mammals, but it does not appear to correspond to events that affected marine invertebrates.

Late Miocene (Tortonian–Messinian)

The late Miocene extinction event amongst tetrapods and amniotes is marked by a decline in family diversity (no. 6: Figs 13.2 and 13.3). Twenty families of amniotes died out, but no reptiles were amongst them. The majority of extinctions affected the mammals, with particular losses amongst primates, artiodactyls, notoungulates, and cetaceans. Extinction rates (Figs 13.5 and 13.6) were not particularly elevated. This event follows the Middle Miocene mass extinction identified by Raup and Sepkoski (1984, 1986) for marine animals.

Pleistocene

The Pleistocene extinctions do not show up on the graphs of non-marine tetrapod and amniote family diversity (Figs 13.2 and 13.3), because of the matching rise in numbers of Pleistocene and Holocene families. Twenty-one families of amniotes died out in the Pleistocene, but amongst reptiles these include only the giant Australian turtles, the Meiolaniidae, and the Euthecodontidae, crocodiles known from Africa and Australasia at that time.

6. *Periodicity of amniote mass extinctions?*

Raup and Sepkoski (1984, 1986), and Sepkoski and Raup (1986) have presented evidence for periodicity in the occurrence of mass extinctions on the basis of several analyses of the fossil record of families and genera of marine animals. In the mid-Permian to Pleistocene span of time, they found a mean spacing of 26 Ma between events, and a considerable literature has already grown up to do with the nature of the periodicity, and the possible extraterrestrial causes (Sepkoski and Raup 1986).

The record of fossil reptiles, and of amniotes as a whole, spans the same time interval and, for the non-marine portion of the data set at least (which represents 90.1 per cent of the numbers of families — 644 non-marine: 58

marine), it provides a partially independent test of the periodicity hypotheses. The main problem, as has been outlined above, is the fact that the fossil record of non-marine amniotes is probably more patchy than that of marine invertebrates.

The mass extinctions identified by Raup and Sepkoski (1984, 1986) on the basis of their analysis of marine animals are listed in Table 13.1, with age dates from Palmer (1983). The mass extinctions amongst amniote

Table 13.1. The extinction events determined in the marine animal fossil record by Raup and Sepkoski (1984, 1986), with age dates from Palmer (1983). Events regarded as 'doubtful' or 'possible' by Raup and Sepkoski (1986) are shown with a question mark. The extinction events that affected non-marine tetrapods, and amniotes in general, are indicated, also with a question mark for the uncertain events. The spacings between the latter events are indicated

Marine events (system and stage)	End of interval (Ma)	Tetrapod and amniote events	Spacing (Ma)
Tertiary, Pliocene?	1.6		
	5.3	Tertiary, late Miocene	
			24.7
Tertiary, Middle Miocene	11.2		
	30.0	Tertiary, early Oligocene	
			6.6
Tertiary, late Eocene	36.6	Tertiary, late Eocene?	
			29.8
Cretaceous, Maastrichtian	66.4	Cretaceous, Maastrichtian	
			21.1
	87.5	Cretaceous, Coniacian?	
			3.5
Cretaceous, Cenomanian	91	Cretaceous, Cenomanian?	
Cretaceous, Aptian	113		
			53
Cretaceous, Hauterivian?	124		
Jurassic, Tithonian	144	Jurassic, Tithonian	
Jurassic, Callovian?	163		
			49
Jurassic, Bajocian?	176		
Jurassic, Pliensbachian	193	Jurassic, Pliensbachian?	
			15
Triassic, 'Rhaetian'	208	Triassic, 'Rhaetian'	
Triassic, Norian	216		17
Triassic, Carnian?	225	Triassic, Carnian	
			15
Triassic, Scythian?	240	Triassic, Scythian?	
			5
Permian, Tatarian	245	Permian, Tatarian	
			18
	263	Permian, Artinskian	

families are listed for comparison. Many of the amniote events correspond to the marine events, but there are large gaps, particularly in the Jurassic and Cretaceous, where there is no clear evidence for mass extinctions. There are also three extinctions amongst amniotes ((?) Coniacian, early Oligocene, late Miocene) that are not matched by marine events, one (Artinskian) that lies outside the time interval studied by Raup and Sepkoski (1984, 1986), and one (Carnian) that seems more certain for amniotes than for marine animals.

In view of the uncertainties involved, it is probably pointless to try to calculate periodicities from the amniote mass extinctions. Nevertheless, the spacings between all probable and possible events are indicated in Table 13.1. Some of the spacings approximate to the 26 Ma period suggested by Raup and Sepkoski (1984, 1986) or to multiples thereof, but the majority do not, particularly those in the Permian and Triassic. If taken at face value, the non-marine tetrapod family data and the amniote family data do not support a model of periodic extinction events.

Acknowledgement

I thank Ms Libby Lawson for drafting the diagrams.

References

Anderson, J. M. and Cruickshank, A. R. I. (1978). The biostratigraphy of the Permian and the Triassic. Part 5. A review of the classification and distribution of Permo-Triassic tetrapods. *Palaeontol. afr.* **21,** 15–44.

Ashlock, P. D. (1971). Monophyly and associated terms. *Syst. Zool.* **20,** 63–9.

Baird, D. (1984). No ichthyosaurs in the Cretaceous of New Jersey. . .or Saskatchewan. *Mosasaur* **2,** 129–33.

Bakker, R. T. (1977). Tetrapod mass extinctions — a model of the regulation of speciation rates and immigration by cycles of topographic diversity. In *Patterns of Evolution as Illustrated by the Fossil Record* (ed. A. Hallam), pp. 439–68. Elsevier, Amsterdam.

Benton, M. J. (1983). Dinosaur success in the Triassic: a noncompetitive ecological model. *Q. Rev. Biol.* **58,** 29–55.

Benton, M. J. (1984). The relationships and early evolution of the Diapsida. *Symp. zool. Soc. Lond.* **52,** 575–96.

Benton, M. J. (1985*a*). Classification and phylogeny of the diapsid reptiles. *Zool. J. Linn. Soc.* **84,** 97–164.

Benton, M. J. (1985*b*). Patterns in the diversification of Mesozoic non-marine tetrapods, and problems in historical diversity analysis. *Spec. Pap. Palaeontol.* **33,** 185–202.

Benton, M. J. (1985*c*). Mass extinction among non-marine tetrapods. *Nature, Lond.* **316,** 811–4.

Benton, M. J. (1986*a*). More than one event in the late Triassic mass extinction. *Nature, Lond.* **321,** 857–61.

Benton, M. J. (1986*b*). The Late Triassic tetrapod extinction events. In *The Beginning of the Age of Dinosaurs,* pp. 303–20. Cambridge University Press, New York.

Benton, M. J. (1987). Mass extinctions among families of non-marine tetrapods: the data. *Mém. Soc. géol. Fr.* **150,** 21–32.

Bray, A. A. (1985). Will impacts become extinct? *Mod. Geol.* **9,** 397–409.

Brown, D. S. (1981). The English Upper Jurassic Pleisiosauroidea (Reptilia) and a review of the phylogeny and classification of the Plesiosauria. *Bull. Br. Mus. nat. Hist.* (*Geol.*) **35,** 253–347.

Buffetaut, E. (1982). Radiation évolutive, paléoecologie et biogéographie des crocodiliens mésosuchiens. *Mém. Soc. géol. Fr.* **142,** 1–88.

Carpenter, K. (1984). Evidence suggesting gradual extinction of latest Cretaceous dinosaurs. *Naturwissenschaften* **70,** 611–2.

Carroll, R. L. (1969). Problems of the origin of reptiles. *Biol. Rev. Camb. philos. Soc.* **44,** 393–432.

Carroll, R. L. (1970). The ancestry of reptiles. *Phil. Trans. Roy. Soc. Lond.* **B257,** 267–308.

Carroll, R. L. (1977). Patterns of amphibian evolution: an extended example of the incompleteness of the fossil record. In *Patterns of Evolution as Illustrated by the Fossil Record* (ed. A. Hallam), pp. 405–37. Elsevier, Amsterdam.

Carroll, R. L. (1982). Early evolution of reptiles. *Ann. Rev. Ecol. Syst.* **13,** 87–109.

Carroll, R. L. (1985). A pleurosaur from the Lower Jurassic and the taxonomic position of the Sphenodontida. *Palaeontographica* **A189,** 1–28.

Charig, A. J. (1973). Kurtèn's theory of ordinal variety and the number of continents. In *Implications of Continental Drift to the Earth Sciences,* Vol. 1 (ed. D. H. Tarling and S. K. Runcorn), 231–45. Academic Press, London.

Colbert, E. H. (1986) Mesozoic tetrapod extinctions: a review. In *Dynamics of Extinction* (ed. D. K. Elliott), pp. 49–62. Wiley, New York.

Cracraft, J. (1981). Pattern and process in paleobiology: The role of cladistic analysis in systematic paleontology. *Paleobiology* **7,** 456–68.

Cutbill, J. L. and Funnell, B. M. (1967) Computer analysis of the fossil record. In *The Fossil Record* (ed. W. B. Harland *et al.*), pp. 791–820. Geological Society of London.

Flessa, K. W. and Jablonski, D. (1985). Declining Phanerozoic background extinction rates; effect of taxonomic structure? *Nature, Lond.* **313,** 216–8.

Fordyce, R. E. (1982). A review of Australian fossil Cetacea. *Mem. nat. Mus. Victoria* **43,** 43–58.

Gaffney, E. S. (1980). Phylogenetic relationships of the major groups of amniotes. In *The Terrestrial Environment and the Origin of Land Vertebrates* (ed. A. L. Panchen), pp. 593–610. Academic Press, London.

Gauthier, J. and Padian, K. (1985). Phylogenetic, functional, and aerodynamic analyses of the origin of birds and their flight. In *The Beginning of Birds* (ed. M. K. Hecht, J. H. Ostrom, G. Viohl, and P. Wellnhofer), pp. 185–97. Freunde des Jura-Museums, Eichstätt.

Harland, W. B., *et al.* (1967). *The Fossil Record.* Geological Society of London.

Harland, W. B., *et al.* (1982). *A Geologic Time Scale.* Cambridge University Press, Cambridge.

Heaton, M. J. (1980). The Cotylosauria: a reconsideration of a group of archaic tetrapods. In *The Terrestrial Environment and the Origin of Land Vertebrates* (ed. A. L. Panchen), pp. 491–551. Academic Press, London.

Heaton, M. J. and Reisz, R. R. (1986). Phylogenetic relationships of captorhinomorph reptiles. *Can. J. Earth Sci.* **23,** 402–18.

Hennig, W. (1966). *Phylogenetic Systematics.* University of Illinois Press, Urbana.

Hoffman, A. (1985). Patterns of family extinction depend on definition and geological time scale. *Nature, Lond.* **315,** 659–62.

Hoffman, A. and Ghiold, J. (1985). Randomness in the pattern of 'mass extinctions' and 'waves of origination'. *Geol. Mag.* **122,** 1–4.

Jablonski, D. (1986). Causes and consequences of mass extinctions: a comparative approach. In *Dynamics of Extinction* (ed. D. K. Elliott), pp. 183–229. Wiley, New York.

Kemp, T. S. (1982). *Mammal-like Reptiles and the Origin of Mammals.* Academic Press, London.

Kitchell, J. A. and Pena, D. (1985). Periodicity of extinctions in the geologic past: deterministic versus stochastic explanations. *Science NY* **226,** 689–92.

Mazin, J-M. (1982). Affinités et phylogénie des Ichthyopterygia. *Géobios Mém. spéc.* **6,** 85–98.

Mazin, J-M. (1984). Les Ichthyopterygia du Trias du Spitsberg. Descriptions complementaires à partir d'un nouveau matériel. *Bull. Mus. nat. Hist. nat., Paris* (4), **6,** Sec. C, 309–20.

Olson, E. C. (1982). Extinctions of Permian and Triassic nonmarine vertebrates. *Spec. Pap. geol. Soc. Am.* **190,** 501–11.

Padian, K. and Clemens, W. A. (1985). Terrestrial vertebrate diversity: episodes and insights. In *Phanerozoic Diversity Patterns* (ed. J. W. Valentine), pp. 41–96. Princeton University Press.

Palmer, A. R. (1983). The decade of North American Geology 1983 time scale. *Geology* **11,** 503–4.

Paul, C. R. C. (1982). The adequacy of the fossil record. In *Problems of Phylogenetic Reconstruction* (ed. K. A. Joysey and A. E. Friday), pp. 75–117. Academic Press, London.

Pitrat, C. W. (1973). Vertebrates and the Permo–Triassic extinctions. *Palaeogeogr. Palaeoclimatol. Palaeoecol.* **14,** 249–64.

Prothero, D. R. (1985). Mid-Oligocene extinction events in North American land mammals. *Science NY* **229,** 550–1.

Rage, J.-C. (1984). La 'Grande Coupure' éocène/oligocène et les herpetofaunas (amphibiens et reptiles): problèmes du synchronisme des évenements paléobiogéographiques. *Bull. Soc. géol. Fr.* 1984 (7) **26,** 1251–7.

Raup, D. M. and Sepkoski, J. J., Jr. (1982). Mass extinctions in the marine fossil record. *Science NY* **215,** 1501–3.

Raup, D. M. and Sepkoski, J. J., Jr. (1984). Periodicity of extinctions in the geologic past. *Proc. nat. Acad. Sci. USA* **81,** 801–5.

Raup, D. M. and Sepkoski, J. J., Jr. (1986). Periodic extinctions of families and genera. *Science NY* **231,** 833–6.

Reisz, R. R., Berman, D. S., and Scott, D. (1984). The anatomy and relationships of the Lower Permian reptile. *Araeoscelis. J. vertebr. Paleontol.* **4,** 57–67.

Romer, A. S. (1966). *Vertebrate Paleontology,* 3rd edn. University of Chicago Press.

Russell, D. A. (1984). The gradual decline of the dinosaurs: fact or fallacy? *Nature, Lond.* **307,** 360–1.
Savage, D. E. and Russell, D. E. (1983). *Mammalian Paleofaunas of the World.* Addison-Wesley, Reading, Mass.
Schopf, J. M. (1982). Extinction of dinosaurs: a 1982 understanding. *Spec. Pap. geol. Soc. Am.* **190,** 415–22.
Sepkoski, J. J., Jr (1978). A kinetic model of Phanerozoic taxonomic diversity. I. Analysis of marine orders. *Paleobiology* **4,** 223–51.
Sepkoski, J. J., Jr. (1982). A compendium of fossil marine families. *Contr. Biol. Geol., Milwaukee Publ. Mus.* **51,** 1–125.
Sepkoski, J. J., Jr and Raup, D. M. (1986). Periodicity in marine extinction events.
Simpson, G. G. (1952). Periodicity in vertebrate evolution. *J. Paleontol.* **26,** 359–70.
Sloan, R. E., Rigby, J. K., Jr, Van Valen, L. M., and Gabriel, D. (1986). Gradual dinosaur extinction and simultaneous ungulate radiation in the Hell Creek Formation. *Science NY* **232,** 629–33.
Thomson, K. S. (1977). The pattern of diversification among fishes. In *Patterns of Evolution as Illustrated by the Fossil Record* (ed. A. Hallam), pp. 377–404. Elsevier, Amsterdam.
Tucker, M. E. and Benton, M. J. (1982). Triassic environments, climates and reptile evolution. *Palaeogeogr. Palaeoclimatol. Palaeoecol.* **40,** 361–79.
Van Valen, L. M. (1984). A resetting of Phanerozoic community evolution. *Nature, Lond.* **307,** 50–2.
Van Valen, L. M. and Maiorana, V. C. (1985). Patterns of origination. *Evolut. Theory* **7,** 107–25.

Appendix

The distributions in geological time of families of marine amniotes. Supplement to the listing of non-marine amniotes in Benton (1987), based largely on Sepkoski (1982; and revisions dated August 1985), with additional sources of information indicated in the relevant places. Families that contain marine and non-marine members have been listed in Benton (1987), and they are not repeated here. Stratigraphic abbreviations are the standard stage abbreviations given in Harland *et al.* (1982).

'Reptilia': *Incertae sedis* (Anderson and Cruickshank 1978)	
Mesosauridae	ART
'Reptilia': *Testudines*	
Chelonioidea	
Protostegidae	CEN–CHT
Toxochelyidae	ALB–YPR
Dermochelyidae	YPR–REC
Cheloniidae	ALB–REC
'Reptilia': *Diapsida*: *Archosauria*	
Crocodylomorpha (Buffetaut 1982)	
Teleosauridae	TOA-VAL
Metriorhynchidae	BTH-HAU
Dyrosauridae	MAA-PRB

'Reptilia': Diapsida: Lepidosauria	
Sauria	
Aigialosauridae	TTH–ALB?
Dolichosauridae	?ALB–CMP?
Mosasauridae	CEN–MAA
'Reptilia': 'Euryapsida'	
Nothosauria	
Nothosauridae	SCY–CRN
Pachypleurosauridae	ANS–LAD (Carroll 1985)
Simosauridae	?ANS–CRN
Plesiosauria (Brown 1981)	
Pistosauridae	LAD
Cymatosauridae	?LAD–CRN
Plesiosauridae	HET–TOA
Leptocleididae	HET–HAU
Pliosauridae	HET–CON?
Elasmosauridae	TOA–MAA
Cryptocleididae	CAL–MAA
Polycotylidae	APT–MAA
Placodontia	
Placodontidae	SCY–LAD
Helveticosauridae	ANS–LAD
Placochelyidae	ANS–RHT
Henodontidae	CRN
Ichthyosauria (Mazin 1982, 1984)	
Unnamed family	SCY–LAD
Omphalosauridae	ANS
Mixosauridae	ANS–LAD
Shastasauridae	LAD–NOR
Ichthyosauridae	HET–TTH
Stenopterygiidae	SIN–PLB
Leptopterygiidae	SIN–CEN
Mammalia (Savage and Russell 1983)	
Cetacea (Fordyce 1982)	
Protocetidae	LUT
Basilosauridae	BRT–RUP
Dorudontidae	PRB–RUP
Squalodontidae	RUP–UMI
Cetotheriidae	RUP–PLI
Patriocetidae	CHT
Agorophiidae	CHT
Eurhinodelphidae	LMI–UMI
Kentriodontidae	LMI–UMI

Acrodelphidae	LMI–PLI
Platanistidae	LMI–REC
Ziphiidae	LMI–REC
Physeteridae	LMI–REC
Phocaenidae	LMI–REC
Delphinidae	LMI–REC
Balaenopteridae	LMI–REC
Balaenidae	MMI–REC
Monodontidae	UMI–REC
Sirenia	
Dugongidae	YPR–REC
Manatidae	LMI–REC
Carnivora: Pinnipedia	
Enaliarctidae	LMI
Desmatophocidae	?LMI–UMI
Odobenidae	LMI–REC
Phocidae	MMI–REC
Otariidae	UMI–REC

14. Extinction and survival in birds

D. M. UNWIN

Department of Zoology, University of Reading, Reading UK

Abstract

A data base for the fossil record of birds is compiled from the distribution of families of birds through time, resolved to the level of the stage. Patterns of diversity and rates of origin and extinction are derived from calculations performed upon the data base. These are investigated with respect to artefacts produced by the parameters used in the compilation, bias in the fossil record, and problems associated with avian taxonomy. It is suggested that the avian fossil record is so badly distorted by these factors that identification of short-term events and patterns, lasting less than 5 Ma, is largely precluded. In contrast, it is proposed that long-term patterns, spanning many millions of years, can be identified. One such pattern is formed by the relative diversity of birds and pterosaurs during the Mesozoic, which corresponds to the pattern produced by the differential survival model of faunal replacement.

Introduction

In a seminal paper published in 1980, Walter Alvarez and a team of associates proposed that the end-Cretaceous mass extinction was caused by an asteroid impact. This and later papers, (Alvarez 1983; Alvarez *et al.* 1980, 1982, 1984) provoked a major controversy which 6 years on has become slighly acrimonious, but still shows little sign of abating. On the positive side, this debate has stimulated a great deal of new research (much of it still in progress), in a variety of disciplines ranging from geochemistry to stratigraphy, palaeontology, and even astronomy.

Extinction and Survival in the Fossil Record (ed. G. P. Larwood), Systematics Association Special Volume No. 34, pp. 295–318. Clarendon Press, Oxford, 1988.

The impact hypothesis contains much of interest to palaeontologists. The existence of an end-Cretaceous mass extinction, postulated on the basis of palaeontological data, has been used as evidence for an impact event (Alvarez 1983; Alvarez *et al.* 1984), which in turn has been cited as a possible mechanism for producing this, and other mass extinctions (Alvarez 1983; Alvarez *et al.* 1982). Clearly, the element of circularity in these arguments lends itself to some criticism, but this is only a minor point and will not be discussed further here. More importantly, this component of the debate, the fossil evidence for mass extinctions, falls within the purview of palaeontology. One of the main aims of palaeontologists so far, has been to try and establish whether mass extinctions have occurred, and if they did, what, if anything, their nature or distribution through time reveals about their causes. The implication being, of course, that they represent distinct phenomena and not an extreme form of background extinction (Raup 1986).

The vast majority of investigations have consisted of quantitative analyses of the fossil record, building upon the pioneering works of Newell (1967) and Valentine (1969). Foremost among recent publications are those of Raup and Sepkoski, whose proposal of a 26 Ma periodicity in mass extinctions (1984, 1986) sparked off another storm of controversy, (e.g. Hallam 1984), and led to some wild speculations concerning companion stars, the Oort cloud and even Planet X (Raup 1986 for main references). Incidentally, the proposal of a 32-Ma periodicity in extinction events, 3 years prior to the Alvarez asteroid (Fischer and Arthur 1977) made relatively little impact.

Back on earth, many of the quantitative analyses, including those of Raup and Sepkoski, have been based on the invertebrate fossil record. This is not surprising as it is much more complete than that of the vertebrates, and owing to its relative importance in stratigraphy, has been much more thoroughly studied. Relatively few recent analyses are based upon, or even include data from the vertebrate fossil record (Newell 1963; Valentine 1973; Bakker 1977), and those that do often focus attention on particular groups of vertebrates such as the dinosaurs (Benton 1983; Russell 1984) or on particular periods such as the Cretaceous–Tertiary (K–T) boundary (Russell 1982) or the Triassic–Jurassic boundary (Benton 1986). Even fewer analyses include data from the fossil record of birds (Fisher 1967; Cutbill and Funnell 1967; Brodkorb 1971), the subject to be discussed here, and only those by Benton (1985*b*, 1987) and Vuilleumier (1984), whose study only concerned fossil South American birds, are relatively up to date.

For this paper a data base was compiled from publications on the fossil record of birds, various calculations performed upon it, and the results analysed. However, unlike most previous investigations, the primary aim was not to establish the existence or otherwise of particular patterns and their relation to extinction events, but to examine the way in which data bases and the results of analysing them can be distorted by various factors, these being the degree of completeness of the fossil record, the effects of preservational bias, the effect of variation within the time intervals chosen upon the resolvability of temporal patterns, and the degree to which current taxonomy reflects true phylogeny and hence true diversity.

It was concluded that the quality of data currently available from the fossil record of birds does not permit events or patterns shorter than 'a few million years' to be distinguished with any degree of confidence. On the other hand, long-term patterns, spread over tens of millions of years, can be identified. One such is provided by the relative diversity of birds and pterosaurs during the Mesozoic, which conforms to the pattern predicted by the differential survival model of faunal replacement (Benton 1983). On the basis of this pattern it is suggested that due to characteristics such as a better terrestrial ability, a more robust wing, and less restricted latitudinal distribution, birds were able to survive into the Tertiary while pterosaurs become extinct at the end of the Cretaceous.

The data, their quality, and associated problems

1. The data base

The data base (Appendix), was compiled from the first and last known occurrence of families of birds, as detailed in the most recently available literature. I was particularly fortunate in that a comprehensive review of fossil birds was published very recently (Olson 1985*a*). This publication is most timely in two respects. To begin with, it contains a wealth of new information on the systematics and stratigraphic ranges of bird families, superseding all other publications to date. Secondly, Olson's review emphasizes the degree to which older reviews (e.g. Fisher 1967; Brodkorb 1971) and catalogues (e.g. Brodkorb 1963) have become outdated. This is particularly important, with respect to quantitative analyses, all of which so far, have been based on these previous reviews. To take but one example, 47 per cent of the ranges of bird families, (97 corrections in total) given by Benton (1987) and correct up to 1985, require emendation in the light of more recent publications.

The family level of taxonomic category was chosen because it represents the most acceptable compromise between known stratigraphic ranges and taxonomic resolution. A higher resolution at the generic, or even the specific level is possible based on the catalogues published by Brodkorb (e.g. 1963), but a significant proportion of the stratigraphic and taxonomic information contained in these works is obsolete. Thus, quantitative assessments based on them would be of little value. Families represented by a single specimen from one horizon were included in the calculations, if based on relatively complete remains, whose higher systematics are generally accepted. Those based on relatively incomplete or fragmentary remains and/or whose higher level systematics are uncertain were excluded from the calculations.

The maximum degree of temporal resolution available for the stratigraphic ranges of most bird families, is that of the stage, though even these rather crude divisions could not be used in the Miocene, which is divided solely into Lower, Middle, and Upper. The Pliocene, Pleistocene, and Holocene are each dealt with as separate units.

The taxonomy of Olson (1985*a*) was adopted, with some minor modifications, partly because it reflects the work of a single researcher and is likely to be more consistent than taxonomies based on a variety of sources, and partly because the best available stratigraphic ranges are largely to be found in Olson's (1985*a*) review.

The basic data were used to compile a simple standing diversity curve. Total rates of origin and extinction were calculated by dividing the length of the stage, in millions of years, by the numbers of families originating or becoming extinct in that particular stage (Sepkoski 1978). Per-taxon origin and extinction rates were also calculated, by dividing the total diversity multiplied by the length of the stage into the number of originations and extinctions which took place in that stage. The results obtained, however, were so variable (as a result of the low quality of the original data) as to not merit further consideration.

2. *The quality of the data*

Ideally, a data base intended for use in identification or testing of events and patterns, short or long term, should contain, or approach as closely as possible, the following criteria: (1) a reliable estimate of the stratigraphic ranges of the taxa involved, (2) ranges based on the lowest resolvable time units available, such as the zone or subzone, (3) a true phylogeny, reflecting true diversity, thus permitting distinction of pseudo-extinction from true extinction and (4) analysis at the lowest (i.e. specific) taxonomic level (Paul, this volume). The degree to which the data approach these criteria gives a rough idea of their quality.

Clearly, the data and the parameters chosen in the compilation of the fossil bird data base, fall far short of the ideal. It is very difficult to assess whether known ranges approach real ranges of particular families. It is usually assumed that continued research will result in known ranges asymptotically approaching real ranges, but how can we assess the current position? There are a variety of quantitative and qualitative assessments that could be made (Paul 1982). One of each is described here.

From a purely qualitative point of view, it would seem that many known ranges do not at present approach even closely the real ranges. Considerable evidence for this is provided not only by the frequent discovery of bird taxa in much earlier periods than those from which they were previously known, but also by the steady discovery of new orders and even subclasses of birds. The avifauna from the Phosphorites du Quercy (Eocene–Oligocene of France) provides a good example of the first case, in that many taxa now known to occur in these deposits (Mourer-Chauvire 1982) were, until recently, only known from the Miocene or Pliocene. Furthermore, the first appearance at this time of many new forms is most probably a result of the paucity of the fossil record immediately preceding the Priabonian (Benton 1987, fig. 2), rather than any real radiation event. There are plenty of examples of new taxa. Orders proposed recently include, Ambiortiformes, Limnornithiformes and Gansuiiformes, though the latter two are not generally accepted. In addition to this, the proposal of a new subclass of birds, the Enantiornithes (Walker 1981) highlights the dangers of assuming the avian fossil record to be reasonably complete even at the supra-family level.

Incidentally, the work of Sibley and Ahlquist (1986) on DNA-DNA hybridization suggests that many taxa appeared much earlier than is currently known. For example, their work predicts that the first passerine appeared about 90 million years ago (Sibley and Ahlquist 1986, fig. 6), but the earliest known passerine is about 30 Ma old (Olson 1985*a*). However, the dating of their splitting events depends on accepting the regularity of the molecular clock and its calibration based on biogeographic dispersal data. The current calibration depends on the assumption that the common ancestor of the rhea (*Rhea americana*) and the ostrich (*Struthio camelus*) (it being assumed that the ratites are monophyletic) was present on Gondwanaland prior to its separation 80 million years ago. However, the monophyly of the ratites now appears extremely doubtful (Feduccia 1980; Houde 1986; Olson 1985*a*), nor is there complete certainty concerning the dating of the disintegration of Gondwanaland, nor is it known whether or not this would have prevented ratite dispersal at some later date. Similar criticisms have been levelled at the other events used for calibration (Houde 1987), thus it would seem that much more precisely determined events are needed before accurate calibration can be achieved.

Benton has carried out the most recent quantitative assessment of the fossil record of birds (1987). Using a simple completeness matrix he demonstrated that the record is 56.9 per cent complete. I would accept that relative to other groups of terrestrial tetrapods, the record of birds is somewhat better than that of lepidosaurs and only a little poorer than that of the Diapsida (Benton 1987), but this analysis only takes into account incidences of taxa missing between their first and last known occurrences, so called 'Lazarus taxa'. It is assumed that the first and last occurrences approach reasonably closely, those of the real ranges. However, the qualitative assessment discussed earlier suggests that they do not. Thus, the real degree of completeness is probably somewhat lower and the same may well hold true for other groups of terrestrial tetrapods.

The resolution of stratigraphic ranges to the level of the stage is clearly very unsatisfactory. The average length of a stage for the time period under consideration is 5 Ma, but they range in length from 1 to 15.5 Ma. Furthermore, these figures vary depending on the timescale used, there being little general agreement as yet, on the length of many of the stages used in this compilation (Benton 1987). There is also the severe problem of worldwide correlation of stages without which global events cannot be demonstrated to have occurred. This problem is further magnified when dealing with terrestrial fossils, since many stages are based on marine sequences whose correlation with terrestrial sequences is only poorly developed. A bird family may originate or become extinct at any time during a stage yet it is assumed in many analyses, as it is here, that all such events took place at one particular point in time, in this case, at the end of the stage. This undoubtedly distorts the reality of the past, and could produce point events, such as extinctions, where none such existed before (cf. Kauffman 1979). Calculating the rate of extinction or origination can produce the opposite effect, smearing what may originally have been point events. Quite clearly, the utilization of such a crude time unit as the stage severely inhibits the distinction of 'real' events with a time period shorter than that of the stage, from artefacts produced by ordering data into such time units.

It is very difficult, if not impossible, to assess whether the current phylogeny of birds actually reflects their true phylogeny. Comparison of the latest review (Olson 1985*a*) with earlier works shows, as discussed below, that a stable and satisfactory classification is still far from being reached. For example, Olson (1985*a*) no longer recognizes the following families: Elopterygidae, Cyphornithidae, Telmabatidae; Odontopterygidae, Pseudodontornithidae, Plegadornithidae, Palaeospizidae, and Scaniornithidae. If correct the invalidation of these families will help to reduce the current degree of pseudo-extinction; but as regards avian taxonomy in general, this is only the tip of the iceberg.

Considerable problems exist, particularly with respect to Mesozoic and early Tertiary taxa, the affinities of many of which are still unclear. These problems are exacerbated by the attempts of some palaeornithologists to place all fossil forms in modern taxa, while others are (rightly) placing distinct fossil forms into separate taxa.

The basic taxonomic unit employed in this analysis, the family, represents a necessary, though unsatisfactory compromise. Its use introduces the possibility of further distortion of past events, primarily because origination and extinction function at the species level, and not at the arbitrarily defined family level. For example, species within a family may become extinct over a long period, yet analyses based at the family level will only recognize this as a point event, thus creating a mini-extinction. Compound these distortions with those produced by using the stage, discussed earlier, and mass extinctions seem almost inevitable.

3. *Associated problems*

There are a number of problems peculiar to the fossil record of birds. To begin with, there is plenty of evidence to suggest that the record is biased in a variety of ways (Fisher 1967). The severest form is undoubtedly taphonomic, but collecting and geographic bias also play prominent roles.

Approximately one in five families of extant birds are marine or littoral forms, the rest are terrestrial. If passerines are excluded, the ratio falls to about two in five. The Mesozoic record provides a sharp contrast, two out of three families known, appear to be marine or littoral, even though a terrestrial origin for birds is generally accepted (Fedducia 1980). That many more terrestrial birds existed, is suggested by the recent dicovery of forms such as *Ambiortus* (Kurochkin 1982), and the Enantiornithes (Walker 1981). The Tertiary record seems less biased in this respect, the ratios (if passerines are excluded) approaching those of the extant fauna. The relatively small size of most bird skeletons when compared with those of mammals, for example, has reduced both their likelihood of preservation and of collection. Small terrestrial forms such as the passerines are particularly poorly represented, partly through the fragility of their skeletons, partly because many collectors tend to overlook small elements (Harrison 1979), but mainly due to the (understandable) reluctance of palaeornithologists to attempt identification of passerine material.

Though the record is somewhat better than many realize (Olson 1985*a*), much of the data stems from a few 'lagerstatten' such as the Niobrara Chalk (Coniacian–Santonian) of Kansas, the Green River Formation (Lower Eocene) of Wyoming and Utah, the Phosphorites du Quercy (Eocene–Oligocene) of France, and the Rancho La Brea tar pits (Pleistocene) of California. Not only do these faunas distort the true picture of bird distribution through time, but also in space. Most well known avifaunas are located in Europe and North America, whilst those of Asia, Australia, and Southern America are, with a few exceptions, relatively poorly known. This reflects to a large degree the current distribution of palaeornithologists and, most importantly, it hinders the establishment of patterns or events on a world-wide basis.

Avian taxonomy is beset with difficulties. To begin with, the skeletal structure of birds presents the taxonomist with some serious, though not insurmountable, problems. The skull bones, for example, of great taxonomic value in living and extinct reptiles, are fused together, and their extent and relationships very difficult to determine. As a result the systematics of extant birds are based either on soft parts or other non-fossilizable characters such as song. The lack of information on the comparative osteoanatomy of extant birds has long held back palaeornithology, though the situation is improving.

In addition to the problems already noted, the skeletal morphology of birds is closely linked to their mode of life. This has led to a high degree of similarity across broad taxonomic groups, and produced some remarkable examples of convergence. The case of the Hesperornithiformes is one of the most celebrated. For many years, the Hesperornithiformes, medium to large flightless divers from the Upper Cretaceous of North America, were considered to be closely related to the grebes (Podicipedidae) and loons (Gaviidae), with which they have many characters in common (e.g. Brodkorb 1963; Fisher 1967). Recently, it has been recognized that all of these characters represent convergences, and that the groups share no close relationship (e.g. Martin 1983). Other examples are legion; the convergence of petrels (Procellariidae), penguins (Sphenisciformes), and auks (Alcidae), all wing-propelled divers, is well known. More important perhaps, is the recent recognition that herons (Ardeidae) and flamingos (Phoenicopteridae) are not closely related to each other or to

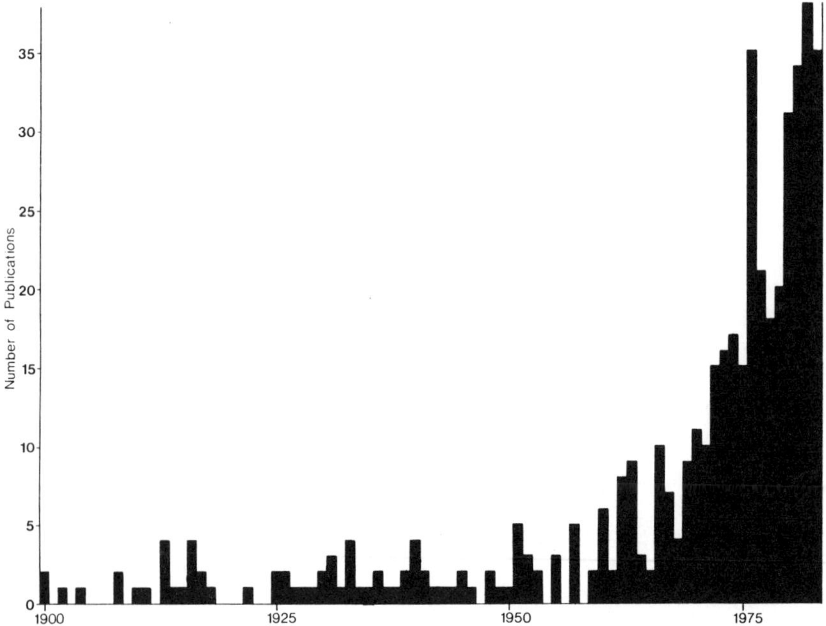

Fig. 14.1. Volume of papers published in *Palaeornithology* 1900–1983, based on publications cited by Lambrecht (1933), Brodkorb (1963, 1978), and Olson (1985*a*).

ibises and storks (Ciconiiformes), but merely share convergent characters such as large size, long legs, and a 'desmognathous' palate (Olson 1979).

The widely, though somewhat incorrectly held belief, that the fossil record of birds is particularly poor, together with the problems of identification, has had two effects. On the one hand there has been a tendency to name any and all material no matter how fragmentary, on the basis that 'bird material is so rare it must be something new'. This has produced considerable congestion in avian taxonomy, perhaps best highlighted by the long list of *Nomina Non Avium* recently published by Brodkorb (1978). On the other hand, the association of a supposedly poor record, with the problems of identification have not surprisingly, rendered avain palaeontology rather unpopular (Brodkorb 1971). It is reasonably fair to say that it lags some 15–20 years behind other disciplines such as mammalian and reptilian palaeontology. The recent advent of more powerful systematic methods such as cladistics, together with the realization that the record is no worse than many others, has led to a great deal of renewed interest, as evidenced by the veritable explosion in numbers of papers being published in this field (Fig. 14.1). Obviously, this situation is to be greatly welcomed, but it does bring some problems. Many new taxa have been proposed, and until they have been reviewed, and their status confirmed or rejected, avian taxonomy will remain in a state of flux. As a result the utility of data bases compiled during this period is likely to be short lived.

The pattern of avian diversity

A simple standing diversity curve, (Fig. 14.2) reveals little more than is currently known about the pattern of avian diversity. Taken at its simplest, it shows that apart from two apparent drops in diversity, one at the end of the Cretaceous, and a small one in the middle Eocene, the trend has been that of steady increase, reaching a high point in the Quaternary. The curve gives little evidence of having been affected by the mass extinction events recently proposed by Raup and Sepkoski (1984, 1986). The only decline in diversity to correlate with one of these extinctions occurs at the K–T boundary, but as discussed below, this is probably not symptomatic of an extinction event.

At present the earliest known bird is *Archaeopteryx,* from the Upper Jurassic of Bavaria. Other contenders for this title are mostly based on fragmentary material of doubtful validity. The claim by Chatterjee (Beardsley 1986) to have found bird remains in the late Triassic Dockum Formation of Texas, has yet to be substantiated. Bird diversity appears to have remained low throughout the Lower Cretaceous, though the discovery of small carinate forms such as *Limnornis* (Kessler 1984) and *Ambiortus*

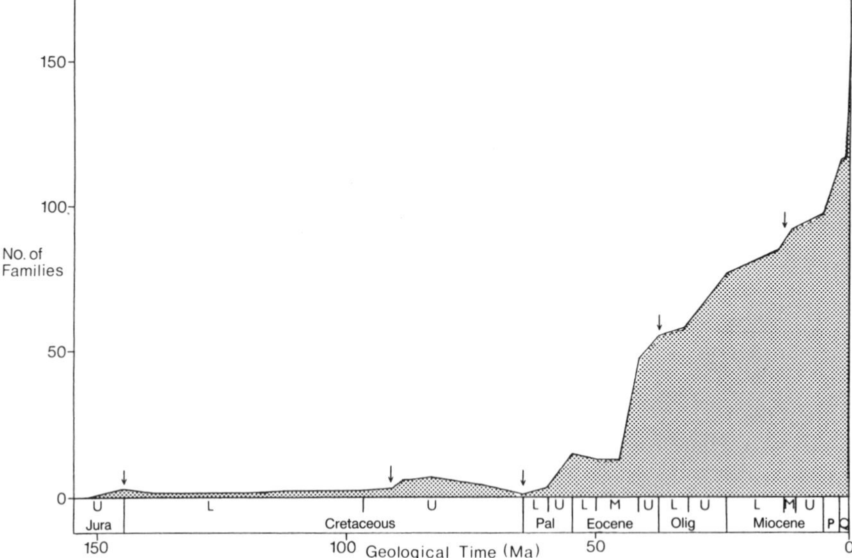

Fig. 14.2. Standing diversity with time for families of birds. Arrows denote 'significant' mass extinctions proposed by Raup and Sepkoski (1986). The time-scale is based on that of Harland *et al* (1982). Abbreviations: Jura, Jurassic; L, Lower; M, Middle; Olig. Oligocene; P, Pliocene; Pal, Palaeocene; Q, Quaternary; U, Upper.

(Kurochkin 1982) together with feather impressions from Mongolia (Kurochkin 1985), Lebanon (Schlee 1973) and Australia (Talent *et al.* 1966) provide substantial evidence for a world wide radiation of small flying forms at this time (Elzanowski 1983). These few and fragmentary remains emphasize the paucity of the Lower Cretaceous avian fossil record.

A number of groups radiated in the Upper Cretaceous, including the Hesperornithiformes and other flightless divers (Martin 1980, 1983), the Enantiornithes, and the 'transitional' Charadriiformes (Olson 1985*a*). The apparent rapid increase in the origination of taxa in the Coniacian and Santonian (Fig. 14.3) is largely an artefact produced by the 'bunching' effect of well preserved faunas from the Niobrara chalk. Fragmentary remains of these birds are known from earlier sediments, and suggest that the radiation of Hesperornithiformes and Ichthyornithiformes began in the Turonian or even earlier.

A recent publication by Molnar (1986) indicates that the Enantiornithes, known only from the late Campanian and Maastrichtian, also originated much earlier and achieved a worldwide distribution during the Upper Cretaceous. The recognition of already described taxa, such as *Alexornis* (and possibly also *Gobipteryx*) as probably being Enantiornithiform (Martin 1983) adds considerable weight to this view.

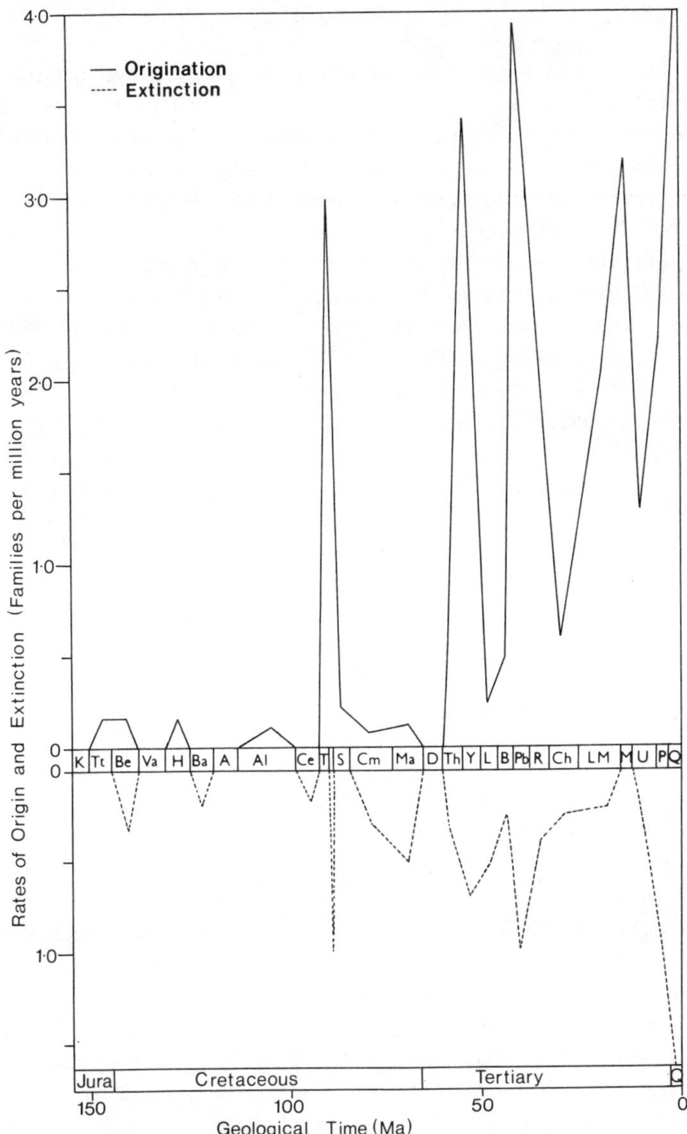

Fig. 14.3. Total rates of origin and extinction of bird families. Pre-Tertiary rates must be treated with great caution as the number of taxa per stage are very low ($n < 8$). The large fluctuations might represent 'real' radiation or extinction events, or may result from the biased nature of the avian fossil record. It is not yet possible to determine which. Abbreviations: A, Aptian; Al, Albian; B, Bartonian; Ba, Barremian; Be, Berriasian; Ce, Cenomanian; Ch Chattian; Cm, Campanian; D, Danian, H, Hauterivian; Jura, Jurassic; K, Kimmeridgian; L, Lutetian, LM, Lower Miocene; M, Middle Miocene; Ma, Maastrichtian; P, Pliocene; Pb, Priabonian; Q, Quaternary; R, Rupelian; S, Santonian; T, Turonian; Th, Thanetian; Tt, Tithonian; UM, Upper Miocene; Va, Valanginian; Y, Ypresian.

Both the standing diversity curve and the plot of total origination and extinction rates (Figs 14.2 and 14.3) seem to indicate that bird diversity reached very low levels at the K–T boundary. This is quite misleading and two factors are operating to produce this artefact. There are, as yet, no well preserved early and middle Maastrichtian avifaunas. As a result, most of the flightless forms seem to become extinct in the Campanian or earlier. It is quite likely, however, that they survived much later than this, as remains such as *Neogaeornis wetzeli* from the latest Cretaceous (Maastrichtian?) of Chile has now been recognized as hesperornithiform (Martin and Tate 1976) though the precise age of this material is not yet established. A number of late Maastrichtian avifaunas are well known, notably those of the Hornerstown Formation, New Jersey, and the Lance Formation of Wyoming. These were originally thought to contain cormorants, flamingos, loons, rails, and shorebirds, but an outline of the revision being undertaken by Olson and Parris (1987) suggests that many of these forms are in fact 'transitional' Charadriiformes and early Procellariiformes. Unfortunately, the higher level systematics of these groups, which began to radiate in the late Cretaceous, are not yet clear; hence, few families were available for this analysis, producing an abnormally low level of diversity across the K–T boundary.

There was a large increase in total diversity, matched by a peak in origination rates, during the Ypresian (Figs 14.2 and 14.3), with the appearance of many water birds, and non-passerine forest birds (Brodkorb 1971). There may have been a real radiation here, but this peak may also reflect a 'bunching' effect as the Thanetian and Danian vertebrate record is very poor (Benton 1987). This is probably because the recovery of bird remains is strongly related to mammal collecting, relatively little of which has been done in Palaeocene strata (Martin 1983). Until the faunal content of these stages is better known, the reality of the Ypresian radiation cannot be judged. A similar though much larger radiation appears to have occurred in the Priabonian. This is almost certainly an artefact produced by the heavily researched fauna of the Phosphorite du Quercy discussed earlier. Again this has led to a bunching of first appearances many of which may have occurred in the poorly represented Bartonian (Benton 1987, fig. 14.2).

A third radiation beginning in the Lower Miocene, and peaking in the Middle Miocene (Fig. 14.3) may be a real event, reflecting the radiation of passerines and land birds exploiting alpine and xeric habitats. A number of passerine families are first known from this period, though Feduccia and Olson (1982) have suggested that they originated somewhat earlier in the Southern hemisphere.

The rapid increase in diversity in the Late Tertiary and Quaternary (Figs 14.2 and 14.3) is almost certainly due to the increased likelihood of preservation, and the 'pull of the Recent'. It is widely recognized that many birds became extinct in the late Pleistocene and Holocene (Feduccia 1980). This demonstrates the degree to which the taxonomic level chosen, can distort events as this extinction led to a decrease of 18 per cent in the number of species, but only produced a drop of 5 per cent in the number of families. A similar effect was noted by Raup and Sepkoski (1986) who found that 'small' events at the family level, become much more pronounced when determined at the generic level.

Available data do not permit the recognition of rapid extinction events, but it is known that many taxa have become extinct. No particular temporal patterns have yet been recognized but many 'doomed' taxa, for example, the pseudodontorns, phorusrhacoids, and teratorns, seem to consist of large and specialized forms. This may constitute one of a number of patterns of 'background extinction', but much more research is needed to establish details of these patterns.

Pterosaurs and birds: an example of differential survival?

It has already been suggested that the fossil record of birds is at present too patchy for detailed quantitative analysis, and that short-term events or patterns based upon them, such as proposed by Vuilleumier (1984) and criticized by Olson (1985*b*) cannot be identified or tested with any degree of accuracy. However, this kind of analysis is not completely redundant as long-term patterns can be recognized. Analysis at the species level demonstrates that birds were at low levels of diversity in the Lower Cretaceous, and increased in diversity in the Upper Cretaceous, reaching their highest levels during the Maastrichtian (Unwin 1987). This data, corrected in the light of recent publications, agrees with the suspected levels of Mesozoic bird diversity outlined earlier.

It has frequently been asserted (e.g. Colbert 1980) that pterosaurs were out-competed by birds, culminating in their demise at the end of the Cretaceous. This can now be tested more rigourously using the amended bird data (Appendix) and data on pterosaur diversity (Unwin 1986). A simple comparison of the percentages of the aerial fauna occupied by birds and pterosaurs (Fig. 14.4), demonstrates that during the Cretaceous, as bird diversity rose, that of the pterosaurs steadily declined. This apparent relationship conforms very closely to the double wedge pattern of the differential survival model of faunal replacement (Benton 1983; Charig 1984). Assuming the diversity levels are related a number of reasons for the replacement of pterosaurs by birds can be postulated. Most importantly, unlike birds, pterosaurs had a poor terrestrial ability (Pennycuick 1986; Unwin 1987; Wellnhofer 1982), and were incapable of exploiting many terrestrial and some surface marine niches. Furthermore, the wing construction of pterosaurs was less flexible than that of birds and more prone

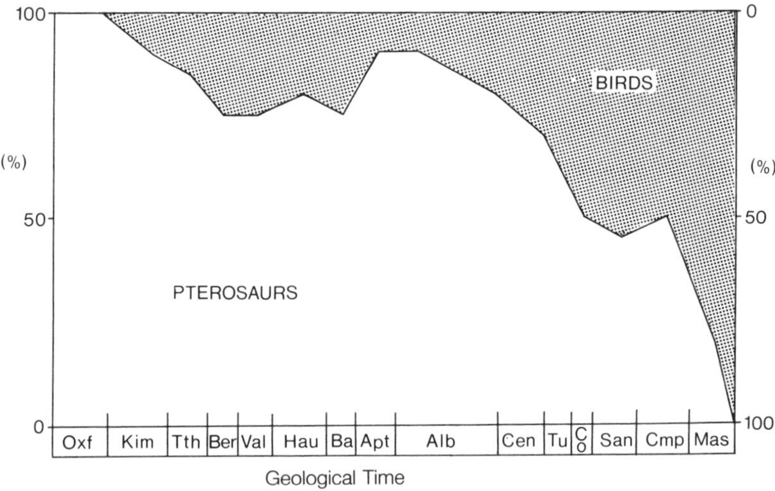

Fig. 14.4. Relative proportions of the aerial fauna occupied by pterosaurs and birds from the Upper Jurassic to the end of the Cretaceous. Abbreviations: Alb, Albian; Apt, Aptian; Ba, Barremian; Be, Berriasian; Cen, Cenomanian; Cmp, Campanian; Co, Coniacian; Hau, Hauterivian; Kim, Kimmeridgian; Mas, Maastrichtian; Oxf, Oxfordian; San, Santonian; Tth, Tithonian; Tu, Turonian; Val, Valanginian.

to damage. Finally, there are both empirical data and theoretical arguments (Bramwell and Whitfield 1974; Wellnhofer 1982) to suggest that pterosaurs were restricted to warm climatic zones. However, the recent discovery of pterosaur remains in the Campanian–Maastrichtian of New Zealand (R. E. Molnar, *pers. comm.*), which was situated at a latitude of about 60° south at that time, casts some doubt on this.

A more clearly defined scenario can now be proposed. Sometime during the Upper Jurassic early birds invaded niches such as those now occupied by shore birds and waders foraging in littoral and marshy environments. These niches were previously unoccupied, as pterosaurs, impeded by their poor terrestrial ability, were unable to exploit them. The pterosaur bauplan also prevented them from exploiting the sub-aqueous environment into which birds radiated in the Lower Cretaceous. Birds probably began to compete with pterosaurs for niches at about this time. Following the extinction of a bird or pterosaur, individuals in both groups could compete for the vacant niche. It would appear that, owing to their more flexible bauplan, birds were more successful, and by the late Cretaceous they appear to have filled many niches, except for those where their avian bauplan was of no particular advantage. These were the niches occupied by large ocean going forms such as *Pteranodon* and possibly *Quetzalcoatlus*, a giant soaring form which has been found in both terrestrial (Lawson 1975) and marine sediments (Monteillet *et al.* 1982). Like petrels

(Matthews 1951), *Pteranodon* at least had little need of a good terrestrial ability. A variety of factors may have brought about the extinction of these last pterosaurs, of which climatic changes seem the most likely candidate (Unwin 1986). Interestingly, the Pseudodontorns, large ocean-going birds which had many characters in common with pterosaurs such as *Pteranodon* and *Nyctosaurus*, first appeared in the early Eocene (S. L. Olson, *pers. comm.*) and filled the niches vacated by pterosaurs at the end of the Cretaceous.

The merit of the hypothesis upon which this scenario is based lies in the predictions which can be made from it, and the ease with which it can be corroborated or refuted by new discoveries of Mesozoic birds and pterosaurs. The predictions are: (1) large soaring birds did not exist at any time during the Mesozoic; (2) small pterosaurs which inhabited terrestrial and coastal environments became extinct by the end of the Lower Cretaceous; and (3) fully aerial birds first appeared in the late Middle or Upper Jurassic.

Conclusions

(1) The fossil record of birds, though better than many realize, is still very patchy and recent analyses of its degree of completeness err on the high side.

(2) Bias in the preservation and collection of avian fossils, together with bias in the geographic location of palaeornithologists, has produced a rather distorted picture of avian history. This distortion has been compounded by the considerable problems posed by avian taxonomy and, until recently, the relative lack of interest in palaeornithology.

(3) The relative crudity of the parameters used in the compilation of the data base, together with the problems mentioned in (2) largely preclude the use of this data in identification or testing of short-term events such as mass extinctions.

(4) It is hypothesized that pterosaurs and birds represent a case of differential survival, since the pattern of their relative diversity during the Mesozoic conforms to the double wedge pattern typical of the differential survival model of faunal replacement. This hypothesis formulates, in a more robust and testable form, previous explanations for the extinction of pterosaurs and survival of birds at the end of the Mesozoic. Furthermore, each new discovery of bird, or pterosaur, will serve to corroborate or refute this hypothesis and the predictions made from it.

Acknowledgements

I am much indebted to Gilbert Larwood for inviting me to take part in the Symposium. I have benefited greatly from discussions with Cyril Walker,

Kevin Padian, Beverley Halstead, and in particular, Mike Benton. Many thanks go to Andrew Kitchener, Charles Deeming, Penelope Milner, and in particular Storrs Olson, who were all kind enough to read this manuscript, and whose comments and criticisms have added much in conciseness and clarity. The remaining errors are, of course, entirely my own. This work was supported by the Wilkie Calvert Postgraduate Scholarship (A/C 030087) generously awarded by the Research Board of Reading University.

References

Alvarez, L. W., Alvarez, W., Asaro, F., and Michel, H. V. (1980). Extraterrestrial cause for the Cretaceous–Tertiary extinction. *Science* **208,** 1095–108.

Alvarez, W. (1983). Experimental evidence that an asteroid impact led to the extinction of many species 65 million years ago. *Proc. nat. Acad. Sci. USA* **80,** 627–42.

Alvarez, W., Asaro, F., Michel, H. V., and Alvarez, L. W. (1982). Iridium anomaly approximately synchronous with terminal Eocene extinctions. *Science* **216,** 886–888.

Alvarez, W., Kauffman, E. G., Surlyk, F., Alvarez, L. W., Asaro, F., and Michel, H. V. (1984). Impact theory of mass extinctions, and the invertebrate fossil record. *Science* **223,** 1135–41.

Bakker, R. T. (1977). Tetrapod mass extinctions. A model of the regulation of speciation rates and immigration by cycles of topographic diversity. In *Patterns of Evolution as Illustrated by the Fossil Record* (ed. A. Hallam), pp. 439–68 Elsevier, Amsterdam.

Beardsley, T. (1986). Fossil bird shakes evolutionary hypothesis. *Nature* **322,** 677.

Benton, M. J. (1983). Dinosaur success in the Triassic: a noncompetitive ecological model. *Q. Rev. Biol.* **58,** 29–55.

Benton, M. J. (1985*a*). Mass extinction among non-marine tetrapods. *Nature, Lond.* **316,** 811–4.

Benton, M. J. (1985*b*). Patterns in the diversification of Mesozoic non-marine tetrapods and problems in historical diversity analysis *Spec. Pap. Palaeont.* **33,** 185–202.

Benton, M. J. (1986). More than one event in the late Triassic mass extinction. *Nature, Lond.* **321,** 858–61.

Benton, M. J. (1987). Mass extinctions among families of non-marine tetrapods: the data. *Mém. Soc. géol. Fr.* **150,** 21–32.

Bramwell, C. D. and Whitfield, G. R. (1974). Biomechanics of *Pteranodon*. *Phil. Trans. Roy. Soc. London.* **B267,** 503–81.

Brodkorb, P. (1963). Catalogue of fossil birds. Part 2 (Archaeopterygiformes through Ardeiformes). *Bull. Fla. State. Mus. Biol. Sci.* **7,** 179–293.

Brodkorb, P. (1971). Origin and evolution of birds. In *Avian Biology,* Vol. 1 (eds D. S. Farner and J. R. King), pp. 19–54. Academic Press, New York.

Brodkorb, P. (1978). Catalogue of fossil birds. Part 5. *Bull. Fla. State. Mus., Biol. Sci.* **23,** 139–228.

Charig, A. J. (1984). Competition between therapsids and archosaurs during the

Triassic period: A review and synthesis of current theories. *Symp. Zool. Soc. Lond.* **52,** 597–628.

Colbert, E. H. (1980). *Evolution of the Vertebrates.* 3rd edn. Wiley Interscience, New York.

Cutbill, J. L. and Funnell, B. M. (1967). Numerical analysis of the Fossil Record. In *The Fossil Record* (ed. W. B. Harland *et al.*), pp. 791–820. Geological Society of London.

Elzanowski, A. (1983). Birds in Cretaceous ecosystems. *Acta palaeontol. polon.* **28,** 75–92.

Fedducia, A. (1980). *The Age of Birds.* Harvard University Press, Cambridge, Mass.

Fedducia, A. and Olson, S. L. (1982). Morphological similarities between the Menurae and Rhinocryptidae, relict passerine birds of the Southern Hemisphere. *Smithson. Contrib. Zool.* **366,** 1–22.

Fischer, A. G. and Arthur, M. A. (1977). Secular variations in the pelagic realm. *S.E.P.M. Spec. Publ.* **25,** 19–50.

Fisher, J. (1967). Fossil birds and their adaptive radiation. In *The Fossil Record* (ed. W. B. Harland *et al.*), pp. 133–154. Geological Society of London.

Hallam, A. (1984). The center of mass extinctions. *Nature, Lond.* **308,** 686–7.

Harland, W. B., Cox, A. V., Llewellyn, P. G., Picton, C. A. G., Smith, A. G., and Walters, R. (1982). *A Geologic Time Scale.* Cambridge University Press, Cambridge.

Harrison, C. J. O. (1979). Small non-passerine birds of the Lower Tertiary as exploiters of ecological niches now occupied by passerines. *Nature, Lond.* **281,** 562–3.

Houde, P. (1986). Ostrich ancestors found in the Northern Hemisphere suggest new hypothesis of ratite origins. *Nature, Lond.* **324,** 536–5.

Houde, P. (1987). *Auk* (In press).

Kauffman, E. G. (1979). The ecology and biogeography of the Cretaceous–Tertiary extinction event. In *Cretaceous–Tertiary Boundary Events,* Vol. 2 Proceedings (eds W. K. Christensen and T. Birkelund), pp. 29–37. University of Copenhagen.

Kessler, E. (1984). Lower Cretaceous birds from Cornet, Roumania. In *Third Symposium on Mesozoic Terrestrial Ecosystems* (ed. W. E. Reif and F. Westphal), pp. 119–21. Attempto Verlag, Tübingen.

Kurochkin, E. N. (1982). New order of birds from the Lower Cretaceous of Mongolia. *Dokl. Akad. Nauk. USSR* **262,** 452–5. [In Russian.]

Kurochkin, E. N. (1985). A true carinate bird from Lower Cretaceous deposits in Mongolia and other evidence of early Cretaceous birds in Asia. *Cretaceous Res.* **6,** 271–8.

Lambrecht, K. (1933). *Handbuch der Palaeornithologie.* Gebruder Borntraeger, Berlin.

Lawson, D. A. (1975). Pterosaur from the latest Cretaceous of Texas. Discovery of the largest flying creature. *Science* **187,** 947–8.

Lianhai, H. and Zhicheng, L. (1984). A new fossil bird from the Lower Cretaceous of Gansu and the early evolution of birds. *Scientia Sinica* (B) **27,** 1296–302.

Martin, L. D. (1980). Foot-propelled diving birds of the Mesozoic. *Acta Congr. Int. Ornithol.* **17,** 1237–42.

Martin, L. D. (1983). The origin and early radiation of birds. In *Perspectives in Ornithology* (ed. A. H. Brush and G. A. Clarke Jr.), pp. 291–338. Cambridge University Press.

Martin, L. D. and Tate, J. (1976). The skeleton of Baptornis advenus (Aves: Hesperornithiformes). *Smithson. Contrib. Paleobiol.* **27,** 36–66.

Matthews, L. H. (1951). *Wandering albatross.* MacGibbon & Kee, London.

Molnar, R. E. (1986). An enantiornithine bird from the Lower Cretaceous of Queensland, Australia. *Nature, Lond.* **322,** 736–8.

Monteillet, J., Lappartient, J. R., and Taquet, P. (1982). Un Ptérosaurian géant dans le Crétacé superieur de Paki (Senegal). *C.R. Acad. Sci., Paris* **295,** 409–14.

Mourer-Chauvire, C. (1982). Les oiseaux fossiles des Phosphorites du Quercy (Eocene Superieur a Oligocene Superieur): implications palaeobiographiiques. *Geobios. Mem. Spec.* **6,** 413–26.

Nessov, L. A. (1984). Upper Cretaceous pterosaurs and birds from Central Asia. *Paleont. Zh.* **1984,** 38–49.

Newell, N. D. (1963). Crises in the history of life. *Scient. Am.* **28,** 76–92.

Newell, N. D. (1967). Revolutions in the history of life. *Geol. Soc. Amer. Spec. Pap.* **89,** 63–91.

Olson, S. L. (1979). Multiple origins of the Ciconiiformes. *Proc. Colonial Waterbird Group* **1978,** 165–70.

Olsen, S. L. (1985*a*). The fossil record of birds. In *Avian Biology* Vol VIII (ed. D. S. Farner, J. R. King, and K. C. Parkes), pp. 79–252. Academic Press, New York.

Olson, S. L. (1985*b*). Faunal turnover in South American fossil avifaunas: the insufficiences of the fossil record. *Evolution* **39,** 1174–7.

Olson, S. L. and Parris, D. (1987). The Cretaceous birds of New Jersey. *Smithsonian Contributions to Palaeobiology.* (In press.)

Paul, C. R. C. (1982). The adequacy of the fossil record. In *Problems of Phylogenetic Reconstruction* Vol. 22 (ed. K. A. Joysey and A. E. Friday), pp. 75–117. Academic Press, London.

Pennycuick, C. J. (1986). Mechanical constraints on the evolution of flight. *Mem. Calif. Acad. Sci.* **8,** 83–98.

Raup, D. M. (1986). Biological extinction in earth history. *Science* **231,** 1528–33.

Raup, D. M. and Sepkoski, J. Jr (1984). Periodicities of extinctions in the geological past. *Proc. nat. Acad. Sci. USA* **81,** 801–5.

Raup, D. M. and Sepkoski, J., Jr (1986). Periodic extinction of families and genera. *Science* **231,** 833–6.

Russell, D. A. (1982). The mass extinctions of the late Mesozoic. *Scient. Am.* **246,** 58–65.

Russell, D. A. (1984). The gradual decline of the dinosaurs–fact or fallacy. *Nature, Lond.* **307,** 360–1.

Schlee, D. (1973). Harzkonservierte fossile Vogelfedern aus der untersten Kreide. *J. Ornithol.* **114,** 207–19.

Sepkoski, J., Jr (1978). A kinetic model of Phanerozoic taxonomic diversity. I. Analysis of marine orders. *Paleobiology* **4,** 223–51.

Sibley, C. G. and Ahlquist, J. E. (1986). Reconstructing bird phylogeny by comparing DNA's. *Sci. Am.* **(4),** 68–78.

Talent, J. A., Duncan, P. M., and Handby, P. L. (1966). Early Cretaceous feather from Victoria. *Emu* **66,** 81–6.

Unwin, D. M. (1987). Pterosaur extinction: nature and causes. *Bull. Soc. géol. France.* (In press).

Unwin, D. M. (1987). Pterosaur locomotion, joggers or waddlers? *Nature, Lond.* **327,** 13–14.
Valentine, J. W. (1969). Patterns of taxonomic and ecological structure of the shelf benthos during Phanerozoic time. *Palaeontology* **12,** 684–709.
Valentine, J. W. (1973). Phanerozoic taxonomic diversity: a test of alternate models. *Science* **180,** 1078–9.
Vuilleumier, F. (1984). Faunal turnover and development of fossil avifaunas in South America. *Evolution* **38,** 1384–96.
Walker, C. A. (1981). New subclass of birds from the Cretaceous of South America. *Nature, Lond.* **292,** 51–3.
Wellnhofer, P. (1982). Zur Biologie der Flugsaurier. *Natur. und Museum* **112,** 278–91.

Appendix

First and last occurrences of families of birds

* Families based on fragmentary remains of dubious taxonomic status: omitted from calculations.

Families still living are designated 'REC'. Those that survived in to historic times, but are now extinct are termed 'HOL'. Stratigraphic abbreviations: ALB, Albian; APT, Aptian; BAR, Barremian; BER, Berriasian; BRT, Bartonian; CEN, Cenomanian; CHT, Chattian; CMP, Campanian; CON, Coniacian; DAN, Danian; EOC, Eocene; HAU, Hauterivian; KIM, Kimmeridgian; LCR, Lower Cretaceous; LMI, Lower Miocene; LUT, Lutetian; MAS, Maastrichtian; MMI, Middle Miocene; OXF, Oxfordian; PLE, Pleistocene; PLI, Pliocene; PRB, Priabonian; RUP, Rupelian; SAN, Santonian; THA, Thanetian; TTH, Tithonian; TUR, Tu onian; UMI, Upper Miocene; VAL, Valanginian; YPR, Ypresian.

Aves

Archaeopterygiformes	
Archaeopterygidae	TTH–BER (Kessler 1984)
Limnornithiformes	
*Limnornithidae	BER (Kessler 1984)
Hesperornithiformes	
Enaliornithidae	ALB–CEN
Baptornithidae	CON–?MAS
Hesperornithidae	CON–CMP
Ichthyornithiformes	
Ichthyornithidae	TUR–CMP
*Zhyraornithidae	TUR–CON (Nessov 1984)
Ambiortiformes	
Ambiortidae	HAU–?BAR (Kurochkin 1982)
Enantiornithiformes	
Enantiornithidae	ALB–MAS (Molnar 1986)
?Alexornithidae	CMP
Gobipterygiformes	
Gobipterygidae	SAN–CMP (Elzanowski 1983)

Gansuiiformes	
*Gansuiidae	LCR (Lianhai and Zhicheng 1984)
Ratitae	
Aepyornithidae	PLE–HOL
Dromiceiidae	PLE–HOL
Casuariidae	UMI–REC
Emeidae	UMI–HOL
Dinornithidae	PLE–HOL
Anomalopterygidae	PLE–REC
Apterygidae	!PLE–REC
Rheidae	PLI–REC
Tinamidae	PLI–REC
Gaviiformes	
Gaviidae	PRB–REC
Procellariiformes	
Diomedeidae	CHT–REC
Procellariidae	RUP–REC
Oceanitidae	UMI–PLE
Pelecanoididae	PLI–REC
Sphenisciformes	
Spheniscidae	PRB–REC
Pelecaniformes	
Prophaethontidae	YPR
Phaethontidae	MMI–REC
Pelecanidae	LMI–REC
Pelagornithidae	YPR–UMI
Sulidae	RUP–REC
Plotopteridae (*S. L. Olson, personal communication*)	RUP–MMI
Phalacrocoracidae	PRB–REC
Anhingidae	LMI–REC
Fregatidae	YPR–REC
Ciconiiformes	
Scopidae	PLI–REC
Balaenicipitidae	PRB–REC
Ciconiidae	PRB–REC
Vulturidae	PRB–REC
Teratornithidae	UMI–HOL
Anseriformes	
Anhimidae	REC
Anatidae	RUP–REC
Galliformes	
Cracidae	YPR–REC
Megapodiidae	PRB–REC
Phasianidae	CHT–REC
Numididae	PRB–REC
Diatrymiformes	
Gastornithidae	YPR–PRB
Diatrymidae (= Gastornithidae?)	YPR–LUT
Gruiformes	
Mesitornithidae	REC

Geranoididae	YPR–BRT
Eogruidae	PRB–UMI
Gruidae	PRB–REC
Struthionidae	UMI–REC
Psophiidae	REC
Ergilornithidae	RUP–PLI
* Orthocnemidae	?PRB
Rallidae	CHT–REC
Heliornithidae	REC
Rhynochetidae	REC
Apterornithidae	PLE–REC
Eurypygidae	YPR–REC
(*S. L. Olson, personal communication*)	
Bathornithidae	RUP–LMI
Cariamidae	?RUP–REC
Phorusrhacidae	PRB–PLI
Idiornithidae	PRB–RUP
* Cunampaiidae	EOC
Podicipedidae	LMI–REC
Charadriiformes	
Jacanidae	CHA–REC
Rostratulidae	REC
Haematopodidae	PLI–REC
Pedionomidae	REC
Graculavidae	MAS
Presbyornithidae	THA–YPR
Charadriidae	PLI–REC
Scolopacidae	PRB–REC
Recurvirostridae	?PRB–REC
Phalaropodidae	PLE–REC
Dromadidae	REC
Burhinidae	LMI–REC
Glareolidae	LMI–REC
Otididae	PRB–REC
Thinocoridae	REC
Chionididae	REC
Stercorariidae	LMI–REC
Laridae	PRB–REC
Rynchopidae	REC
Alcidae	MIO–REC
Plataleidae	LUT–REC
Phoenicopteridae	YPR–REC
(*S. L. Olson, personal communication*)	
Columbiformes	
Pteroclidae	PRB–REC
Columbidae	LMI–REC
Psittaciformes	
Psittacidae	PRB–REC
Strigiformes	
Tytonidae	LMI–REC
Protostrigidae	YPR–PRB

Strigidae	LMI–REC
Ogygoptyngidae	THA
Caprimulgiformes	
Steatornithidae	YPR–REC
Aegothelidae	?PRY–REC
Podargidae	PRB–REC
Caprimulgidae	PRB–REC
Nyctibiidae	PLE–REC
Apodiformes	
Aegialornithidae	YPR–RUP
Hemiprocnidae	?LUT–REC
Apodidae	LMI–REC
Coliiformes	
Coliidae	PRB–REC
Coraciiformes	
Halcyonidae	PRB–REC
Todidae	PRB–REC
Momotidae	RUP–REC
Meropidae	PRB–REC
Coraciidae	PRB–REC
Atelornithidae	REC
Leptosomidae	REC
Archaeotrogonidae	PRB–CHT
Trogonidae	RUP–REC
Bucconidae	YPR–REC
* Primobucconidae	YPR
Galbulidae	REC
Bucerotiformes	
Upupidae	PRB–REC
Phoeniculidae	LMI–REC
Bucerotidae	MMI–REC
Piciformes	
Capitonidae	LMI–REC
Indicatoridae	PLI–REC
Rhamphastidae	PLE–REC
Picidae	MMI–REC
Passeriformes	
Eurylaimidae	LIM–REC
Furnariidae	PLE–REC
Formicariidae	PLE–REC
Conopophagidae	REC
Rhinocryptidae	?LMI–REC
Pittidae	REC
Philepittidae	REC
Acanthisittidae	REC
Tyrannidae	PLE–REC
Pipridae	REC
Cotingidae	REC
Phytotomidae	REC
Menuridae	REC
Atrichornithidae	REC
Alaudidae	LMI–REC

Hirundinidae	PLE–REC
Motacillidae	LMI–REC
Campephagidae	REC
Pycnonotidae	PLE–REC
Irenidae	REC
Laniidae	LMI–REC
Vangidae	REC
Bombycillidae	PLE–REC
Dulidae	REC
Cinclidae	PLE–REC
Palaeoscinidae	MMI–REC
Troglodytidae	PLE–REC
Mimidae	PLE–REC
Prunellidae	PLE–REC
Muscicapidae	LMI–REC
Paridae	PRB–REC
Sittidae	?MMI–REC
Certhiidae	PLE–REC
Dicaeidae	REC
Nectariniidae	REC
Zosteropidae	REC
Meliphagidae	HOL–REC
Emberizidae	UMI–REC
Parulidae	PLE–REC
Drepanididae	PLE–REC
Vireonidae	PLE–REC
Icteridae	PLE–REC
Fringillidae	?MMI–REC
Estrildidae	REC
Viduidae	REC
Ploceidae	?LMI–REC
Sturnidae	PRB–REC
Oriolidae	PLE–REC
Dicruridae	PLE–REC
Sylviidae	UMI–REC
Calleidae	HOL–REC
Grallinidae	REC
Artamiidae	REC
Cracticidae	REC
Ptilinorhynchidae	REC
Paradisaeidae	REC
Corvidae	?MMI–REC
Family *incertae sedis*	
Ardeidae	PRB–REC
* Agnopteridae	PRB–CHT
Turnicidae	PLI–REC
Tronchilidae	PLE–REC
Zygodactylidae	LMI–UMI
Dromornithidae	MMI–PLE
Falconidae	PRB–REC
Sagittariidae	PRB–REC
Accipitridae	PRB–REC

Pandionidae	CHT–REC
Opisthocomidae	UMI–REC
Cuculidae	PRB–REC
Musophagidae	CHT–REC
* Dakotornithidae	THA
Apatornithidae	CON–MAS

15. Extinction and the fossil mammal record

R. J. G. SAVAGE

Department of Geology, University of Bristol, Bristol, UK

Abstract

Extinction is seen as the inevitable and inseparable complement to evolution. The quest is to identify patterns and processes of extinction, by testing models and by examining the data base. Mass extinctions (as distinct from background noise) may be periodic in time, worldwide in distribution, and selective in the taxa extinguished. Analyses of these patterns may suggest the influence of physical processes.

The fossil mammal record is most profitably examined at generic and familial levels. The data base is relatively good for North America and Europe throughout the Cenozoic. The stratigraphic record at these peak extinction events is examined for evidence of environmental stress or for coincidence with supposed cyclic extinction events. Patterns are identifiable in the episodes, but each episode has its own individual stamp and they share little in common. It is proposed that environmental influences are the dominant trigger mechanism which act on the biota to produce mass extinction of terrestrial mammals at local, regional, or global level.

Introduction

Extinction and evolution are as inseparable as night and day, as life and death, as Yin and Yang. Without extinction there would be no fossil record worth studying; only remains of living species. Yet while a vast amount of time and effort has been devoted to evolution studies, the problems of extinction have until recently been relatively neglected. What we seek to

Extinction and Survival in the Fossil Record (ed. G. P. Larwood), Systematics Association Special Volume No. 34, pp. 319–34. Clarendon Press, Oxford, 1988.

identify in extinction records is pattern(s) and if possible offer an explanation in terms of causal processes.

We know the chronostratigraphical record to be incomplete in varying degree, yet at every known level there are last appearances of species. From this we may reasonably conclude that extinction *per se* is essentially continuous. Our problem is to identify clusters or peaks amongst the background noise, the often quoted 'mass extinctions'. The peaks may effect many stocks simultaneously and be recorded over one or more continents. They may be very brief events or may plateau out over several million years. In examining the record of peaks through the geological column, we can look for a pattern such as cyclicity.

An alternative (or indeed sequential) approach is to examine processes and deduce how these might influence the extinction patterns. The processes may conveniently be seen as physical, environmental, and biological. Movements associated with plate tectonics, both horizontal and vertical, have dramatic effects on the distribution of land and sea, on global sea-levels, on watertables, temperatures, and climatic regimes. These physical changes drastically effect the environment and its vegetation cover, the trophic pyramids, and the competition elements which operate in natural selection. The biotic response is either extinction in the face of changes or adaptation to them by natural selection.

Our first task must be to look at the data base. For mammals this is uneven. Stratigraphically, it is poor in the Mesozoic, fair in the Palaeogene, and progressively better through the Neogene toward the Pleistocene; in short, the more recent the record, the better it is. Geographically, the mammal fossil record is good for North America and Europe, but patchy elsewhere. This reflects the amount of work done by palaeontologists on these two continents and our relative ignorance of other continents. Taxonomically, the record is good for megafaunas and weaker for microfaunas; it is easy to spot elephants, but difficult to recover rodent teeth. There is a bias toward primates which are over-represented in the record due to the enormous amount of time, effort, and money devoted to their recovery — though there are some spin offs for other stocks. Because of these constraints, this paper will largely, though not exclusively, concentrate on the Cenozoic mammal faunas of North America and Europe. It is felt that to include the Mesozoic record or the Cenozoic record on other continents would introduce so much inadequate data that no meaningful interpretation could result.

The chronostratigraphic base for terrestrial mammals is the sequence of land mammal ages (Table 15.1). This is well established for the Cenozoic in North America, reasonably good for Europe and South America. It is much less developed for other continents. The assignment of land mammal ages to classical epochs (e.g. Miocene) poses major problems and relies heavily on interdigitations of marine and continental sediments (which are rare)

Table 15.1. Cenozoic land mammal ages in North America and Europe.

		Land mammal ages	
Age (Ma)	Epochs	North American	European
	Holocene		
	———	Rancholabrean	Oldenburgian
	Pleistocene		
		Irvingtonian	Biharian
2.5	———		
			Villafranchian
	Pliocene	Blancan	
5	———		Ruscinian
		Hemphillian	Turolian
		Clarendonian	Vallesian
	Miocene	Barstovian	Astaracian
		Hemingfordian	Orleanian
			Agenian
24	———	Arikareean	
		Geringian	Arvernian
	Oligocene		
		Whitneyan	
		Orellan	Suevian
37	———	Chadronian	
		Duchesnian	
			Headonian
		Uintan	
	Eocene	Bridgerian	Geiseltalian
		Wasatchian	Cuisian
54	———	Clarkforkian	
			Cernaysian
		Tiffanian	
	Palaeocene		
		Torrejonian	
		Puercan	
65			

and on radiometrically dated beds, usually ash bands; it follows that the correlation of land mammal ages between continents often lacks precision. The mean duration of land mammal ages is around 3.4 Ma; this is the major limiting factor in the resolution of taxa duration, especially at generic and specific levels.

The choice of taxonomic level of enquiry is important Table 15.2. We recognize 44 orders of mammals in total, of which 26 are extinct. Orders are probably well represented in the fossil record and it is unlikely that the

Table 15.2. Numbers of extinct and living mammalian orders, families, genera, and species

Mammals	Extinct	Living	Total	Mean duration
Orders	26	18	44	—
Families	180	130	310	23 Ma
Genera	*c.* 3000	1000	*c.* 4000	10 Ma
Species	*c.* 8000	4000	*c.* 12 000	1.5 Ma

number of extinct orders will be doubled. They are also for the most part long-lived. Of the 26 extinct orders, no less than eight disappeared during early Oligocene times (discussed in detail below). The others show no concentration and their extinctions are scattered through the stratigraphical column, in ones, twos, and a three.

Looking next at mammalian families, some 130 families of living mammals are generally recognized and there are currently about 180 extinct families known. The number of living families is unlikely to change much in the future, but the number of extinct families could increase very considerably as more faunas become known. The family record is useful for overviews, and for geographic and stratigraphic range analyses. It is less useful for analyses of extinction as most families are relatively long lived. Non-therian families are very long lived; the mean for 12 extinct families is 45 Ma. Marsupial families are rather short lived; the mean for 10 extinct families is 14 Ma. Therian mammal families have a mean duration of around 23 Ma, and the figure holds whether the living families are included or excluded. The mean duration figure for all mammal families is 23 Ma; this figure results because the therians dominate with only a handful of marsupial and non-therian families.

Mammalian genera are in many ways the most suitable taxonomic units to utilize in extinction studies. There is good agreement on the recognition of genera, they are usually widely distributed in space and have limited time ranges. In round figures, there are about 1000 living genera and about 3000 extinct genera known. Few if any genera have a duration less than 5 Ma, few exceed 20 Ma, and only a handful exceed 30 Ma. The mean duration of extinct genera appears to be about 10 Ma, and on average they will range through two or three mammal ages.

The number of living species of mammal is around 4000. The number of fossil species named is two or three times that figure; it certainly includes many synonyms and excludes many as yet undescribed species. The guess of 8000 fossil species is a very conservative estimate; the figure might be nearer 10 000 or even 12 000. The problems of parity between fossil and living taxa is greater at the species level than at higher levels; the diagnosis of living species often relies on data unavailable in the fossil record (e.g. soft parts, behaviour, distribution). Only in the Pleistocene do we have a sufficiently detailed chronology to make estimates of species longevity and it is around 1.5 Ma. This short duration, combined with the limited geographical range of most species, does not make them the best units for extinction studies. In the analysis that follows it is primarily genera, and to a lesser degree families, which will be examined.

The mammal extinction record

Table 15.3 summarizes data for land mammal extinctions throughout the Cenozoic in North America and Europe. Savage and Russell (1983) provide the indispensable data base for this study with their tabulation (chapter 9, tables 1 and 2) of the disappearance rate for genera at each land mammal age on the two continents. Their 'disappearance rate' is perhaps rather freely interpreted here as 'extinction rate'. In other words no attempt has been made to weed out Lazarus taxa; at generic level, this is believed to be a relatively trivial problem. The tables in Savage and Russell show that the mean extinction rate for genera on both continents is 16, that is the number of genera that become extinct per million years. Taking 16 extinctions per million years as the mean background level of extinctions, attention is here focused on stratigraphic intervals where the extinction rate is equal to or greater than the mean rate. These extinction events are shown in Table 15.3 on either side of the central time scale, the figures in each case representing the actual rates. In the columns lateral to these are figures of the duration of each extinction peak and the length of the intervening gaps between peaks. Raup and Sepkoski (1984) recognized a cyclicity in extinction peaks in marine faunas at 26 Ma intervals; their Cenozoic extinction peaks are at 11.3, 38, and 65 Ma. Similarly, Fischer and Arthur (1977) recognized a cyclicity in marine faunal extinctions which peaked at 32 Ma intervals, with Cenozoic maxima at 2, 30, and 62 Ma.

Before undertaking a detailed examination of this data a number of preliminary observations can be made. The peaks of extinction are of variable height, ranging here from the mean base of 16 to 73 genera per Ma. The duration of these peak events varies from 5 to less than 1 Ma. The peaks are not always separate and followed by a trough; throughout most of the Miocene a series of extinction peaks follow each other without an interval, both in North America and in Europe. At other stratigraphic levels where the time resolution is less precise, these would be rolled into one major event. There is some degree of correlation of the peaks between North America and Europe, but it is not complete. The gaps between the peaks are very variable, ranging from 2 to 18 Ma in duration.

Table 15.3. Comparative table for North American and European land mammal ages in which the extinction rates for genera are equal to or greater than the mean; for details see text

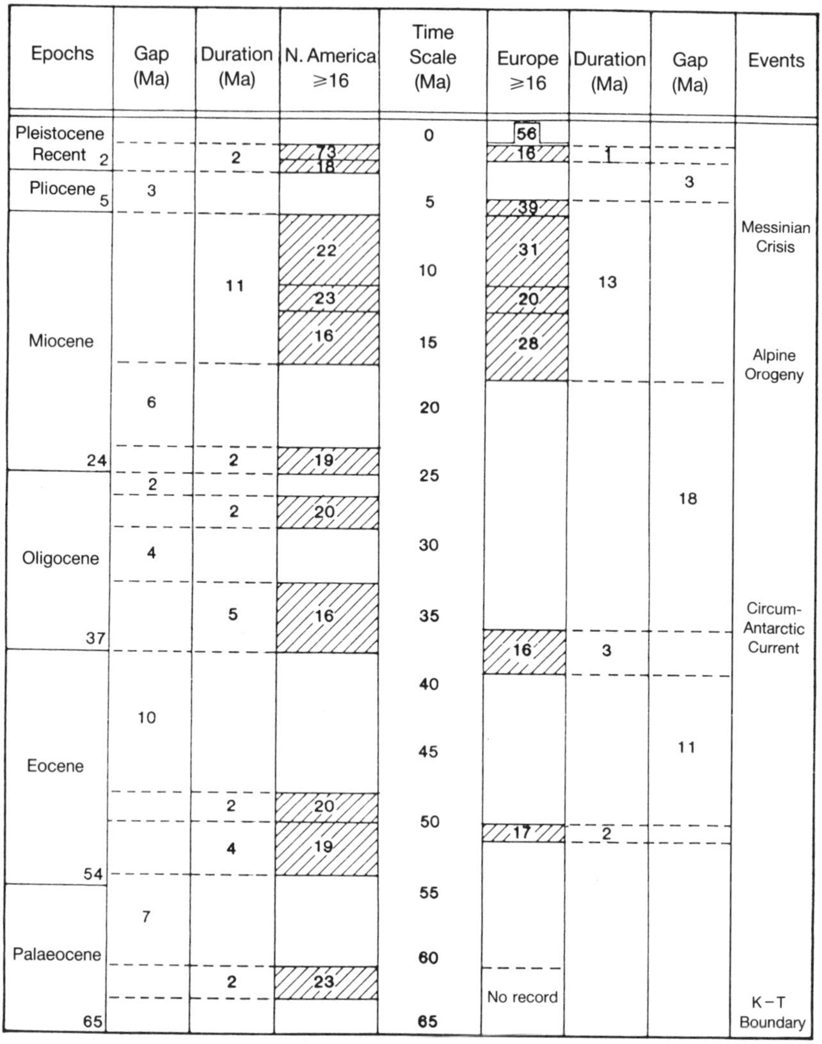

Epochs	Gap (Ma)	Duration (Ma)	N. America ≥16	Time Scale (Ma)	Europe ≥16	Duration (Ma)	Gap (Ma)	Events
Pleistocene				0	56			
Recent 2		2	73 18		16	1		
Pliocene 5	3			5	39		3	
Miocene		11	22	10	31	13		Messinian Crisis
			23		20			
			16	15	28			Alpine Orogeny
	6			20				
24		2	19	25			18	
Oligocene	2							
		2	20					
	4			30				
37		5	16	35				Circum-Antarctic Current
Eocene				40	16	3		
	10			45			11	
		2	20					
		4	19	50	17	2		
54								
Palaeocene	7			55				
		2	23	60	No record			
65				65				K–T Boundary

The K–T boundary (65 Ma)

The K–T boundary is one of the best known extinction peaks in the fossil record, but the data base for mammals is very poor. Only in western North America do we know a continual succession of fossiliferous strata through the boundary. The land mammal ages on either side of the boundary are the Lance in the latest Cretaceous and the Puercan in the earliest

Table 15.4. Mammal faunas in western North America around the K–T boundary

K–T Boundary N. America	Lance 5 Ma		Puercan 3 Ma	
	G.	sp.	G.	sp.
Multituberculates	10	34	10	12
carryovers			5	1
Marsupials	5	21	1	1
carryovers			1	?
Placentals	7	15	22	38
carryovers			4	0
Totals	22	70	33	51
carryovers			10	1

Palaeocene (Table 15.4). The same three groups of mammals are known in each level, but the proportions differ. The multituberculates were small herbivorous mammals, filling an ecological niche similar to living voles; there are 10 genera in each level, five of those in the Lance carrying over into the Puercan, though with fewer species. Both the marsupial and placental mammals were small, shrew-like, and essentially insect feeders. The marsupials suffered a dramatic decline at the boundary, almost becoming extinct, while the placentals experienced an explosive radiation in the Puercan. The total number of mammalian species is actually fewer in the Puercan than in the Lance. The absence of a record across the boundary in Europe makes comparison impossible. With almost half the genera carrying on across the boundary, it is only the marsupials which appear to have been overstressed. Note also that these two mammal ages span 8 million years; measured from the present, this time lapse would extend back well into the Miocene, and in that interval we currently recognize four, not two land mammal ages.

The middle Palaeocene (60 Ma)

In North America the Torrejonian land mammal age in the Middle Palaeocene has a high extinction rate (23 genera per Ma over a 2 Ma interval). Unfortunately, no European mammal fauna of this age is known. The Torrejonian mammal age is believed to have lasted about 2 Ma and from it are recorded about 88 genera in 29 families. Fifty per cent of these genera become extinct before the next mammal age (Tiffanian), but all the families continue into later faunas. It is instructive to look at the succeeding Tiffanian age, where there are 94 genera in 38 families; 56 per cent of the

Table 15.5. Data on early Eocene mammal faunas from North America and Europe

Mammals late early Eocene (*c.* 50 Ma)	Europe Cuisian 2 Ma	N. America Wasatchian 4 Ma
Total genera	58	113
Carryovers	39 (67%)	36 (32%)
Disappear	34 (59%)	75 (66%)
Survive	24 (41%)	38 (34%)
Total families	35	39
Disappear	6 (17%)	10 (26%)

genera and 18 per cent of the families become extinct before the next age (Clarkforkian), yet this turnover does not make a peak above the noise level because the Tiffanian age is reckoned to have been about 4 Ma long. The most probable pattern of events in the Palaeocene is of mean or slightly greater than mean turnover rates.

The early Eocene (50 Ma)

There is a marked extinction event in early Eocene times in both North America and Europe, with an apparent overlap in time, though it seems the North American event lasted longer and extended into mid-Eocene times. In Europe, the Cuisan mammal age in France spans about 2 Ma between 49 and 51 Ma. Its time equivalent in North America is probably the late Wasatchian fauna; the whole Wasatchian spans about 4 Ma and it contains about double the number of genera known from the Cuisian (see Table 15.5). In both faunas around 60 per cent of their genera become extinct, and the loss of families is equally impressive. In the Cuisian six families disappear and in the Wasatchian 10 families suffer a similar fate. At generic level these figures do represent major losses and the succeeding Bridgerian in particular is marked by a very high generic appearance rate. The familial extinctions are, however, less impressive when we note that by early Eocene times all these families had been reduced to a single species or at most a few species. The early Eocene sees the beginning of the end for the first great wave of Cenozoic mammalian radiations.

The early Oligocene (37 Ma)

In 1909 Stehlin introduced the term *La grande coupure* to describe the great faunal break in Europe at the Eocene–Oligocene junction. There was the loss of tropical and semi-tropical elements in the flora and fauna, evidence of cooling with a change to dry/humid seasonality and the development of open savannah with gallery forests lining the rivers. The changes in the mammals were quite dramatic; almost all the European endemics disappeared, all save the moles and dormice, and their place was taken by invaders from central Asia. For the first time we see hamsters, beavers, rabbits, and rhinoceroses in European faunas. The end of the Eocene saw the end of the tapir-like lophiodonts but the horse-like palaeotheres which continued into the Oligocene adapted to browsing on tough tree foliage by evolving high crowned cheek teeth. Nine mammal families did not survive the end of the Eocene in Europe and in the early Oligocene we witness the appearance of no less than 17 families new to the continent. This impressive invasion of faunas came from central Asia and was made possible by the drying up of the Turgai Straits, a seaway which had in earlier times extended from the Arctic ocean across central Asia to the Tethys.

In North America there are also major changes in the mammal faunas across the Eocene–Oligocene boundary. Prothero (1985) has analysed the changes in the faunas through three land mammal ages, the late Eocene Duchesnian (41–38 Ma), the early Oligocene Chadronian (38–32 Ma), and the mid-Oligocene Orellan (32–30 Ma). Prothero has been able to use the magnetic polarity time scale to correlate many of the North American faunas through these mammal ages; this gives a precision of dating not hitherto available at these stratigraphic levels. The Eocene–Oligocene transition sees the extinction of no less than 15 archaic families of mammals which had been dominant during earlier Palaeogene times, while another six families cling on before becoming extinct at the next junction. Also near the Eocene–Oligocene transition we see the arrival of many new stocks with an Asian ancestry, having migrated along the new dry Beringia landbridge. The second major turnover is at the early – mid Oligocene transition (Chadronian–Orellan junction) which can be dated at 32 Ma. Within a very short time (probably less than 200 000 years) toward the end of Chadronian times there was a second wave of extinction; in this disappeared many of the remaining archaic groups, including the great titanotheres, pantolestids, mole-like epoicotheres, paramyid and cylindrodont rodents. This level also saw the appearance of new stocks such as leptauchenine oreodonts and cricetid rodents.

Palaeontologists have long recognized major faunal changes at or near the Eocene–Oligocene boundary. As we get more data, the pattern becomes clearer, though it does not show up as a high extinction peak on the plot; Table 15.3 shows it to merely equal the mean extinction rate in both Europe and North America. In contrast, the appearance rate in Europe is high in early Oligocene and just above the mean in North America. Another point that comes out vividly is the double pulsed changeover in western North America, with the disappearances of archaic stocks and the arrival of new stocks from Asia in two waves.

Turning briefly from the family level to that of orders, the early Oligocene saw the disappearance of no less than eight mammalian orders. Multituberculates, having persisted for nearly 150 Ma — longer than any other mammalian order, and having dispersed through all the northern continents — finally died out in North America in the early Oligocene. The little known pantolestids disappear from North America, Europe and north Africa. Also from Africa, the last rhinoceros-like embrithopods vanish. Elephant-like pyrotheres vanish from the South American scene. Asia witnesses the last anagalids, pantodonts and mesonychids. In North America the apatemyids die out and condylarths disappear from South America.

On a global scale, early Oligocene times witnessed major tectonic readjustments. The separation of Antarctica from both Australia and South America during the Oligocene led to the decoupling of the polar and equatorial currents, the establishment of the psychrosphere and the circum-Antarctic current. Other associated phenomena included atmospheric cooling, development of seasonal climates, and vegetational changes. While the Antarctic continent became isolated and cooler, the associated lower sea-levels allowed land migrations across the holarctic continental masses. This period (about 38 Ma ago) is also recognized by Raup and Sepkoski as one of their extinction peaks.

Oligocene – Miocene boundary (24 Ma)

The boundary is not marked by any major change in the European mammal faunas, but the North American faunas display extinction peaks in two mammal ages on either side of the boundary. The faunas of the Whitneyan land mammal age in late Oligocene times contain about 70 genera, of which 50 are carryovers from the previous Orellan age, and about 30 carry on into the succeeding Geringian age. About 20 genera make their first appearance and some 40 disappear during the Whitneyan age. There is little evidence of immigrant taxa and most of the replacements appear to have evolved locally; all the orders and half the families are still living. The Whitneyan lacks primates (no more in North America until Man arrives) and the fauna is dominated by browsers adapted to open woodland and bush savannah. Although on the table the Whitneyan stands out as a peak with below mean extinction rate ages on either side, all three ages have similar sized faunas; the Geringian has double the number of new genera and the Orellan almost half the number of extinctions.

The Arikareean faunas of the early Miocene are known from an abundance of taxa, some 128 genera of mammals of which 75 become extinct during the age and some 67 new genera make their first appearance. During early Miocene times the Cordilleras witnessed the outpourings of voluminous volcanic lava flows and ash falls. These activities were associated with plate tectonic activity which built up during Miocene times into the Alpine orogeny and resulted in the great mountain chains of the Rockies and Alps. Although tectonic activity was also building up in Europe and we have good faunas over the same period, they do not appear to demonstrate any peaks of extinction in these early stages.

Miocene extinctions

Mammal faunas are very well known throughout the Miocene and Pliocene in both North America and Europe. On each continent we recognize five mammal ages within the Miocene epoch, though each has not an exact temporal equivalent on the other continent. In all these ages there are abundant and varied mammal faunas. The middle and late Miocene ages of North America, and the middle Miocene to early Pliocene ages of Europe undergo successive changes with high rates of disappearance accompanied by rates of appearance which are almost all close to the mean (only the Ruscinian has an exceptionally high appearance rate, due in large part to it being a very short age of 1.5 Ma). Thus, the picture is that of a succession of faunal turnovers rather than one large and short extinction episode. The turnovers span about 11 Ma (16–5 Ma) in North America and about 13 Ma (17–4 Ma) in Europe. The extinction peak recognized by Raup and Sepkoski at 11.3 Ma falls in the midst of these.

European mammal faunas in mid-Miocene times had an abundance of ursids (bears), mustelids, hyaenas, felids (cats), shrews, suids (pigs), cervids (deer), giraffids, bovids, rhinoceroses, equids (horses), and mastodonts (the latter had recently arrived from Africa). By late Miocene times the families present were largely the same, though the genera had in many cases changed; bovids were becoming increasingly dominant among the herbivores as the grasses spread and they adapted to feeding on them. Among other groups to extensively colonize territories in late Miocene times were cricetids and murids (voles and mice). In North America in mid-Miocene times the faunas were especially rich in canids (dogs), merycoidodonts, camels, cervids (deer), antilocaprids (pronghorns), equids (horses), cricetids (voles), and geomyids (pocket gophers), with among other stocks, some mastodonts. In late Miocene times most of these families are still abundant, but with important changes. The equids changed from being leaf browsers to being grass grazers and, thus, successfully exploit the open prairies. The mastodonts greatly diversified and a few ground sloths make their appearance from South America.

The early Miocene was a period of submergent continents with seaways through Beringia and Panama isolating North America, and an extensive Tethys and Paratethys in Europe. During mid- and late Miocene times there is evidence of the climate becoming cooler and drier as the continental masses emerged with the uplifting of the Alpine orogenic chains. Land-bridges became established across Beringia and Panama, and island chains probably developed between Madagascar and Africa, and between Australia and southeast Asia, and perhaps even between west Africa and South America. In the Tethyan area at the close of Miocene times we witness the Messinian Crisis with the seaway becoming reduced to a series of saline basins. The accompaniment of lavas and especially volcanic ashes with lacustrine sediments in cratonic basin sequences helps greatly to date and preserve the faunas. The relatively rapid environmental changes taking place throughout the Miocene may be an important factor in establishment of patterns of extinction.

Plio–Pleistocene extinctions

The Pleistocene fossil record differs from the earlier record in several ways. It is much more extensive geographically, more abundant in individuals and species, and much shorter in duration — only 2 Ma. Yet within this short epoch can be recognized a succession of mammal faunas in both Europe and North America. Extinction rates are higher than at any other time in the past.

Kurten and Anderson (1980) have analysed the North American Plio–Pleistocene faunas in great detail (Table 15.6). The mammal species

Table 15.6. Species extinction record for North American late Pliocene and Pleistocene mammals (after Kurten and Anderson 1980).

		Species extinction	Subtotals
Rancholabrean	Late	77	
	Middle	21	105
	Early	7	
Irvingtonian	Late	33	
	Middle	14	86
	Early	39	
Blancan	Late	49	
	Middle	27	137
	Early	49	
	Very early	12	
Total number of species extinctions			328

extinction record is given for three mammal ages; the Blancan may be seen as spanning the Pliocene epoch and the extinction rate for the age as a whole is high, with two peaks separated by two troughs. The Irvingtonian age represents the early and mid-Pleistocene time span, the Rancholabrean equating with the late Pleistocene; both ages show a similar pattern of peaks with high extinction rates separated by troughs with low extinction rates. Particularly dramatic is the massive extinction during the Wisconsinian, the last glaciation event. These Plio–Pleistocene extinctions affected mammals of all sizes; in the Blancan 72 per cent of the extinctions were of mammals under 1 kg. In the Wisconsinian the large mammals make up some 66 per cent of the extinctions.

Table 15.7. North American and European Plio-Pleistocene mammal extinctions at species level

Mammal Species			Europe			N. America		
	Time-span	Length	Total	Extinct.	Rate	Total	Extinct.	Rate
Late Pleistocene	0.5–0.1	0.5	151	22	30	338	108	64
Early Pleistocene	1.9–0.5	1.4	204	90	31	217	89	29
Late Blancan	3.5–1.9	1.6	120	33	17	102	49	31

The European record can also be examined in detail (Kurten 1968); Table 15.7 records the best comparison that can be made with the North American faunas. The Villafranchian mammal age is taken as equivalent to the late Blancan, the two Pleistocene ages being the Biharian and Oldenburgian. Note that the species totals are comparable for both continents in the first two ages, but in the late Pleistocene the North American fauna is more than double that known from Europe. In the late Pliocene the extinction rate in North America is nearly double that in Europe; in the early Pleistocene both continents have similar extinction rates and in the late Pleistocene the North American rate is again more than double that in Europe.

The Plio–Pleistocene epochs were dominated by global climatic fluctuations, with both Europe and North America experiencing a succession of glacial periods, each separated by intervals with temperate regimes. The glaciations brought about lowered sea-levels (in some cases over 100 m), thus allowing migrations across extended continental masses. Conversely, the interglacial periods enabled island faunas to flourish. With the latitudinal movements of temperature, so did the vegetation belts move, and their dependent mammal faunas. The Plio–Pleistocene is remarkable as much for the evolution of new genera and species as for the extinction of existing taxa. It was an epoch of rapid faunal turnovers.

One of the most remarkable features of the Pleistocene mammal faunas is the rise and radiation of three great herbivore stocks — elephants, horses, and bovines; the horses are wholly grazers, the bovines very largely grazers, and the elephants mix grazing with browsing. The elephants had their ancestry in the browsing mastodonts of the Miocene and Pliocene. These mastodonts declined as the elephantids radiated from their origin in Africa, forming mammoth lineages in Eurasia and North America. The end of the Pleistocene saw a sharp decline in elephantid numbers, with the loss of all mammoths and the demise of the last mastodonts in North America. The one-toed grazing horses had originated in North America in the Pliocene and during the Pleistocene the new genus *Equus* spread into Eurasia and Africa, replacing the earlier three-toed *Hipparion* which finally became extinct at the end of the Pleistocene. From the Pliocene *Leptobos* of Eurasia arose a series of bovine genera that were to dominate the grazing faunas of both the Old and New Worlds during the Pleistocene. These include *Bison, Bos* (cattle), and *Bubalus* (water buffalo). These and other grazing herbivores were to dominate the great prairies of the northern hemisphere, at the expense of the earlier browsing stocks, many of which suffered great reductions or became extinct. Deinotheres, mastodonts, and anthracotheres hung on through the Pleistocene to disappear at the end. Camels abounded in North America during much of the Pleistocene, but disappeared from the continent as the epoch closed. Tapirs which had been diverse in Eurasia in early Pleistocene times, declined and vanished, to survive only in the tropics.

Much has been written about the loss of the late Pleistocene megamammals — giant deer, aurochs, woolly rhinoceros, giant ground sloth, cave bear, cave hyaena, cave lion, sabretooth cats, giant beavers, giant pigs, giant camels, and many others. However, as shown above it is not only the giants which vanish; many medium and small sized mammals also became extinct. Nor can it be shown that these extinctions are simultaneous on all continents; the resolution is lacking to establish a clear picture. Man has often been blamed for 'overkill', exterminating the large mammals by his hunting success; again an unproven case. In the holarctic picture. Man has often been blamed for 'overkill', exterminating the large mammals by his hunting success; again an unproven case. In the holarctic realm bison and mammoth seem to be the only large mammals which were hunted extensively, and one of these is still extant. The North American Pleistocene faunas contain about 200 mammalian genera, which is almost exactly double the number living there today. The corresponding figures for Europe are 140 Pleistocene genera and 70 living genera. Few large mammals survive on either continent today and the faunas appear impoverished compared with the Pleistocene as a whole. However, lumping all Pleistocene genera for comparison with the Recent means comparing almost 2 million years with a few thousand years. Table 15.8 shows a

Table 15.8. Numbers of mammalian genera in the North American and European Plio–Pleistocene and Recent faunas

	Numbers of mammalian genera	
	Europe	N. America
Living	69	99
Late Pleistocene	94	134
Early Pleistocene	98	119
Late Pliocene	123	69

distribution breakdown for each subdivision on both continents. The late Pliocene fauna of North America is smaller than any succeeding fauna, while in Europe it is larger than any succeeding fauna. Early and late Pleistocene faunas maintain high diversity on both continents. The living faunas on both continents retain about 74 per cent of the late Pleistocene diversity; certainly a drop, but not a disastrous impoverishment.

Conclusions

The search for mass extinction events affecting Cainozoic mammals in North America and Europe has revealed a series of extinction episodes which surpass the mean extinction rate, regarded as background noise. These events are shown to differ in severity and duration. Some are single short events, others multiple and prolonged events. There is some correlation between the events on the two continents, but the histories do not run in parallel. There is no evidence of cyclicity in these extinction events and only the slenderest coincidence with the Raup and Sepkoski extinction peaks.

While it is possible to construct scenarios of physical, environmental, and biological influences which could produce stress conditions in the faunas, it is impossible to prove any direct causal connection. Nevertheless, different combinations of these factors seem very probably to have been in each case at least in part responsible for the triggering of natural selection pressures which periodically culminated in a mass extinction event.

References

Fischer, A. G. and Arthur, M. A. (1977). Secular variations in the pelagic realm. *Soc. econ. Paleont. Miner. Spec. Publn.* **25,** 19–50.

Kurten, B. (1968). *Pleistocene Mammals of Europe.* Weidenfeld & Nicolson, London.

Kurten, B. and Anderson, E. (1980). *Pleistocene Mammals of North America.* Columbia University Press, New York.

Prothero, D. R. (1985). Mid-Oligocene extinction event in North American land mammals. *Science* **229,** 550–1.

Raup, D. M. and Sepkoski, J. J. (1984). Periodicity of extinctions in the geologic past. *Proc. nat. Acad. Sci.* **81,** 801–5.

Savage, D. E. and Russell, D. E. (1983). *Mammalian Paleofaunas of the World.* Addison-Wesley, Reading, Mass.

Stehlin, H. G. (1909). Remarques sur les faunules de mammiferes des couches eocenes et oligocenes du Bassin de Paris. *Bull. Soc. géol. Fr.* **9,** 488–520.

Systematics Association Publications

1. Bibliography of key works for the identification of the British fauna and flora *3rd edition* (1967)
   Edited by G. J. Kerrich, R. D. Meikle and N. Tebble
2. Function and taxonomic importance (1959)
   Edited by A. J. Cain
3. The species concept in palaeontology (1956)
   Edited by P. C. Sylvester-Bradley
4. Taxonomy and geography (1962)
   Edited by D. Nichols
5. Speciation in the sea (1963)
   Edited by J. P. Harding and N. Tebble
6. Phenetic and phylogenetic classification (1964)
   Edited by V. H. Heywood and J. McNeil
   Out of print
7. Aspects of Tethyan biogeography (1967)
   Edited by C. G. Adams and D. V. Ager
8. The soil ecosystem (1969)
9. Organisms and continents through time (1973)†
   Edited by N. F. Hughes
   Published by the Association (out of print)

Systematics Association Special Volumes

1. The new systematics (1940)
   Edited by Julian Huxley (Reprinted 1971)
2. Chemotaxonomy and serotaxonomy (1968)*
   Edited by J. G. Hawkes

3. Data processing in biology and geology (1971)*
Edited by J. L. Cutbill
4. Scanning electron microscopy (1971)*
Edited by V. H. Heywood
(out of print)
5. Taxonomy and ecology (1973)*
Edited by V. H. Heywood
6. The changing flora and fauna of Britain (1974)*
Edited by D. L. Hawksworth
7. Biological identification with computers (1975)*
Edited by R. J. Pankhurst
8. Lichenology: progress and problems (1976)*
Edited by D. H. Brown, D. L. Hawksworth and R. H. Bailey
9. Key works to the fauna and flora of the British Isles and north-western Europe (1978)*
Edited by G. J. Kerrich, D. L. Hawksworth and R. W. Sims
10. Modern approaches to the taxonomy of red and brown algae (1978)*
Edited by D. E. G. Irvine and J. H. Price
11. Biology and systematics of colonial organisms (1979)*
Edited by G. Larwood and B. R. Rosen
12. The origin of major invertebrate groups (1979)*
Edited by M. R. House
13. Advances in Bryozoology (1979)*
Edited by G. P. Larwood and M. B. Abbot
14. Bryophyte systematics (1979)*
Edited by G. C. S. Clarke and J. G. Duckett
15. The terrestrial environment and the origin of land vertebrates (1980)*
Edited by A. L. Panchen
16. Chemosystematics: principles and practice (1980)*
Edited by F. A. Bisby, J. G. Vaughan and C. A. Wright
17. The shore environment: methods and ecosystems (2 Volumes) (1980)*
Edited by J. H. Price, D. E. G. Irvine and W. F. Farnham
18. The Ammonoidea (1981)*
Edited by M. R. House and J. R. Senior
19. Biosystematics of social insects (1981)*
Edited by P. E. Howse and J.-L. Clément
20. Genome evolution (1982)*
Edited by G. A. Dover and R. B. Flavell
21. Problems of phylogenetic reconstruction (1982)*
Edited by K. A. Joysey and A. E. Friday
22. Concepts in nematode systematics (1983)*
Edited by A. R. Stone, H. M. Platt and L. F. Khalil

23. Evolution, time and space: the emergence of the biosphere (1983)*
Edited by R. W. Sims, J. H. Price and P. E. S. Whalley
24. Protein polymorphism: adaptive and taxonomic significance (1983)*
Edited by G. S. Oxford and D. Rollinson
25. Current concepts in plant taxonomy (1983)*
Edited by V. H. Heywood and D. M. Moore
26. Databases in systematics (1984)*
Edited by R. Allkin and F. A. Bisby
27. Systematics of the green algae (1984)*
Edited by D. E. G. Irvine and D. M. John
28. The origins and relationships of lower invertebrates (1985)‡
Edited by S. Conway Morris, J. D. George, R. Gibson, and H. M. Platt
29. Infraspecific classification of wild and cultivated plants (1986)‡
Edited by B. T. Styles
30. Biomineralization in lower plants and animals (1986)‡
Edited by B. S. C. Leadbeater and R. Riding
31. Systematic and taxonomic approaches in palaeobotany (1986)‡
Edited by R. A. Spicer and B. A. Thomas
32. Coevolution and systematics (1986)‡
Edited by A. R. Stone and D. L. Hawksworth
33. Key works to the fauna and flora of the British Isles and northwestern Europe 5*th edition* (1987)‡
Edited by R. W. Sims, P. Freeman, and D. L. Hawksworth
34. Extinction and survival in the fossil record (1988)‡
Edited by G. P. Larwood

*Published by Academic Press for the Systematics Association
†Published by the Palaeontological Association in conjunction with the Systematics Association
‡Published by the Oxford University Press for the Systematics Association

Author index

(*Bold numbers refer to the reference sections*)

Systematic index

Subject and stratigraphical index